Basics for Numerical and Symbolic Computation

Basics for Numerical and Symbolic Computation

Edited by:

Jovan Pehcevski

ARCLER PRESS

www.arclerpress.com

Basics for Numerical and Symbolic Computation

Jovan Pehcevski

Arcler Press
224 Shoreacres Road
Burlington, ON L7L 2H2
Canada
www.arclerpress.com
Email: orders@arclereducation.com

ISBN: 978-1-77469-445-9 (Hardcover)

DECLARATION

Some content or chapters in this book are open access copyright free published research work, which is published under Creative Commons License and are indicated with the citation. We are thankful to the publishers and authors of the content and chapters as without them this book wouldn't have been possible.

ABOUT THE EDITOR

Jovan currently works as a presales Technology Consultant at Dell Technologies. He is a result-oriented technology leader with demonstrated subject matter expertise in planning, architecting and managing ICT solutions to reflect business objectives and achieve operational excellence. Jovan has broad deep technical knowledge in the fields of data center and big data technologies, combined with consultative selling approach and exceptional client-facing presentation skills. Before joining Dell Technologies in 2017, Jovan spent nearly a decade as a researcher, university professor and IT business consultant. In these capacities, he served as a trusted advisor to a multitude of customers in financial services, health care, retail, and academic sectors. He holds a PhD in Computer Science from RMIT University in Australia and worked as a post-doctoral visiting scientist at the renowned INRIA research institute in France. He is a proud father of two, an aspiring tennis player, and an avid Science Fiction/Fantasy book reader."

TABLE OF CONTENTS

LIST OF CONTRIBUTORS

Ioan R. Ciric
The University of Manitoba, Winnipeg, Canada

Riccardo Bernardini
University of Udine, Udine, Italy

Cheon Seoung Ryoo
Department of Mathematics, Hannam University, Daejeon, Korea

A. A. Obayomi
Department of Mathematics, Faculty of Science, Ekiti State University, Ado-Ekiti, Nigeria

S. O. Ayinde
Department of Mathematics, Faculty of Science, Ekiti State University, Ado-Ekiti, Nigeria

O. M. Ogunmiloro
Department of Mathematics, Faculty of Science, Ekiti State University, Ado-Ekiti, Nigeria

Xuefeng Li
Department of Mathematical Sciences, Loyola University, New Orleans, LA, USA

J.J. Miralles
University of Castilla-La Mancha Spain

J.A. Martinez
University of Castilla-La Mancha Spain

I. Marin
University of Castilla-La Mancha Spain

M.M. Artigao
University of Castilla-La Mancha Spain

Barnabé Tang Ahanda
Department de Mathematics, Faculty of Sciences, University of Yaounde 1, Yaounde, Cameroon

Jean Gérard Aghoukeng Jiofack
Department de Mathematics, Faculty of Sciences, University of Yaounde 1, Yaounde, Cameroon

Romain Germain Nzangué
Department de Mathematics, Faculty of Sciences, University of Yaounde 1, Yaounde, Cameroon

Gilbert Hapi Mbiakop
Department de Mathematics, Faculty of Sciences, University of Yaounde 1, Yaounde, Cameroon

Inès Ferrer-Mallorquí
Institut d'Informàtica i Aplicacions, Universitat de Girona, Girona, Spain

Josep Vehí
Institut d'Informàtica i Aplicacions, Universitat de Girona, Girona, Spain

Mohammed A. Sharaf
Department of Astronomy, Faculty of Science, King Abdulaziz University, Jeddah, Saudi Arabia

Abdel-naby S. Saad
Department of Astronomy, National Research Institute of Astronomy and Geophysics, Cairo, Egypt

Zainab A. Mominkhan
Department of Mathematics, College of Science for Girls, King Abdulaziz University, Jeddah, Saudi Arabia

Yuan Yuan
Hydroforce (Beijing) Technology Co., Ltd., Beijing, China

A. Chaudhuri
CORIA-UMR CNRS 6614, Avenue de l'Universite, 76801 St Etienne du Rouvray France

A. Hadjadj
CORIA-UMR CNRS 6614, Avenue de l'Universite, 76801 St Etienne du Rouvray France

C. Guha
Chemical Engineering Department, Jadavpur University, Kolkata-700032 India

T. K. Dutta
Chemical Engineering Department, Jadavpur University, Kolkata-700032 India

Youssef Belhamadia
University of Alberta, Campus Saint-Jean Edmonton, Alberta, Canada

Leonardo Di G. Sigalotti
Area of Physics of Irreversible Processes, Department of Basic Sciences, Autonomous Metropolitan University-Azcapotzalco (UAM-A), Mexico City, Mexico

Jaime Klapp
Physics Department, National Institute of Nuclear Research (ININ), Ocoyoacac, Mexico

Karla Pedroza
Los Cabos H+ Hospital, San José del Cabo, Mexico

Edgar Nathal
National Institute of Neurology and Neurosurgery "Manuel Velasco Suárez", Mexico City, Mexico

Carlos E. Alvarado-Rodríguez
Departament of Chemical Engineering, DCNyE, University of Guanajuato, Guanajuato, Mexico

Ali Erfani Agah
Department of Water Engineering, Faculty of Agriculture, Shahrood University of Technology, Shahrood, Iran

Wiston Adrián Risso
Institute of Economics (IECON), University of the Republic, Montevideo, Uruguay

Fuding Xie
Department of Computer Science, Liaoning Normal University, Liaoning, Dalian 116081, China
Key Laboratory of Mathematics and Mechanization (KLMM), Academy of Mathematics and Systems Science, Chinese Academy of Science, Beijing 100080, China

Zhen Wang
School of Physics and Electronic Technology, Liaoning Normal University, Liaoning, Dalian 116029, China

Min Ji
Department of Computer Science, Liaoning Normal University, Liaoning, Dalian 116081, China

David Stirling
The University of Wollongong Australia

Amir Hesami
The University of Wollongong Australia

Christian Ritz
The University of Wollongong Australia

Kevin Adistambha
The University of Wollongong Australia

Fazel Naghdy
The University of Wollongong Australia

Gehan Abouelseoud
Alexandria University

Amin Shoukry
Computer Science and Eng. Dept, Egypt-Japan University of Science and Technology (EJUST), Alexandria, Egypt

LIST OF ABBREVIATIONS

AALN	Ada Augusta Learning Network
AE	Average Error
ANNs	Artificial Neural Networks
ARIMA	Autoregressive Integrated Moving Average
AUSM	Advection Upwind Splitting Methods
BFGS	Broyden–Fletcher– Goldfarb–Shanno
BMM	Balanced Milstein Method
BZ	Belousov-Zhabotinsky
CIR	Cox-Ingersoll-Ross
CNS	Central Nervous System
CRM	Coefficient of Residual Mass
CUDA	Compute Unified Device Architecture
DDT	Deflagration-To-Detonation Transition
DFP	Dynamic Finger Print
DOF	Degrees Of Freedom
DSMC	Direct Solution Monte Carlo
EPE	Engineering Problem Encryption
FDM	Finite Difference Method
FEEP	Field Emission Electric Propulsion
FEM	Finite Element Method
FFT	Fast Fourier Transform
FHE	Fully Homomorphic Encryption
FRS	Fuzzy Rule Systems
FVM	Finite Volume Method
GARCH	Generalized Autoregressive Conditional Heteroscedasticity
GPU	Graphics Processor Unit
HCI	Human Computer Interfaces
HPC	High-Performance Computing

HRR	Heat Recirculation Reactor
IRCAD	Interval Robust Control framework for Analysis and Design
ISUIA	International Study of Unruptured Intracranial Aneurysms
JSPs	JavaServer Pages
LN	Learning Network
MEMS	Micro-Electro-Mechanical-Systems
MST	Minimal Spanning Tree
NDDE	Nonlinear Differential-Difference Equation
NEMS	Nano-Electro Mechanical-Systems
NL	Network Learning
NLAR	Nonlinear Autoregressive
NLMA	Nonlinear Moving Average
NPE	Numerical Problem Encryption
NS	Navier-Stokes
NVC	Non-Verbal Communication
PCA	Principle Components Analysis
PCP	Parallel Coordinate Plot
PDF	Probability Density Function
QE	Quadratic-Exponential
QFT	Quantitative Feedback Theory
RCBD	Random Complete Block Design
RHS	Right Hand Side
RMSE	Root Mean Square Error
SDB	Symbolic Data Base
SDEs	Stochastic Differential Equations
SMPC	Secure Multiparty Computation
SNA	Social Network Analysis
SPH	Smoothed Particle Hydrodynamics
STSA	Symbolic Time Series Analysis
TG	Truncated Gaussian
TMAC	Tangential Momentum Accommodation Coefficient
VAR	Vector Autoregressive
WENO	Weighted- Essentially Non-Oscillatory

PREFACE

Symbolic calculus or algebraic calculus is a scientific field that deals with the study and development of algorithms and software for handling and solving mathematical expressions. Unlike the numerical calculus, which deals with specific numerical values of variables, the symbolic calculus manipulates expressions in which variables are treated as symbols without a specific value. Symbolic computing, on the other hand, emphasizes exact computation with expressions containing variables that have values, which are not given and are thus manipulated as symbols (hence the name symbolic calculation).

Software applications that perform symbolic calculations are called computing algebra systems, with a goal system adjusted to the complexity of major applications that include, at least: a method for representing mathematical data in a computer; a user programming language (usually different from the language used for implementation); memory manager; user interface for input / output of mathematical expressions; and large set of templates for performing common tasks, such as simplification of expressions, differentiation using chain rules, polynomial factorization, and indeterminate integrations.

At the beginning of computing algebra, around 1970, when well-known algorithms were firstly used on computers, symbolic calculations turned out to be very inefficient. Therefore, much of the field research work consisted of revisiting classical algebra to make it more efficient and discovering efficient algorithms to meet this efficiency. A typical example of this work is the calculation of the polynomials of the greatest common divisors, which was needed to simplify fractions. Surprisingly, it turned out that the classical Euclidean algorithm was inefficient for polynomials in infinite fields, so it was necessary to develop a new algorithm. The same was valid for classical algorithms in the linear algebra.

Apart from numbers and variables, each mathematical expression can be viewed as an operator: a symbol followed by a series of operands. In computing algebra software, expressions are usually represented this way. This kind of representation is very flexible, and many things, which do not seem to be mathematical expressions at first glance, can be represented and manipulated in this way. For example, an equation is an expression with "=" as an operator, a matrix can be represented as an expression with "Matrix" as an operator, and its rows as an operand.

This edition covers different topics from numerical and symbolic computing, including the main numerical and symbolic methods, schemes and applications.

Section 1 focuses on numerical computing methods, describing: matrix differential equations for solving systems of linear algebraic equations, numerical problem encryption for high-performance computing applications, numerical verification method of solutions for elliptic variational inequalities, trigonometric numerical integrator for solving first order ordinary differential equation, and augmented Lagrangian methods for numerical solutions to higher order differential equations.

Section 2 focuses on symbolic computing methods, describing learning network assisted by means of symbolic computation, multiple factorial analysis of symbolic data, combining symbolic tools with interval analysis, and an application to solve robust control problems, symbolic and graphical computations of a class of slightly perturbed equations.

Section 3 focuses on numerical computing applications, describing non-negativity preserving numerical algorithms for problems in mathematical finance, numerical simulations of microscale gas flows: continuum approach, recent numerical methods in electro-cardiology, numerical simulation of the blood flow through a brain vascular aneurysm with an artificial stent using the SPH method, and numerical modeling of soil water flow and nitrogen dynamics in a tomato field irrigated with municipal wastewater.

Section 4 focuses on symbolic computing applications, describing symbolic time series analysis and its application in social sciences, application of symbolic computation in nonlinear differential-difference equations, symbolic modelling of dynamic human motions, and a framework for bridging the gap between symbolic and non-symbolic AI.

SECTION 1: NUMERICAL COMPUTING METHODS

Chapter

1

USING MATRIX DIFFERENTIAL EQUATIONS FOR SOLVING SYSTEMS OF LINEAR ALGEBRAIC EQUATIONS

Ioan R. Ciric

The University of Manitoba, Winnipeg, Canada

ABSTRACT

Various ordinary differential equations of the first order have recently been used by the author for the solution of general, large linear systems of algebraic equations. Exact solutions were derived in terms of a new kind of infinite series of matrices which are truncated and applied repeatedly to approximate the solution. In these matrix series, each new term is obtained from the preceding one by multiplication with a matrix which becomes better and better conditioned tending to the identity matrix. Obviously, this helps the numerical computations. For a more efficient computation of approximate solutions of the algebraic systems, we consider new differential

Citation: Ioan R. Ciric, "Using Matrix Differential Equations for Solving Systems of Linear Algebraic Equations", IntechOpen Matrix Theory - Clasics and Advances, 2022, DOI: 10.5772/intechopen.104209.

equations which are solved by simple techniques of numerical integration. The solution procedure allows to easily control and monitor the magnitude of the residual vector at each step of integration. A related iterative method is also proposed. The solution methods are flexible, permitting various intervening parameters to be changed whenever necessary in order to increase their efficiency. Efficient computation of a rough approximation of the solution, applicable even to poorly conditioned systems, is also performed based on the alternate application of two different types of minimization of associated functionals. A smaller amount of computation is needed to obtain an approximate solution of large linear systems as compared to existing methods.

INTRODUCTION

Exact analytic expressions in the form of infinite series of matrices for the solution of linear systems of algebraic equations were derived in [1] by integrating associated ordinary differential equations. These differential equations were obtained using a quadratic functional related to the system of algebraic equations and describe the orthogonal trajectories of the family of hypersurfaces representing the functional. More convergent matrix series were presented in [2] which can be applied to approximate the solution of the system of equations. Solution of linear systems based on the numerical integration of differential equations has originally been formulated in [3].

In Section 2 of the present book chapter, we use recently derived highly convergent series formulae for matrix exponentials [3] in order to construct improved iterative methods for solving approximately large systems of algebraic equations. In Section 3, we use novel functionals that allow to formulate differential equations which lead to a substantial increase in the efficiency of the solution process [4]. Independently of the starting point, the paths of integration of these equations converge all to the solution point of the system considered. At each step of the numerical solution, one passes, in fact, from one path to a slightly different one due to computation errors. The procedure does not require to find accurately an entire path but only the solution point which is common to all the paths. This is why we apply the simple Euler method [5] to integrate the differential equations. The computation errors are now determined by the magnitude of the second derivative of the position vector with respect to the parameter defining the location along the path. A related iterative method [6] is also described. In Section 4, two different kinds of minimization of the system functionals are applied alternately for quick computation of a rough solution of large linear systems [7].

MATRIX SERIES FORMULAE FOR THE SOLUTION OF LINEAR ALGEBRAIC SYSTEMS

Consider a system of equations written in matrix form as

$$Ax - b = 0 \tag{1}$$

where $A \in R^{n \times n}$ is an arbitrary nonsingular matrix, $b \in R^n$ is a given *n*-dimensional vector and $x=(x_1,x_2,\ldots,x_n)^T$ is the unknown *n*-dimensional vector, with *T* indicating the transpose. Assume that x is a continuous function of the real variable *v* over a certain interval and associate to (1) the vector differential equation of the first order

$$\frac{dx}{dv} = f(v)(Ax - b) \tag{2}$$

where f(v) is a continuous function to be chosen.

Exact Analytic Expressions for the Solution of (2)

Imposing the condition $x(v_0)=x_0$, being a chosen position vector, (2) has a unique solution over a specified interval [4], namely,

$$x(v) = x_o + e^{-Ag(v_o)} \sum_{k=0}^{\infty} \frac{A^k}{(k+1)!} \left[(g(v))^{k+1} - (g(v_o))^{k+1} \right] (Ax_o - b) \tag{3}$$

where $g(v) \equiv \int f(v)dv$ is a primitive of f(v), i.e., f(v)=dg/dv. If f(v) is taken to be f(v)=1/v, then g(v)=lnv. Choosing $v_0=1$ gives $g(v_0)=0$ and (3) becomes now

$$x(v) = x_o + (\ln v) \sum_{k=0}^{\infty} \frac{(A \ln v)^k}{(k+1)!} (Ax_o - b) \tag{4}$$

Over the interval $v \in (0,2)$,x(v) can also be expressed in the form [3]

$$x(v) = x_o - (1-v) \left[I + \sum_{k=1}^{\infty} \frac{(1-v)^k}{k+1} (I - A)\left(I - \frac{A}{2}\right) \cdots \left(I - \frac{A}{k}\right) \right] (Ax_o - b) \tag{5}$$

I denoting the identity matrix of order *n*. The solution of (1) is theoretically obtained for v=0, the series being in general extremely low convergent. Since the rate of convergence of the series in (5) is very small for *v* very close to the value corresponding to the solution of (1), i.e., v=0, this formula should be applied repeatedly for a *v* not too close to zero, $v \in (0,1)$, until a satisfactory small distance to the solution x(0) is reached.

Highly Convergent Series Iteration Formulae

Practical formulae for an approximate solution of (1) can be derived by using matrix series that are much more convergent than the one in (5). This can be done by writing (4) in the form

$$x(v) = x_o + A^{-1}\left(e^{A\ln v} - I\right)(Ax_o - b) \tag{6}$$

and by expressing the matrix exponential in terms of series of superior convergence given recently in ref. [3]. Very close to the solution v=0, say for $v=e^{-N}, N \gg 1$, we have $e^{A\ln v}=e^{-NA}$ and with [3]

$$e^{-NA} - I = -10^{-q} c_q NA\left[I + \sum_{k=1}^{\infty}\frac{(-1)^k 10^{-qk}}{k+1}(I + c_q NA)\times\left(I + \frac{c_q NA}{2}\right)\cdots\left(I + \frac{c_q NA}{k}\right)\right] \tag{7}$$

where $q > 0$ and $c_q \equiv 1/\ln(1+10^{-q})$, we get

$$x\left(e^{-N}\right) = x_o - 10^{-q} c_q N\left[I + \sum_{k=1}^{\infty}\frac{(-1)^k 10^{-qk}}{k+1}(I + c_q NA)\times\left(I + \frac{c_q NA}{2}\right)\cdots\left(I + \frac{c_q NA}{k}\right)\right](Ax_o - b) \tag{8}$$

In order to perform numerical computations, the number of terms in the series has to be appropriately chosen. To have a small c_q in (8) one has to take a small q. For $q = 0.5$, e.g., $c_q=3.639409$. One may start with $N = 15$ which would require to retain about 50 terms in this alternating series to get a rough approximation of $x(e^{-15})$. The computation is repeated with the new x_o taken to be the preceding $x(e^{-N})$ until an acceptable accuracy of the solution of (1) is reached.

Much more convergent formulae are constructed if we apply in (6) the expansion [3]

$$e^{-NA} = (1+10^{-q})^{-p}\Big\{I + p!\sum_{k=1}^{p}\frac{10^{-qk}}{k!(p-k)!}\left(I - \frac{c_q NA}{p}\right)\left(I - \frac{c_q NA}{p-1}\right)\cdots\left(I - \frac{c_q NA}{p-k+1}\right) - 10^{-(p+1)q} c_q NA(I - c_q NA)\left(I - \frac{c_q NA}{2}\right)\cdots\left(I - \frac{c_q NA}{p}\right)\left[\frac{1}{p+1}I + p!\sum_{k=1}^{\infty}\frac{(-1)^k 10^{-qk}}{(k+1)(k+2)\cdots(k+p+1)}(I + c_q NA)\left(I + \frac{c_q NA}{2}\right)\cdots\left(I + \frac{c_q NA}{k}\right)\right]\Big\}, \tag{9}$$

p=1,2,…

The multiplication with A^{-1} from (6) is avoided by arranging the first summation in (9) in terms of powers of *A* and by taking into account that

$$p!\sum_{k=1}^{p}\frac{10^{qk}}{k!(p-k)!}=(1+10^{-q})^{p}-1 \quad (10)$$

It is obvious that using (9) leads to computations with a much smaller number of terms retained in the series for a given *N*. With the same amount of computation we obtain an $x(e^{-N})$ which is much closer to the exact solution $x^{(0)}$ of the system of equations (1).

METHODS OF NUMERICAL INTEGRATION

In this Section, we use special kinds of functionals that lead to the construction of differential equations which allow a substantial increase in the efficiency of the solution of large linear algebraic systems.

Vector Differential Equations and Their Application to the Solution of (1)

Consider a functional of the form

$$F(x)=\|Ax-b\|^{\alpha} \quad (11)$$

associated to (1) where α is a positive real number to be chosen. A real variable $v, v>0, v\in(v_o,v_S)$, is now defined by

$$F(x(v))=F(x(v_o))\,h(v) \quad (12)$$

where h(v) is an appropriately selected real function with definite first and second derivatives in (v_o,v_S),v_o corresponding to a starting point $x(v_o)\equiv x^{(0)}$, , with $h(v_o)=1$, and v_S corresponding to the solution vector of (1)$x(v_S)\equiv x_S$,, with $h(v_S)=0$. Denoting $F(x^{(0)})\equiv F_o$, we have

$$F_o\frac{dh}{dv}=\alpha\|Ax-b\|^{\alpha-2}\left(A^T(Ax-b)\right)^T\frac{dx}{dv} \quad (13)$$

Thus,

$$\frac{dx}{dv}=\frac{F_o}{\alpha}\frac{dh}{dv}\,\frac{A^T(Ax-b)}{\|Ax-b\|^{\alpha-2}\|A^T(Ax-b)\|^2} \quad (14)$$

which is the differential equation to be integrated from $v=v_o$ to $v=v_S$. The second derivative of x is

$$\frac{d^2x}{dv^2} = \frac{F_o}{\alpha}\frac{d^2h}{dv^2}\frac{A^T(Ax-b)}{\|Ax-b\|^{\alpha-2}\|A^T(Ax-b)\|^2} + \left(\frac{F_o}{\alpha}\right)^2\left(\frac{dh}{dv}\right)^2\frac{1}{\|Ax-b\|^{2(\alpha-1)}\|A^T(Ax-b)\|^2} \times\left\{\frac{\|Ax-b\|^2}{\|A^T(Ax-b)\|^2}\left[A^TA-2\frac{\|AA^T(Ax-b)\|^2}{\|A^T(Ax-b)\|^2}I\right]+(2-\alpha)I\right\}A^T(Ax-b) \tag{15}$$

From (11) and (12) we get a useful relationship,

$$\|Ax-b\| = (F_o\,h(v))^{1/\alpha} \tag{16}$$

that allows to simply monitor the magnitude of the residual vector of (1) during the computation process.

As explained in the Introduction we apply the Euler method for the solution of (14) and compute successively

$$x^{(i+1)} = x^{(i)} + \eta\left(\frac{dx}{dv}\right)_{x=x^{(i)}}, \quad i = 0, 1, 2, \ldots \tag{17}$$

where η is the step size. In the absence of any hint about a good starting point $x^{(0)}$ corresponding to $v=v_o$, we have used the point along the normal from the origin x=0 to the surface F(x)=const in (11) which is the closest to the solution point x_S [8], i.e.,

$$x^{(0)} = \frac{\|b\|^2}{\|A^Tb\|^2}A^Tb \tag{18}$$

The function h(v) and the parameter α are chosen such that the first and the second derivatives of x in (14) and (15) remain finite when $v \to v_S$, i.e., when the residual $\|Ax-b\| \to 0$, while the errors evaluated with the second derivative are kept reasonably small along the interval of integration $v \in (v_o, v_S)$. For each h(v), α is determined by imposing the condition that the second derivative in (15) tends to zero as $\|Ax-b\| \to 0$. To decide on the value of α for a given h(v), we require to have a good rate of decrease of $\|Ax-b\|$ at the beginning of the computation process, for instance to have (see (11) and (12))

$$\frac{\|Ax^{(1)}-b\|}{\|Ax^{(0)}-b\|} = (h(\eta))^{1/\alpha} \cong 0.8 \tag{19}$$

when $\eta=0.1$. Finally, to solve (1), we determine the actual step size to be used for the numerical solution such that the errors at the beginning of the computation process are small, say

$$\frac{\eta^2}{2}\frac{1}{\|x^{(1)}\|}\left\|\frac{d^2x}{dv^2}\right\|_{x=x^{(0)}} < 0.01 \quad (20)$$

If the magnitude of the residual vector, which is computed at each step, does not decrease anymore the computation is continued with a new cycle of integration by applying (17) with $x^{(0)}$ replaced by a new starting point, i.e.,

$$x^{(new)} = x^{(last)} - \frac{\left\|Ax^{(last)} - b\right\|^2}{\left\|A^T\left(Ax^{(last)} - b\right)\right\|^2} \times A^T\left(Ax^{(last)} - b\right) \quad (21)$$

where $x^{(last)}$ is the position vector from the preceding step. $x^{(new)}$ is the closest point to x_S along the normal to the surface F(x)=const taken at x(last)[8]. Subsequent cycles of integration are performed in the same way until a satisfactory approximate solution of (1) is obtained. If the difference between $x^{(new)}$ and $x^{(last)}$ is insignificant we find the point along the normal to F(x)=const, taken at $x^{(new)}$, where *F* has a minimum and, then, apply (21) again. It should be noted that as one approaches the solution point the direction of the gradient $A^T(Ax-b)$ tends to become more and more perpendicular to the direction of x_S-x and, thus, the residual $\|Ax-b\|$ and the relative error $\|x-x_S\|/\|x_S\|$ will not decrease any more as expected. This is why the computation has to be continued by opening a new cycle of integration. For systems with higher condition numbers, this happens more quickly.

Numerical experiments obtained using $h(v)=1-v$ with $\alpha=0.45$ and also using $h(v)=(1-v)^2$ with $\alpha=0.9$ shows that only two up to five integration cycles with a step size of 0.1 are needed in order to get an accuracy of about 1% for the solution of systems with condition numbers up to 100.

A Related Iterative Method

The basic idea of this method is to find, starting from a point $x^{(0)}$, a point $x^{(1)}$ along the gradient of a functional (11) associated with the general system (1), such that the magnitude of the new residual vector is an established fraction of its initial value, $\left\|Ax^{(1)} - b\right\| = \tau\left\|Ax^{(0)} - b\right\|, \tau < 1$. Instead of performing an integration as in Section 3.1, one proceeds iteratively, i.e.,

$$\left\|Ax^{(i+1)} - b\right\| = \tau\left\|Ax^{(i)} - b\right\|, \quad i = 0, 1, 2,\ldots \quad (22)$$

for each iteration the starting point being the point found in the preceding iteration, with τ maintained as tight as possible at the same value. To do this, we impose that F(x) in (11) varies from $x^{(i)}$ to $x^{(i+1)}$ in the same way for each iteration, namely,

$$F(x) = F\left(x^{(i)}\right) h(v) \tag{23}$$

where h(v) is now a real function of a real variable, monotone decreasing in the interval $v \in [v_o, v_o + \eta], v_o \geq 0, \eta > 0$, with definite first and second derivatives, and with $h(v_o)=1$ such that

$$F\left(x^{(i+1)}\right) = F\left(x^{(i)}\right) h(v_o + \eta), \quad i = 0, 1, 2, \ldots \tag{24}$$

Then, from (11),

$$\frac{F\left(x^{(i+1)}\right)}{F\left(x^{(i)}\right)} = \frac{\left\|Ax^{(i+1)} - b\right\|^{\alpha}}{\left\|Ax^{(i)} - b\right\|^{\alpha}} = \tau^{\alpha} \tag{25}$$

and

$$\tau = \left(h(v_o + \eta)\right)^{1/\alpha} \tag{26}$$

$x^{(i+1)}$ is computed as

$$x^{(i+1)} = x^{(i)} + \eta\left(\frac{dx}{dv}\right)_{v=v_o} \tag{27}$$

i.e., with (14) where F_0 is replaced with $F(x^{(i)})$,

$$x^{(i+1)} = x^{(i)} + \frac{\eta}{\alpha}\left(\frac{dh}{dv}\right)_{v=v_o} \frac{\left\|Ax^{(i)} - b\right\|^{2}}{\left\|A^{T}\left(Ax^{(i)} - b\right)\right\|^{2}}$$

$$\times A^{T}\left(Ax^{(i)} - b\right), \quad i = 0, 1, 2, \ldots \tag{28}$$

in which, taking into account (21), one has to have $\left|(\eta/\alpha)(dh/dv)_{v=v_o}\right| < 1$. This expression corresponds to the first step in the numerical integration by Euler's method of the differential Eq. (14), starting from $x^{(i)}$ with a step η, or to the first two terms of the Taylor series expansion of $x(v_o+\eta)$.

The starting value $x^{(0)}$ and the function h(v) are chosen in the same way as in Section 3.1. For selected ratios η/α the iteration cycle continues as long as the residual $\|\|Ax^{(i)}-b\|\|$ decreases at a proper rate. Theoretically, to make

$\left\|Ax^{(i)}-b\right\|=\varepsilon\left\|Ax^{(0)}-b\right\|$ with $\varepsilon \ll 1$ one would need to conduct $(\ln\varepsilon)/\ln\tau$ iterations. Since computation errors are introduced at each iteration (as in the Euler method), the initially chosen value of τ cannot be maintained the same as the iterative process continues. An approximate solution of (1) is obtained at the end of the iteration cycle as before, applying (21). Subsequent iteration cycles are performed with the starting point in each cycle being the point determined in the preceding cycle.

Numerical results were generated using, as in the method presented in Section 3.1, $h(v)=1-v$ but with $\eta/\alpha=0.2$ and $h(v)=(1-v)^2$ with $\eta/\alpha=0.1$. Now, in both cases, $v_o=0$, $(\eta/\alpha)(dh/dv)_{v=0}=-0.2$ in (28), and $\tau\cong 0.8$. A substantial increase in τ, approaching $\tau=1$, or oscillations of its value during the first few iterations show that the computation errors evaluated from (15) (with F_o replaced by $\|Ax-b\|^{\alpha}$) are too big. In such a situation, one has to decrease the factor $(\eta/\alpha)(dh/dv)_{v=0}$ in (28), i.e., to decrease η/α for a given $h(v)$, which leads to an increase of τ and a decrease of $(1/2)\eta^2\left\|d^2x/dv^2\right\|_{v=v_o}$. For accuracy of 1% for the solution of (1), one needs a number of three to six short iteration cycles, with about eight iterations per cycle, for systems with condition numbers below 100. Of course, as in the previous method, an increased number of iteration cycles is required to reach the same accuracy for the solution of poorly conditioned systems.

Remarks. One can easily see that $x^{(i+1)}$ in (28) can be expressed in the form

$$x^{(i+1)}=x^{(0)}+\sum_{\ell=0}^{i}x_d^{(\ell)},\quad i=0,1,2,\ldots \tag{29}$$

where

$$x_d^{(\ell)}\equiv x^{(\ell+1)}-x^{(\ell)}=\frac{\eta}{\alpha}\left(\frac{dh}{dv}\right)_{v=v_o}\frac{\left\|Ax^{(\ell)}-b\right\|^2}{\left\|A^T\left(Ax^{(\ell)}-b\right)\right\|^2}\times A^T\left(Ax^{(\ell)}-b\right) \tag{30}$$

with the solution of (1) given by the infinite series

$$x_S=x^{(0)}+x_d^{(0)}+x_d^{(1)}+x_d^{(2)}+\cdots \tag{31}$$

$x_d^{(\ell)}$ in (30) can also be written as

$$x_d^{(\ell)}=\frac{\eta}{\alpha}\left(\frac{dh}{dv}\right)_{v=v_o}\frac{\left\|b^{(\ell)}\right\|^2}{\left\|A^Tb^{(\ell)}\right\|^2}A^Tb^{(\ell)} \tag{32}$$

where

$$b^{(\ell)} = Ax_d^{(\ell-1)} - b^{(\ell-1)}, \quad \ell = 1, 2, \ldots;$$
$$b^{(0)} = Ax^{(0)} - b \tag{33}$$

This shows that the difference $x_d^{(\ell)}$ is a "rough approximation" to the solution of a system (1) whose right-hand side is, at each iteration, just the residual vector of the system in the preceding iteration, which decreases in magnitude from one iteration to the next. Thus, the method presented in this Section represents a practical, concrete implementation of the well-known idea of successive approximations/perturbations [9].

To search for a possible increase in efficiency, more general functionals of the form

$$F(x) = \mathscr{F}(\|Ax - b\|) \tag{34}$$

could be tested, where F and its first derivative are finite and continuous at all the points within the interval of integration. Now the corresponding (14) is

$$\frac{dx}{dv} = F_o \frac{dh}{dv}\left(\frac{d\mathscr{F}}{d\|Ax - b\|}\right)^{-1} \frac{\|Ax - b\|}{\|A^T(Ax - b)\|^2} \times A^T(Ax - b) \tag{35}$$

Note. The paths of integration in the methods presented can be looked at as being the field lines of a Poissonian electrostatic field in a homogeneous region bounded by a surface of constant potential $F(x)=\|Ax_0-b\|^{\alpha}$ and with a zero potential at the solution point, $F(x_s)=0$. In the particular case of $\alpha=2$ the ratio of the volume density of charge within the region to the material permittivity is constant, namely, $-2\sum_{i=1}^{n}\sum_{k=1}^{n} a_{ik}^2$ where a_{ik} are the entries of *A*. By altering this electrostatic field one could eventually make quicker the approach to the solution point along the integration path.

METHOD OF ALTERNATE MINIMIZATIONS

This simple method is based on the property of a functional of the form (34) associated with a general system of equations (1) to allow not only to minimize the value of the functional but also the distance to the solution point of (1). Using only minimizations along ordinary gradients of the functional is not efficient unless the system is very well-conditioned.

For computing efficiently an approximate solution of general, large linear systems of algebraic equations, we propose in this Section the alternate application of minimizations of a functional and of the distance to the solution point, along the direction of the gradient of the functional.

Consider the functional in (34) where F is a real function defined for all xx from a chosen starting point x(s) to the solution point x_S of (1), monotone decreasing with ||Ax−b||,F(0)=F(x_S). The gradient of F(x) is

$$\nabla F\left(x\right) = \frac{d\mathscr{F}(\|Ax-b\|)}{d\|Ax-b\|}\frac{A^T(Ax-b)}{\|Ax-b\|} \tag{36}$$

i.e., in the direction of the vector AT(Ax−b).

The minimum of F(x) along a straight line through x(s) in the direction defined by an *n*-dimensional vector *d* is found from the condition that x−x(s)=λd, where λ is a scalar to be determined, and AT(Ax−b) are perpendicular. This gives λ and x for the minimum of F(x), namely,

$$x^{(d)}_{\min F} = x^{(s)} - \frac{\left[(Ad)^T\left(Ax^{(s)}-b\right)\right]}{\|Ad\|^2}d \tag{37}$$

The point at which F(x) is minimum along the normal taken at x(s) is determined by replacing *d* with A^T(Ax(s)−b) in (37),

$$x_{\min F} = x^{(s)} - \frac{\left\|A^T\left(Ax^{(s)}-b\right)\right\|^2}{\left\|AA^T\left(Ax^{(s)}-b\right)\right\|^2}A^T\left(Ax^{(s)}-b\right) \tag{38}$$

The minimum distance between the solution point x^S and a point x along the direction of the gradient of F(x) taken at $x^{(s)}$ is at

$$x = x^{(s)} + \mu A^T\left(Ax - b\right) \tag{39}$$

where the scalar μ is determined by requiring the distance $\|x-x^S\|$ to be minimum. This gives μ and x for this minimum, namely,

$$x_{\min D} = x^{(s)} - \frac{\left\|Ax^{(s)}-b\right\|^2}{\left\|A^T\left(Ax^{(s)}-b\right)\right\|^2}A^T\left(Ax^{(s)}-b\right) \tag{40}$$

which depends only on the residual vector and on AT(Ax−b) at the point $x^{(s)}$, independently of the form of F.

As already mentioned, repeated minimizations using only (38) are not efficient for solving a general system (1). The same is true when using only (40). To obtain an approximate solution of (1), we apply the formulae (40) and (38) alternately, the starting point x(s) being each time the point determined in the preceding minimization. As in Section 3, when there is no indication about a convenient first starting point x(s), one can use the origin $x^{(s)}=0$. Only a few iterations, up to ten, are needed for a solution with a relative error $\|x\sim-x_S\|/\|x_S\|$ of about 1% for systems with condition numbers of up to 100.

The procedure is surprisingly efficient for a rough solution even for very ill-conditioned systems. For example, for the system Hx=b where *H* is the Hilbert matrix of order eight and whose solution is xS=[1,1,…,1] ^{T} [10], a solution of 6% accuracy is obtained with a starting point $x^{(s)}=0$ by performing only seven alternate minimizations (40), (38), with no equilibration or regularization preoperated on the system. By comparison [10], for the Gauss elimination method, the accuracy is only 40.6%, for the Gauss elimination with equilibration 9.15%, and for the Householder method with equilibration 5.6%.

Experimental results show that, as the new point x given by (40), (38) becomes closer to the solution point x_S, the direction AT(Ax−b) of the gradient in (36), i.e., of the correction terms in (40) and (38), becomes more and more orthogonal to the direction of $x-x_S$. This causes the magnitude $\|Ax-b\|$ of the residual vector of (1) and the relative difference $\|x-x_S\|/\|x_S\|$ to not significantly decrease anymore. To progress with the computation one can intercalate minimizations of F(x) along directions that are different from that of AT(Ax−b). By numerical testing, it is observed that $x-x^{(0)}$, where $x^{(0)}$ is the original starting point, has at this stage, in most cases, a significant component along the direction of $x-x_S$. Thus, one can use such a minimization direction to try to improve the solution accuracy.

CONCLUSIONS

Highly convergent iteration formulae for solving general, large linear systems of algebraic equations are derived from exact analytic solutions of particular differential equations based on new, accurate series representations of the matrix exponential recently published in ref. [3]. Specialized differential equations which make it possible to monitor and control the computation errors and the decrease of the magnitude of the residual vector $\|Ax-b\|Ax-b$ of (1) at each stage of the computation procedure are constructed and integrated numerically to approximate the solution of these systems. Two methods of the solution have been presented in this book chapter and the simplest Euler method is applied for the numerical integration of the vector differential equations. In the first method cycles of integration are used, each cycle starting from a convenient value of the unknown and continuing until the rate of decrease of $\|Ax-b\|$ becomes too small. The second method is an iterative method where a fixed rate of decrease of $||Ax-b||$ is imposed at the beginning of the iteration cycles. In this method, only the first step in the Euler method is computed at each iteration and each cycle of iteration is conducted until there is no significant change in the magnitude of the residual vector. These two methods are highly efficient for large systems with condition numbers below 100 since only up to six cycles with less than ten steps per cycle are necessary to get a solution accuracy of 1%, at each step within a cycle having to compute two matrix-vector multiplications. The method in Section 3.2 seems to be more efficient for systems with bigger condition numbers. The number of cycles of integration/iteration increases with the condition number and preconditioning should be done for ill-conditioned systems before attempting to apply the methods presented in this work.

The iterative method of alternate minimizations presented in Section 4 is intended for computing quickly a rough approximation of the solution of linear systems of equations. In this method, preequilibration or preregularization/preconditioning are not required to obtain useful results even for systems with poorly conditioned matrices.

The present Book Chapter has been intended to constitute a review of the work done so far on the subject matter. It describes the proposed new methods for an approximate solution of large linear algebraic systems using appropriately chosen matrix functionals and shows the procedures for constructing the concrete solution algorithms. At this stage, the results presented are only validated by preliminary numerical experiments which indicate the efficiency of the proposed procedures for deriving approximate/rough solutions of large systems. More numerical experiments involving systems with higher condition numbers, as well as theoretical results, will be presented in future work. It is my hope that other researchers will be attracted to this new area and rigorous theoretical results will also be established (theorems, etc.).

REFERENCES

1. Ciric IR. Series solutions of vector differential equations related to linear systems of algebraic equations. In: Proceedings of the 12th International Symposium on Antenna Technology and Applied Electromagnetics and Canadian Radio Sciences Conference (ANTEM/URSI). Montréal, Canada: IEEE; 2006. pp. 317-321. ISBN: 978-0-9638425-1-7
2. Ciric IR. Rapidly convergent matrix series for solving linear systems of equations. In: Proceedings of the 17th International Symposium on Antenna Technology and Applied Electromagnetics (ANTEM). Montréal, Canada: IEEE; 2016. DOI: 10.1109/ANTEM.2016.755022/978-1-4673-8478-0
3. Ciric IR. New matrix series formulae for matrix exponentials and for the solution of linear systems of algebraic equations. In: Shah K, Okutmustur B, editors. Functional Calculus. London: IntechOpen; 2020. DOI: 10.5772/Intechopen.2022.77599/978-1-83880-007-9
4. Ciric IR. Approximate solution of large linear algebraic systems using differential equations. In: Proceedings of the 19th International Symposium on Antenna Technology and Applied Electromagnetics (ANTEM). Winnipeg, Canada: IEEE; 2021. ISBN: 978-1-6654-0335-1
5. Burden RL, Faires JD. Numerical Analysis. 5th ed. Boston: PWS-KENT; 1993. ISBN: 0-534-93219-3
6. Ciric IR. Using selected rates of residual vector decrease for solving iteratively large linear systems. In: Proceedings of the 19th International Symposium on Antenna Technology and Applied Electromagnetics (ANTEM). Winnipeg, Canada: IEEE; 2021. ISBN: 978-1-6654-0335-1
7. Ciric IR. Alternate minimizations for an approximate solution of general systems of linear algebraic equations. In: Proceedings of the 19th International Symposium on Antenna Technology and Applied Electromagnetics (ANTEM). Winnipeg, Canada: IEEE; 2021. ISBN: 978-1-6654-0335-1
8. Ciric IR. A geometric property of the functional associated with general systems of algebraic equations. In: Proceedings of the 12th International Symposium on Antenna Technology and Applied Electromagnetics

and Canadian Radio Sciences Conference (ANTEM/URSI). Montréal, Canada: IEEE; 2006. pp. 311-315. ISBN: 978-0-9638425-1-7

9. Lanczos C. Applied Analysis. New York: Dover Publications, Inc.; 1988. ISBN: 0-486-65656-X (Paperback)

10. Hämmerlin G, Hoffmann K-H. Numerical Mathematics. New York: Springer-Verlag; 1991

Chapter

NUMERICAL PROBLEM ENCRYPTION FOR HIGH-PERFORMANCE COMPUTING APPLICATIONS

2

Riccardo Bernardini

University of Udine, Udine, Italy

ABSTRACT

Recent years witnessed the diffusion of cloud-based services. Cloud services have the interesting advantage that they can provide resources (CPU, disk space, etc.) that would be too expensive to deploy and maintain in-house. A major drawback of cloud-based services is the problem of handling private data and—possibly—intellectual property to a third party. With some service (e.g., data storage), cryptography can provide a solution; however, there are some services that are more difficult to protect. An example of such services is the renting of CPU to carry out numerical computation such as differential

Citation: Riccardo Bernardini, "Numerical Problem Encryption for High-Performance Computing Applications", IntechOpen Modern Cryptography, 2018, DOI: 10.5772/intechopen.85565.

equation solving. In this chapter, we discuss the problem of encrypting numerical problems so that their solution can be safely outsourced. The idea is to transform (encrypt) a given numerical problem into a different one whose solution can be mapped back to the solution of the original problem if the key used at the encryption stage is known.

Keywords: HPC, numerical analysis, security, cloud

INTRODUCTION

The rise of *cloud computing* has made it possible for SMEs to procure, on a pay-per-use basis, resources that until few years ago they had to acquire themselves. Although cloud computing offers interesting opportunities, it has some drawbacks too. One important drawback is the lack of *data privacy*: as soon as the SME hands its data to the cloud provider, there is the risk that the data could be exposed to third parties. Privacy protection in cloud services is indeed one of the key challenges highlighted by the *Public Consultation on Cloud Computing and Software* [1].

In some cases, the SME can take some simple countermeasures to mitigate privacy risks. For example, the SME can send to the provider an encrypted version of the data in the case of storage service (see Figure 1a). This is possible since the cloud provider can store the data just as a "binary blob" without the need to understand them.

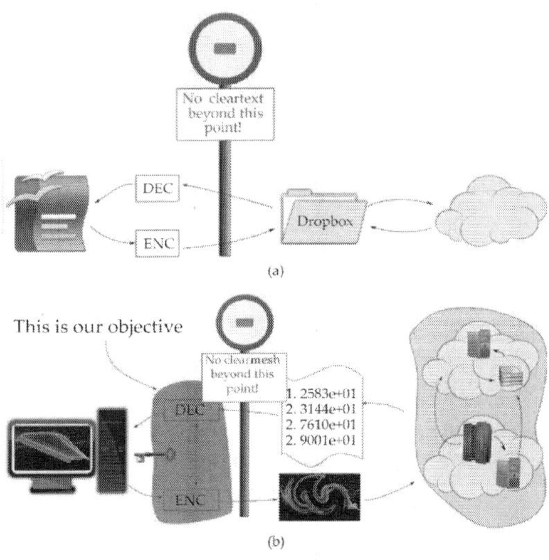

Figure 1: (a) Using safely storage on networks and (b) renting safely CPU on network.

However, there are some services where simple solutions are not feasible, for example, the procurement of high-performance computing (HPC) resources on the cloud. It is a fact that many potential users are not eager to employ cloud HPC services because of the risk of data disclosure. In fact, simple data encryption mechanisms are not sufficient to deal with security and privacy protection in data centric environments, and even "the right to be forgotten" does not cover indirect security and privacy aspects [2]. Ideally, we would like a solution reminiscent of what is done for storage: encrypt the computation before sending it to the cloud and decrypt what we receive from the cloud. Figure 1b shows a graphical representation of this idea: on the left hand side, we see a user that needs to compute the airflow around a new wing in an aerodynamic application. In order to outsource that aerodynamic problem, the user processes it with a secret key in the block ENC with the aim of transforming it into a *different numerical* problem that is uploaded to the cloud. A number of HPC providers collaborating on the cloud may be needed to solve this computation-intensive problem. The encrypted solution is then synthesized and sent back to the user (wing designer) that processes it with DEC in order to get the final answer. Standard cryptography techniques (e.g., RSA, AES, etc.) in this context provide only limited result. The question is if the user uploads their data in encrypted form to the cloud, how can the cloud process it?

At a first glance, this seems an unsolvable problem, but recent developments in the field of theoretical cryptography can address this apparent paradox using a set of secure techniques belonging to the family of secure multiparty computation (SMPC). Some efficient SMPC solutions, such as additive homomorphic encryption or garbled circuit, allow evaluating particular functions on encrypted data, but unfortunately they require interaction with the user, often making the cooperation more expensive than the direct solution. On the other hand, fully homomorphic encryption (FHE) allows evaluating functions on encrypted binary data, without decrypting it nor interacting with the client. This implies that the cloud can compute any function of the data uploaded by the user (without learning anything about the plain-text data) and return the encrypted answer.

Even more recent developments in the field of cryptography show that even cryptographic obfuscation can be achieved (under strong but plausible assumptions): using obfuscation, one can upload a piece of software to the cloud, which now can be run on any input data without learning anything about the proprietary code—think of a researcher who find a new algorithm to diagnose some diseases. Now the researcher can sell diagnosis as an external service on a cloud provider, without fearing that the cloud can steal or leak his proprietary methods.

While SMPC and FHE provide wonderful theoretical results, they will have little or no practical impact in the way users work with the cloud. The computational overhead introduced by FHE is huge, and an obfuscated version of even a simple function (a point function with few bit inputs) requires gigabytes of memory.

THE SETUP

The Basic Setup

Figure 2 shows a more detailed setup with two possible scenarios. We have, with reference to Figure 2a, an original problem that we will call the *engineering problem*; for example, determine drag and lift of a new innovative wing profile or the electronic configuration of a new molecule. The first step in the solution of the engineering problem is its transformation into a *numerical problem* (typically a linear system or an eigenvalue problem) that is solved by means of a well-known numerical algorithm in order to obtain a *numerical solution*. The final step is to use the numerical solution to obtain the *engineering solution*, that is, the values that are of interest for us (e.g., drag and lift in the wing problem). It is worth to discuss in some details the three steps.

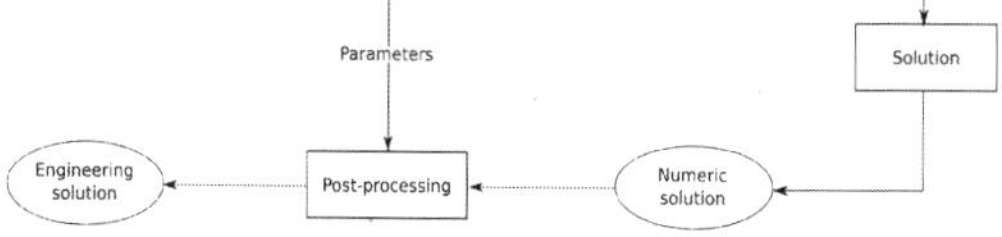

(a)

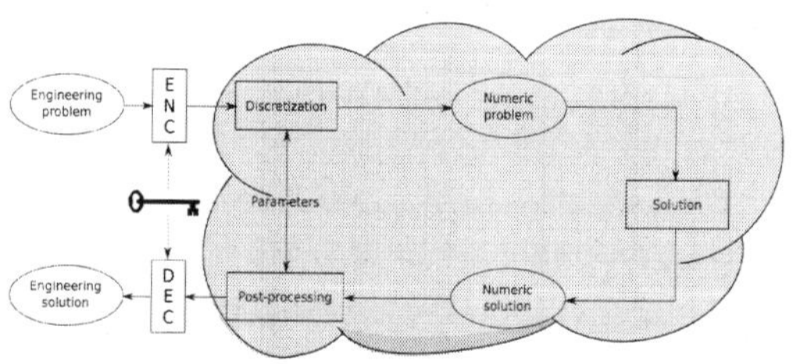

(b)

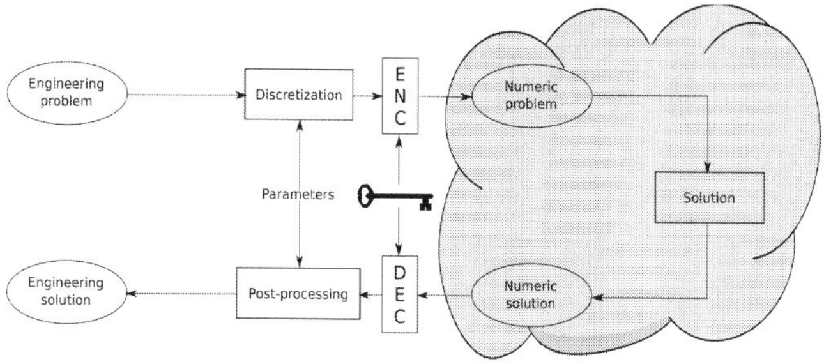

(c)

Figure 2: (a) Typical scenario: an engineering problem is converted to a numeric problem that is solved to derive the desired engineering solution, (b) outsourcing by encrypting the engineering problem and (c) outsourcing by encrypting the numerical problem.

From the engineering problem to the numerical problem

Many engineering problems require the solution of differential equations that when discretized give rise, usually, to a linear system or an eigenvalue problem. The matrices obtained with the discretization can be quite large but (usually) sparse. Note that instead the engineering problem has usually a relatively compact description; in the example at hand, it would be a description of the geometry of the wing (by means of 3D model format) together with a description of the air flow [3, 4, 5]; in the case of the molecule, it could be a description of the configuration of the molecule (e.g., by means of its structural formula).

Although intuitively one can imagine that the solution of the numerical problem is the heaviest part, even the discretization step can be nontrivial. For example, some discretization technique—e.g., the popular finite element method (FEM)—requires to partition the space using a grid whose generation can be fairly complex [6]. Finally, it is worth observing that with large problems, not only the required CPU time can be a problem but also the amount of memory required that can become readily prohibitive because of the "dimension curse." Indeed many engineering problems involve at least four dimensions: one for time and three for space. This suggests that working with very sparse matrices (and preserving their sparse structure) is of utmost importance. This will be important in the following when talking about some proposals found in the literature.

Finally, it is worth observing that there is a good degree of arbitrary in the discretization step: if a grid is used, it is not uniquely determined; in finite difference method (FDM), the ordering used to map the grid points to matrix coordinates is arbitrary; in methods based on function space discretization (e.g., Galerkin-like methods), there is a wide freedom in the choice of basis functions. This implies that while it is known how to go from an engineering problem to its numerical counterpart, it is not clear how (or "if") it is possible to go in the other direction. This is important in our context since we can expect that the opponent will be interested in the engineering problem rather than in the numerical one.

Numerical solution

In this step the numerical problem (typically, a linear system or an eigenvalue problem) is solved to give the numerical solution (typically, a vector). If the problem is very large, this step can be very CPU-consuming. Algorithms used for the solution typically exploit the very sparse structure of the matrices involved. For example, in the solution of a linear system, usually an iterative procedure is preferred, since the iterative procedure exploits more easily any sparsity (the cost of a product matrix-vector is proportional to the number of non-zero elements), while the inverse of a sparse matrix is not necessarily sparse, requiring much more memory and CPU time to be handled.

This suggests that if the encryption is applied at the numerical problem level (see Section 3 in the following), care must be exercised in order not to spoil the sparsity, although this goes against with the requirement—sometimes stated—of hiding the number of zeros (see Section 4.1.1).

From the numerical solution to the engineering solution

This step usually is not as computationally demanding as the previous ones. It amounts to extract from the numerical solution the values of interest (e.g., drag and lift in the wing example) or producing suitable visualizations of the data (e.g., in the electronic distribution of a molecule example).

ENCRYPTING THE PROBLEM

There are two possibilities, shown in Figure 2, of encrypting the problem: in a case (see Figure 2b) the original problem is directly encrypted and outsourced, leaving both discretization and numerical solution to the cloud; we will call this solution the *engineering problem encryption* (*EPE*). In the other case (see in Figure 2c), the discretization step is done on the client computer, and only the numerical problem (e.g., solution of linear system) is encrypted and outsourced; we will call this solution the *numerical problem encryption* (*NPE*). Pros and cons of the two solutions are as follows.

EPE: From the point of view of the work to be done on the client, this solution is preferable since the possibly large cost of the discretization step is outsourced to the cloud. This solution, however, requires a different encryption technique for every problem (e.g., an encryption technique suitable for aerodynamic problems cannot be used for computational chemistry problems). To the best of our knowledge, there is currently no proposal working at the EPE level.

NPE: On the one hand, this solution requires that the client does the discretization step and the computational cost of this step cannot be negligible; on the other hand, many engineering problems reduce to a just few numerical problems when discretized. This means that a procedure to encrypt, say, a linear system can be applied in many engineering problems that require the solution of differential equations.

Note that in both cases the key remains in the client, similarly to what happens in the case of remote storage. This implies that the client can employ very long keys, maybe as long as the data to be encrypted, since there is not a problem of distributing the key.

Numerical Issues

Since we are interested in numerical problems, the issue of the impact of the encryption/decryption on the precision of the result needs to be taken into account. For example, consider the following protocol, similar to many techniques proposed for protecting a linear system [7, 8, 9]:

$$\mathbf{Ax} = \mathbf{b} \qquad (1)$$

with $\mathbf{A} \in \mathbb{R}^{n\times n}$ and $\mathbf{x}, \mathbf{b} \in \mathbb{R}^{n}$. System (1) is protected by randomly generating two permutation matrices $\mathbf{P}, \mathbf{Q} \in \mathbb{R}^{n\times n}$ and a diagonal matrix $\mathbf{D} \in \mathbb{R}^{n\times n}$ and replacing ΛΛ and bb with, respectively,

$$\hat{\mathbf{A}} = \mathbf{PAQD} \qquad (2)$$

$$\hat{\mathbf{b}} = \mathbf{Pb} \qquad (3)$$

and sending the problem

$$\hat{\mathbf{A}}\hat{\mathbf{x}} = \hat{\mathbf{b}} \tag{4}$$

to the cloud. It is immediate to check that the solutions of (1) and (4) are related by

$$\mathbf{x} = \mathbf{QD}\,\hat{\mathbf{x}} \tag{5}$$

Eq. (5) can be considered as the decryption algorithm. This looks fine, but the value returned for $\hat{x}$ usually has some error ε . It is immediate to verify that this induces an error $QD\varepsilon$ on the decrypted value. If D has some large entries, this will amplify the noise that affects the decrypted value. Moreover, right multiplication by D can affect the condition number of A , worsening the conditioning of the system. Figure 3 shows the result of a numerical experiment demonstrating this problem: we generated 5000 symmetric matrices A and 5000 diagonal matrices D of sizes ranging from 10×10 to 200×200 , and for every iteration, we computed the condition number amplification

$$\rho = \text{cond}\,(\mathbf{AD})/\text{cond}\,(\mathbf{A}) \tag{6}$$

where $\text{cond}\,(\mathbf{A}) = |||A|||_2|||\mathbf{A}^{-1}|||_2$ is the 2-norm condition number of A . Entries of A were Gaussian with zero mean and unit variance, while entries of D have been generated using a variety of distributions: Gaussian as the entries of A , uniform in [0,1] , uniform in [1/2,3/2] , and uniform in [1,2].

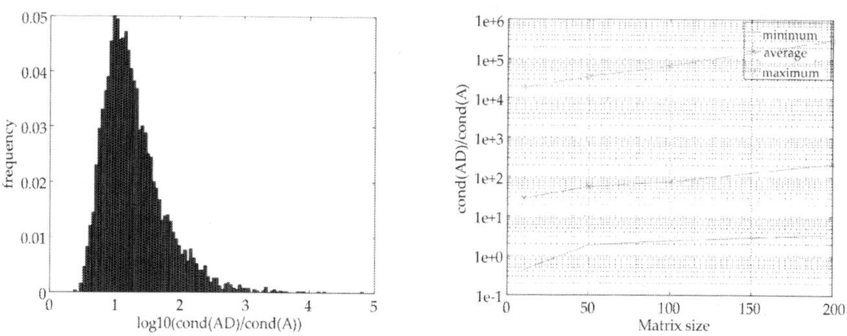

(a) (b)

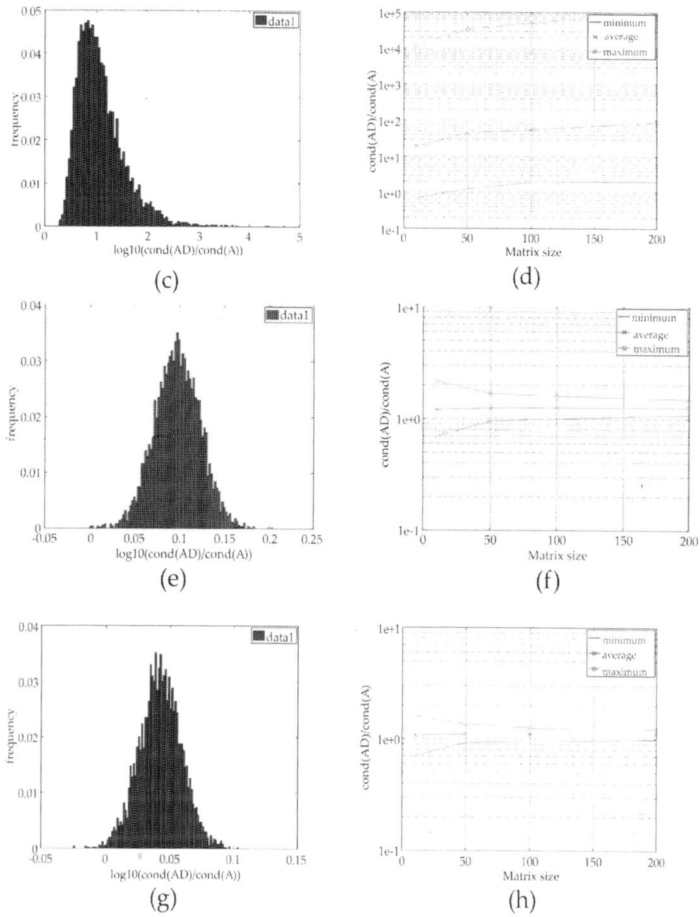

Figure 3: (a) Histogram of the condition number amplification ρ=cond(AD)/cond(A) in logarithmic scale for 100×100 matrices; matrix entries of both A and D are Gaussian with mean zero and unitary variance; (b) minimum, maximum, and average values of ρ as a function of the matrix size; (c) and (d) like (a) and (b), but the entries of D are uniformly distributed in [0,1]; (e) and (f) like (c) and (d), but the entries of D are uniformly distributed in [1/2,3/2] ; (g) and (h) like (c) and (d), but the entries of D are uniformly distributed in [1,2].

The right hand column of Figure 3 shows the histogram of ρ in logarithmic scale for 100 × 100 matrices; Figure 3b shows the minimum, maximum, and average values of ρ as function of the matrix size. Different rows of Figure 3 are relative to different ways of generating D . Although from Figure 3 one can see that sometimes the conditioning can improve (ρ<1), it is clear that there is a non-negligible probability of worsening the condition number of several orders of magnitude, especially if the entries of D can go near to zero (first two rows of Figure 3 relative to the cases Gaussian and uniform in [0,1]).

THE ADVERSARY MODEL

In the literature, two kinds of "bad behavior" can be present together or not in a specific opponent:

Honest but curious: In this case all the parties follow correctly the protocol, and the computed result is correct, but the cloud tries to learn the protected secret. Note that it is reasonable to expect that the opponent is interested in the *engineering problem* or the corresponding solution rather than in the numerical counterparts. The numerical problem/solution can be seen as a mean to find out the engineering problem/solution. Note that the freedom that one has in the discretization step can potentially be used to make it more difficult to invert it.

Malicious: The cloud does not follow correctly the protocol and, in particular, could try to "cut some corners" returning results that are not correct. In many cases, this attack can be easily counteracted by checking the result returned by the cloud; this is possible since many problems can be computationally heavy to solve, but it is fairly easy to check if a solution is correct; think, for example, the case of a linear system.

In some cases, even a full check can be quite demanding; in those cases, a random selection of checks to be done can give a fairly good assurance that the result is correct. For example, in the case of a linear system, a random subset of the equations can be checked. The probability of a false positive (an uncorrected result is accepted) decreases exponentially with the number of checks, while the computational cost grows only linearly. Indeed, many check techniques proposed in the literature can be reduced to this idea.

Remark 4.1

It is worth observing that since we are talking about floating point computation, we need to take into account numerical noise that results from computation. For example, if a numerical system is solved with an iterative approach, the returned solution will differ from the "true" solution by a, hopefully, small error. This needs to be taken into account when checking if the solution is correct.

In a typical context the adversary is interested in getting the original problem or the solution. It is worth observing that while in the classical cryptography setup, the opponent recovers completely the message or nothing at all; in this case, there is the possibility that the adversary recovers an approximate version of the problem/solution. Observe also that the setup of Figure 2c makes the problem for the adversary more difficult since after recovering the numerical problem there is the problem of doing the inverse of discretization.

Finally, it is worth observing that the encryption/decryption steps could amplify the numerical noise introduced by the solution algorithm.

Security Criterion

A major difference between traditional cryptography and cryptography of numerical problems is the information that an opponent can gain about the secret. In the most typical case, in a practical application of classical cryptography, two cases are possible: (i) the opponent succeeds in the attack and learns the whole secret and (ii) the attack is unsuccessful and very little, or nothing is learned about the secret; the case where a partial success can be achieved is quite uncommon. Suitable security criterion in this kind of application is information-theoretical criterion (where the adversary has unlimited computational power) or based on computational indistinguishably (where we admit that the computational power of the adversary can grow only polynomially).

In the context of encryption of numerical problems instead, it is possible that the opponent gains some partial or *low-resolution* information. For example, in the case of a new wing, the adversary is interested in learning the profile of the wing; the adversary has a strong *a priori* information about the profile, and only some detail information to integrate such a priori are needed. Depending on the specific encryption technique, it could be possible for the adversary to find said details but only up to a resolution. The ambiguity could be "unsolvable" even with infinite computational power (more or less the equivalent of information-theoretical security), or maybe it could be that the amount of computation required to improve the resolution increases more than the polynomial with the required resolution (this would be the equivalent of classical computational indistinguishably).

There is also the possibility that the encryption technique leaks some *invariant* of the problem. For example, if linear system Ax=b is encrypted by applying two orthogonal matrices U,V as in

$$\underbrace{\mathbf{UAV}}_{\hat{\mathbf{A}}}\underbrace{\mathbf{V}^t\mathbf{x}}_{\hat{\mathbf{x}}} = \underbrace{\mathbf{Ub}}_{\hat{\mathbf{b}}}, \tag{7}$$

matrix $\hat{A}$ preserves the singular values of A . Is this a problem? Most probably, it depends on the specific underlining engineering problem. As in another example, consider the encryption protocol described in (2). In this case, $\hat{A}$ preserves not the singular values of A but its sparsity (i.e., the number of non-zero entries). Again, if this is a problem or not probably depends on the corresponding engineering problem.

To the best of our knowledge, most of the literature consider the classical cryptography criterion of computational indistinguishably, and more research about criterion specific for numerical problem could prove useful.

About sparsity preservation

As said before, many discretization techniques produce matrices that are very sparse. This is very important from the viewpoint of computational efficiency since there are algorithms that are able to exploit the sparseness, working only with the non-zero entries of the matrix. This suggests that the encryption should preserve the sparseness or, at least, not reduce it in a significant way. However, some researchers raise the concern that even the fraction of non-zero entries can be private information. This would suggest that the encryption step should not preserve matrix sparsity.

It is also worth observing that in some application, the number of non-zero entries is known a priori with good precision. For example, in FDM/FEM discretization methods for differential equations, the fraction of non-zero entries depend on the structure of the grid employed which can be considered known with good precision. A quantitative and objective approach to the importance of preserving or hiding the sparsity can be an interesting field for future research activity.

EXISTING TECHNIQUES

The field of encrypting numerical problems is relatively new, and there is not a huge variety of encryption algorithms proposed; many of them are just variations of some basic scheme. To the best of our knowledge, no algorithm found in the literature tries to encrypt the engineering problem; it rather tackles the numerical problem. Also, it is very difficult to find some discussion about the numerical stability of the proposed scheme.

Basic Techniques

There are few basic techniques used in the literature to encrypt matrices and vectors.

Addition: To matrix A a matrix K is added to obtain $\tilde{A}$ =A+K. This approach is not widely used, probably because sum does not "propagate" nicely in usual linear algebra operations (e.g., linear systems, eigenvalue problems, determinants, singular value decomposition). Moreover, in the case of linear systems, it is not obvious how to choose K in order to have $\tilde{A}$ invertible.

Product by a diagonal matrix: Matrix A is multiplied (on the left, on the right, or both) by a diagonal matrix D . The effect on problems like linear systems is quite easy to describe, and it suffices that all the diagonal entries are not zero in order to guarantee the invertibility of $\tilde{A}$, but, as shown in Section 3.1, if D is randomly chosen, the condition number of A can increase by several orders of magnitude.

Product by a permutation: Matrix AA is multiplied (on the left, on the right, or both) by a permutation matrix PP . This grants that cond($\tilde{A}$)=cond(A) (therefore also invertibility is preserved). A product by a permutation requires no floating point multiplications, although the cost of data movement (especially if A is very large) is not necessarily small. Sparsity is preserved, and this can be seen either as an advantage or as a drawback (see Section 4.1.1). Some characteristic values such as determinant and singular values are preserved.

Product by a unitary matrix: Matrix A is multiplied (on the left, on the right, or both) by a matrix R such that $R^tR=I$. This grants that cond($\tilde{A}$)=cond(A). Product by R does not necessarily preserve sparsity. If R is generated as a sequence of planar rotations, product by R (stored as sequence of rotation, not as a full matrix) can be more efficient than usual matrix product. Singular values and determinant are preserved.

Brief Literature Review

Maybe the numerical problem most frequently addressed is the problem of solving linear systems. Lei et al. [7] proposed a scheme similar to (2) where AA is both left and right multiplied by a diagonal matrix and a permutation. The proposal of Wang et al. [10] is one of the few that proposes to use classical homomorphic encryption together with an iterative algorithm; the approach of Wang et al. requires to use matrices with integer entries (eventually by rescaling the system) and a continuous exchange between the cloud and the user. Chen et al. in [11] pointed out a weakness of [10] and proposed a variation to solve the problem. Another solution based on matrix pre- and post-multiplication is suggested in [8].

A problem similar to linear system is linear regression. Chen et al. in [9] used pre- and post-multiplication approaches with two diagonal matrices. Zhou et al. in [12] pointed out a possible weakness (in the hypothesis of an integer problem). Similar approaches, still based on matrix pre- and post-multiplication, can be found also for matrix product [13, 14], determinant computation [15], and singular value decomposition [16].

CONCLUSIONS

We analyzed the problem of outsourcing safely numerical and engineering problems to the cloud. It turns out that the field is still in an evolving phase. Many approaches are different instances of pre- and post-multiplication masking techniques; security criterion is a reformulation of criterion from classical cryptography and does not address the peculiarities of numerical problem protections; finally, the problem of how the encryption/decryption impacts on issue such as numerical conditioning of the problem is usually not addressed. Future research directions will aim to cover such still open areas.

REFERENCES

1. Griffin D. Report on the public consultation for H2020 Work Programme 2016–17: Cloud computing and software. Tech. Rep., DG CONNECT E.2. 2014
2. Jeffery K, Schubert L. Complete computing: Toward information, incentive and intention. Tech. rep., DG CONNECT E.2. 2014
3. Hsieh A. Direct numerical simulation of complex turbulence [PhD thesis]. University of Colorado; 2017
4. Alfonsi G. On direct numerical simulation of turbulent flows. Applied Mechanics Reviews. 2011;64:0802-0803
5. Landahl MT, Mollo-Christensen E. Turbulence and Random Processes in Fluid Mechanics. 2nd ed. Cambridge, UK: Cambridge University Press; 1992
6. Belytschko T, Rabczuk T, Huerta A, Fernández-Méndez S. Meshfree Methods, Ch. 10. Atlanta, GA, USA: American Cancer Society; 2004
7. Lei X, Liao X, Huang T, Li H, Hu C. Outsourcing large matrix inversion computation to a public cloud. IEEE Transactions on Cloud Computing. 2013;1(1):1
8. Kumar M, Meena J, and Vardhan M, Secure amp;amp; efficient delegation of system of linear equation to a malicious cloud server. In: 2017 4th International Conference on Electronics and Communication Systems (ICECS); Feb 2017. pp. 12-17
9. Chen F, Xiang T, Lei X, Chen J. Highly efficient linear regression outsourcing to a cloud. IEEE Transactions on Cloud Computing. 2014;2:499-508
10. Wang C, Ren K, Wang J, Wang Q. Harnessing the cloud for securely outsourcing large-scale systems of linear equations. IEEE Transactions on Parallel and Distributed Systems. 2013;24:1172-1181
11. Chen F, Xiang T, Yang Y. Privacy-preserving and verifiable protocols for scientific computation outsourcing to the cloud. Journal of Parallel and Distributed Computing. 2014;74(3);2141-2151
12. Zhou L, Zhu Y, Choo K-KR. Efficiently and securely harnessing cloud to solve linear regression and other matrix operations. Future Generation Computer Systems. 2018;81:404-413
13. Lei X, Liao X, Huang T, Heriniaina F. Achieving security, robust cheating resistance, and high-efficiency for outsourcing large matrix

multiplication computation to a malicious cloud. Information Sciences. 2014;280:205-217

14. Kong S, Cai Y, Xue F, Yu H, Ditta A. Cloud outsourcing computing security protocol of matrix multiplication computation based on similarity transformation. International Journal of Wireless and Mobile Computing. 2018;14(1):90-96
15. Lei X, Liao X, Huang T, Li H. Cloud computing service: The case of large matrix determinant computation. IEEE Transactions on Services Computing. 2015;8:688-700
16. Pramkaew C, Ngamsuriyaroj S. Lightweight scheme of secure outsourcing svd of a large matrix on cloud. Journal of Information Security and Applications. 2018;41:92-102

Chapter

NUMERICAL VERIFICATION METHOD OF SOLUTIONS FOR ELLIPTIC VARIATIONAL INEQUALITIES

3

Cheon Seoung Ryoo

Department of Mathematics, Hannam University, Daejeon, Korea

ABSTRACT

In this chapter, we propose numerical techniques which enable us to verify the existence of solutions for the free boundary problems governed by two kinds of elliptic variational inequalities. Based upon the finite element approximations and explicit a priori error estimates for some elliptic variational inequalities, we present effective verification procedures that, through numerical computation, generat a set which includes exact solutions. We describe a survey of the previous works as well as show newly obtained results up to now.

Citation: Cheon Seoung Ryoo, "Numerical Verification Method of Solutions for Elliptic Variational Inequalities", IntechOpen Simulation Modeling, 2021, DOI: 10.5772/intechopen.101357.

Keywords: numerical verification method, variational inequalities, error estimates, fixed point formulation, newton-like method, finite element method

INTRODUCTION

Numerical verification methods of solutions for differential equations have been the subject of extensive study in recent years and much progress have been made both mathematically and computationally [1, 2, 3, 4, 5, 6, 7, 8, 9, 10, 11, 12, 13, 14, 15, 16, 17, 18, 19, 20, 21, 22, 23]. However, for some problems governed by the elliptic variational inequalities, there are very few approaches. As far as we know, it is hard to find any applicable methods except for those of Nakao and Ryoo [13, 24, 25, 26, 27, 28, 29, 30, 31, 32, 33, 34, 35, 36, 37, 38, 39, 40, 41, 42, 43, 44, 45, 46].

The authors have studied for several years the numerical verification method of solutions for elliptic variational inequalities using finite element method and the constructive error estimates combining with Schauder's and Banach's fixed point theorem. Several results in our research are already published in [13, 24, 25, 26, 27, 28, 29, 30, 31, 32, 33, 34, 35, 36, 37, 38, 39, 40, 41, 42, 43, 44, 45, 46]. In this chapter, we briefly overview our resent research results including works not yet published.

The outline of this chapter is as follows. In Section 2, the two types of elliptic variational inequalities are considered. In Subsection 2.1, we describe the elliptic variational inequalities and give a fixed point formulation to prove the existence of solutions. In Subsections 2.2 and 2.3, the main tool of the verification method is explained at an abstract level. In Subsection 2.2, we present a simple iteration method for numerical verification of solutions for the elliptic variational inequalities. We construct the concepts of rounding and rounding error for functions and present a computer algorithm to construct the set satisfying the verification conditions. However, it is difficult to apply the method in Subsection 2.2 to a problem in which an associated operator is not retractive in a neighborhood of the solution, because it is based upon a simple iteration method. In Subsection 2.3, we propose another approach to overcome such a difficulty. This method can be applied to general elliptic variational inequalities without any retraction property of the associated operator. We introduce a Newton-like operator and reformulate the problem using it. Particularly, special emphasis is placed on the way to devise the Newton-like operator for a kind of non-differentiable map which defines the original problem. We introduce a computational verification condition. In order to show a concrete usage of the tool, in Section 3, we present an application to some problems governed by the elliptic variational inequalities. Many difficulties remain to be overcome in the construction of general techniques applicable to a broader range of problems. However, the authors have no doubt that investigation along this line will lead to a new approach employing numerical methods in the field of existence theory of solutions for various variational inequalities that appear in mathematical analysis.

ELLIPTIC VARIATIONAL INEQUALITIES

The theory of elliptic variational inequalities has become a rich source of inspiration in both mathematical and engineering sciences. Elliptic variational inequalities are an effective tool for studying the existence of solutions of constrained problems arising in mechanics, optimization and control, operation research, engineering science, etc. [47, 48, 49, 50, 51, 52]. It is the aim of this chapter to introduce a numerical technique to verify the solutions for elliptic variational inequalities. The basic approach of this technique consists of the fixed point formulation of elliptic variational inequalities and construction of the function set, on computer, satisfying the validation condition of a certain infinite dimensional fixed point theorem. For fixed point formulation, we consider a candidate set which possibly contains a solution. In order to get such a candidate set, we divide the verification procedure into two phases: one is the computation of a projection into a closed convex subset of some finite dimensional subspace (rounding); the other is the estimation of the error for the projection (rounding error). Combining these methods with some iterative technique, the exact solution can be enclosed by sum of rounding parts, which is a subset of finite dimensional space, and the rounding error, which is indicated by a nonnegative real number. These two procedures enable us to treat infinite dimensional problems as finite procedures, thta is, by computer.

Notations

- V: real Hilbert space with scalar product $(\cdot,\cdot)$ and associated norm $\|\cdot\|$,
- V∗V∗: the dual space of V,
- $a(\cdot,\cdot):V\times V\rightarrow R$a··:V×V→R is a bilinear, continuous and V-*elliptic* from on V×V.

A bilinear form a(·,·) is said to be V-*elliptic* if there exists a positive constant α such that a(v,v)≥α‖v‖2,∀v∈V.

In general we do not assume a(·,·) to be symmetric, since in some applications nonsymmetric bilinear forms may occur naturally.

- L:V→R continuous, linear functional,
- K is a closed convex nonempty subset of V,
- j(·):V→R∪{∞} is a convex lower semicontinuous (l.s.c) and proper functional (j(·) is proper if j(v)>−∞,∀v∈V and j≠+∞).

The Two Types of Elliptic Variational Inequalities

We consider two classes of elliptic variational inequalities.

- Elliptic variational inequalities of the first kind: *Find*u∈V *such thatu*u*is a solution of the problem*

$a\left(u,v-u\right)\geq L\left(v-u\right),\forall v\in K,u\in K.$

- Elliptic variational inequalities of the second kind: *Find*u∈V *such thatu*u*is a solution of the problem*

$a\left(u,-u\right)+j\left(v\right)-j\left(u\right)\geq L\left(v-u\right),\forall v\in V,u\in V.$

The Problem and the Fixed Point Formulation

Let us first set a few notations [1, 47, 49, 50, 53, 54, 55, 56, 57, 58, 59, 60, 61]. In what follows we shall make use of the Sobolev spaces $W^{k,p}(\Omega)$ of functions which possess generalized derivatives integrable with the pth power up to and including the kth order. For p=2, we shall write $W^{k,p}\left(\Omega\right)=H^{k}\left(\Omega\right),H^{0}\left(\Omega\right)=L^{2}\left(\Omega\right)$. Further, we introduce the scalar product in $L^2(\Omega)$ by

$\left(f,g\right)=\int_{\Omega}f\left(x\right)g\left(x\right)dx.$

The norm in $H^k(\Omega)$ will be denoted by $\|\cdot\|_{H^{k}(\Omega)}$. The symbol $|\cdot|_{H^{k}(\Omega)}$ will stand for the seminorm,

$$|u|_{H^{k}(\Omega)}=\left(\sum_{|\alpha|=k}\left\|D^{\alpha}u\right\|_{L^{2}(\Omega)}^{2}\right)^{\frac{1}{2}},\quad \|u\|_{H^{k}(\Omega)}=\left(\sum_{j=0}^{k}|u|_{H^{j}(\Omega)}^{2}\right)^{\frac{1}{2}}.$$

Let V be a real Hilbert space with a scalar product $(\cdot,\cdot)_V$ and an associated norm $\|\cdot\|_{V},V^{*}$ its dual space. K denotes a nonempty closed convex subset of $V,a\left(\cdot,\cdot\right):V\times V\rightarrow\mathbf{R}$ is a bilinear, symmetric, continuous and elliptic form of $V, a(\cdot,\cdot) : V \times V \to \mathbf{R}$ is a bilinear, symmetric, continuous and elliptic form of V×V; that is, there exist constants α>0, and β>0 such that $a(u,v) \leq \alpha \| u \|_V \| v \|_V, \forall u, v \in V$ and $a(v,v) \geq \beta \| v \|_V^2, \forall v \in V$. The pairing between V and V* is denoted by <·,·>. Let Λ be a canonical isomorphism from V* onto V defined, for g∈V*, by $< g, v >= (\Lambda g, v)_V, \forall v \in V$. We can easily see that $\|\Lambda \|_{V^*} = \| \Lambda^{-1} \|_V = 1$. Now, let us consider the following variational inequality:

Find $u \in K$ such that $a(u, v-u) \geq < f(u), v-u >, \forall v \in K,$ (1)

where f is a nonlinear operator such that f(u)∈V*.

In order to obtain a fixed point formulation of variational inequality (1) we need the following standard result.

Lemma 1. *Let* K *be a closed convex subset of* V. *Then* u=P_Kω, *the projection of* ω *on* K, *if and only if*

$$u \in K : (u-\omega, v-u)_V \geq 0, \forall v \in K. \quad (2)$$

For some constant ρ>0, let us define a mapping G:V→V by

$$G(u) = P_K \Lambda \Phi(u), \quad (3)$$

where u∈V, Φ(u)∈V* is defined by

$$< \Phi(u), v >= (u,v)_V - \rho a(u,v) + \rho < f(u), v >, \forall v \in V. \quad (4)$$

For some constant ρ>0, problem (1) can be written as

$$(u, v-u)_V - \{(u, v-u)_V - \rho a(u, v-u) + \rho < f(u), v-u >\} \geq 0, \forall v \in K.$$

Using (4) in the above inequality, problem (1) is equivalent to that of finding u∈K such that

$$(u - \Lambda\Phi(u), v-u)_V \geq 0, \forall v \in K. \quad (5)$$

By (2) and (5), we now have the following fixed point problem for the operator G:

$$u = P_K \Lambda \Phi(u) = G(u). \quad (6)$$

Under appropriate conditions on the space V and the operator G:V→V(e.g., continuity, compactness), which usually have to be verified by theoretical means, fixed point theorem yields the existence of a solution u of the problem (1) in some suitable set U⊂V, provided that

$$G(U) \subset U. \quad (7)$$

In order to compute an explicit inclusion, we must therefore construct U explicitly. For the numerical verification of condition (7), we have to use interval analysis on many levels between basic interval arithmetic and functional analysis. For the appropriate and suitable choice of the operator f, the form a(·,·)a··, and the convex set K; one encounters problems governed by the elliptic variational inequality as special cases from the problem (1) [48, 49, 50, 51, 52]. Inbrief, it is clear that the problem (1) is the most common. Up to now, devising a verification technique for the problem (1) is still an open problem. It is an important and interesting area of future research to find the numerical inclusion methods for the problem (1) by using (6). In this paper, we suppose that $V \subset L^2(\Omega)$ and the nonlinear map $f(\cdot):V \rightarrow L^2(\Omega)$ satisfies the following assumptions.

A1.f is a continuous map from V to $L^2(\Omega)$.

A2. For each bounded subset $W \subset V$,f(W) is also bounded in $L^2(\Omega)$.

If we restrict the nonlinear map f as above, then it can be shown that the problem (1) can be characterized by a class of variational inequality of the type,

$$\text{find } u \in K \text{ such that } a(u, v - u) \geq (f(u), v - u), \forall v \in K. \quad (8)$$

The problem (8) has the restricted condition; even so (8) is an important and very useful class of nonlinear problems arising in mathematical physics, mechanics, engineering sciences, etc. In Section 3, we briefly consider a particular example of interest in applications. Another example is given in [13, 24, 25, 26, 27, 28, 29, 30, 31, 32, 33, 34, 35, 36, 37, 38, 39, 40, 41, 42, 43, 44, 45, 46]. In the special case in which K≡V, (8) yields the variational theory of the boundary value problems for partial differential equations. We will discuss existence and inclusion methods for problem (8). These are methods providing the existence of a solution of the problem (8) within explicitly computable bounds. As we have seen before, the transformation of problem (8) into some fixed point formulation (6) can be carried out in the same way. In a conclusion problem (8) is equivalent to the fixed point problem of finding u∈K such that

$$u = S(u), \quad (9)$$

where S denotes a specific operator, not necessarily the same as in (6). In particular for a given problem, we reduced the problem (8) to the fixed point formulation (9) and the continuity and compactness of S is discussed. For this reason, we shall say nothing about this problem for which we refer to [13, 24, 25, 26, 27, 28, 29, 30, 31, 32, 33, 34, 35, 36, 37, 38, 39, 40, 41, 42, 43, 44, 45, 46]. In order to simplify argument we assume that S is a continuous and compact operator. Since S is continuous and compact, as a result of Schauder's fixed point theorem, if there exists a nonempty, bounded, convex, and closed subset U such that S(U)⊂U, then there exists a solution of u=S(u) in U. In Sections 2.2 and 2.3, we describe how to construct U explicitly.

Verification by a Simple Iteration Method

In this subsection, we describe a simple iteration method for numerical verification of solutions for elliptic variational inequalities. In order to treat functions and variational inequalities in the infinite dimensional space V by computer, we introduce two concepts, rounding and rounding error. Now, let V_h be a finite dimensional subspace of V dependent on h(0<h<1) and let K_h be a nonempty closed convex subset of V_h. Usually, V_h is taken to be a finite element subspace with mesh size h. For the sake of simplicity, we shall define K_h, an approximate subset of K, by $K_h=V_h \cap K$.K_h is a closed convex subset of V_h. In practical applications, the construction of K_h is one of the difficulties presented by variational inequalities. For a given problem, several approximations are available. For a general study of the approximation of convex sets, we refer the reader to the work of Mosco [51]. We define the projection PK_h from V into K_h [49, 50]. That is, vh=PKh(u), the projection of u into K_h, is defined as follows:

$$u = S(u), v_h \in K_h : \quad (v_h, \zeta - v_h)_V \geq (u, \zeta - v_h)_V, \quad \forall \zeta \in K_h. \tag{10}$$

To verify the existence of a solution of (9), we determine a set W for a bounded, convex, and closed subset U⊂V as

$$W = \{v \in V : \ v = S(u), u \in U\}.$$

From Schauder's fixed point theorem, if W⊂U holds, then there exists a solution of (8) in the set U. Our goal is to find a set U which includes W. For any subset W⊂V, we define R(W)⊂K_h by the projection of V to K_h, which is called the rounding of W. Additionally, we define RE(W), the rounding error of W, as a subset of V so that W⊂R(W)+RE(W) holds. Using R(W)+RE(W) instead of W, the verification condition becomes

$$R(W) + RE(W) \subset U. \tag{11}$$

Let us describe the procedure more concretely. First, we consider the auxiliary problem: given g∈$L^2(\Omega)$,

$$\text{find } u \in K \text{ such that } a(u, v-u) \geq (g, v-u), \forall v \in K. \tag{12}$$

We note that, by well known result [49], there is a unique element u which satisfies (12).

Secondly, we define the approximate problem corresponding to (12) as

$$a\left(u_h, v_h - u_h\right) \geq \left(g, v_h - u_h\right), \forall v_h \in K_h, u_h \in K_h \tag{13}$$

and (13) admit one and only one solution [49]. Error estimates for the variational inequalities can be found in [48, 49, 52], etc. Now, using (10), (12), (13) and error estimates, we make the following assumption.

A3. For each u∈V, there exists a positive constant CC, independent of u and h, such that

$$\left\| u - P_{K_h} u \right\|_V \leq Ch \| g \|_{L^2(\Omega)}. \tag{14}$$

In order to verify the solutions numerically, it is necessary to determine the constant C that appears in a priori error estimations; this constant will be discussed later.

In order to construct the set U satisfying the verification condition (11) in a computer, we use an iterative procedure, that is, the sequential iteration. We propose a computer algorithm to obtain the set U which satisfies the condition (11).

(1) First, we obtain an approximate solution $v_h^{(0)} \in K_h$ to (8) by an appropriate method. Set $U_h^{(0)} = \left\{ v_h^{(0)} \right\}$ and $\alpha_0 = 0.$

(2) Next we will define $R(W^{(i)})$ and $RE(W^{(i)})$ for i≥0, where $W^{(i)}$ is the set defined as follows:

$$W^{(i)} = \left\{ v^{(i)} \in V : v^{(i)} = S\left(u^{(i)}\right), \quad u^{(i)} \in U^{(i)} \right\}.$$

$R(W^{(i)})$ is defined by the subset of K_h which consists of all the elements $v_h^{(i)} \in K_h$ such that

$$a\left(v_h^{(i)}, \psi - v_h^{(i)}\right) \geq \left(f\left(u^{(i)}\right), \psi - v_h^{(i)}\right), \quad \forall \psi \in K_h, \tag{15}$$

holds for some $u^{(i)} \in U^{(i)}$. Note that $R(W^{(i)})$ can be enclosed by $R\left(W^{(i)}\right) \subset \sum_{j=1}^{M} A_j \phi_j$, where $A_j = \left[\underline{A_j}, \overline{A_j}\right]$ are intervals, $\{\phi_j\}_{j=1}^{M}$ is a basis of V_h, and $M=\dim V_h$. For details of the interval calculation, we refer the reader to Nakao [6, 7, 12]. Next $RE(W^{(i)})$ is defined as

$$RE\left(W^{(i)}\right) = \left\{ v \in V : \| v \|_V \leq Ch \sup_{u^{(i)} \in U^{(i)}} \left\| f\left(u^{(i)}\right) \right\|_{L^2(\Omega)} \right\}. \tag{16}$$

Here, C is the same constant as in (14). Hence,

$W^{(i)} \subset R\left(W^{(i)}\right)+RE\left(W^{(i)}\right)$ holds.

(3) Check the verification condition:

$$R\left(W^{(i)}\right)+RE\left(W^{(i)}\right) \subset U^{(i)}. \tag{17}$$

If the condition is satisfied, then U(i)Ui is the desired set, and a solution to (8) exists in $W^{(i)}$, and hence in $U^{(i)}$.

(4) If the condition is not satisfied, we continue the simple iteration by using δ−inflation; that is, let δ be a certain positive constant given beforehand, and take

$$\alpha_{i+1}=Ch\sup_{u^{(i)}\in U^{(i)}} \| f\left(u^{(i)}\right) \|_{L^2(\Omega)} +\delta,$$

$$[\alpha_{i+1}]=\{v\in V:\| v \|_V\leq \alpha_{i+1}\},$$

$$U_h^{(i+1)}=\sum_{j=1}^{M}\left[\underline{A_j}-\delta,\ \overline{A_j}+\delta\right]\phi_j,$$

$$U^{(i+1)}=U_h^{(i+1)}+[\alpha_{i+1}],$$

and then go back to the second step. The reader may refer to [26, 27, 28, 29, 30, 31, 32, 33, 34, 35, 36, 37, 38, 39, 40, 41, 42, 43, 44, 45, 46] for the details. If the condition (17) is satisfied, in our inclusion method of solutions for (9), the solution uu is enclosed in the set $U^{(i)}$, which we call 'a candidate set' of the form $U^{(i)}=U_h^{(i)}+[\alpha_i]$.

Verification by a Newton-like Method

The significance of a Newton-like operator was already pointed out in [29, 43]. Hence we will not discuss it in detail here. In Subsection 2.1, numerical verification of solutions for elliptic variational inequalities using a finite element method have been discussed only for simple iteration method. The method proposed in Subsection 2.2 is such that $\left\{\left(U_h^{(i)},\alpha_i\right)\right\}$ always converges to the limit value $\{(U_h,\alpha)\}$ from an arbitrary initial value $\left\{\left(U_h^{(0)},\alpha_0\right)\right\}$ if S in (9) is retractive operator (we refer to Zeidler [59, 60, 61] for the definition of retraction), while no convergence can generally be expected if S is not retractive operator. Briefly, for not retractive operator in the neighborhood of the solution, it is difficult to use the previous scheme proposed in Subsection 2.2. To overcome such a difficulty, in this section, we newly formulate a verification method using the Newton-like method. This approach enables us to remove the restriction in Subsection 2.2 to the retraction property of the operator in the neighborhood of the solution. Namely, this technique can be applied to general variational inequalities without any retraction property of the associated operator S. We refer to [29, 43] for a detailed study of the properties of the Newton-like Method.

In this subsection, we use the notation of Section 2.2. We assume that $K_h=V_h \cap K$ is a closed convex cone with vertex at 0 and K_h^* its dual. We note that K_h^* is also a closed convex cone with vertex at 0, which is the only point common to K_h and K_h^*. From (10) it follows that K_h^* is the set of points whose projections into K_h is 0. We need some additional lemma.

Lemma 2. *Any* u∈V *can be uniquely decomposed into the sum of two orthogonal elements. That is,*

$$u = P_{K_h} u \oplus (I - P_{K_h})u = P_{K_h} u \oplus P_{K_h^*} u.$$

Here, ⊕ *denotes the sum of two orthogonal elements in the sense of* V..

Note that (9) can be rewritten as the following decomposed form in K_h and K_h^*:

$$\begin{cases} P_{K_h} u = P_{K_h} S(u), \\ (I - P_{K_h})u = (I - P_{K_h})S(u). \end{cases} \quad (18)$$

In order to formulate a Newton-like verification condition for (18), we need a Fréchet derivative of the operator SS. For most of the variational inequalities, the S in (9) is not Fréchet differentiable at all. Therefore, in order to use a Newton-like type method, a major difficulty in numerically solving the fixed point formulation u=S(u) is the treatment of the non-differentiable operator S. We need a suitable modification of the Fréchet derivative of S. Using some techniques, we can devise the approximate Fréchet derivative of S. Hence we shall assume that $\widetilde{D}S(u)$ is the approximate Fréchet derivative of the S(u) at uu as the linear operator. Let $\widetilde{D}S(u)$ be designated as the Fréchet-like derivative of S at u.

To consider the Newton-like operator for (18), we define the nonlinear operator N_h:V→V_h as

$$N_h(u) \equiv P_{K_h} u - \left[I - \widetilde{D}S(u_h)\right]_h^{-1} (P_{K_h} - P_{K_h} S)(u).$$

Here II is the identity operator and $\left[I - \widetilde{D}S(u_h)\right]_h^{-1}$ denotes the inverse on V_h of the restriction operator $\left[I - \widetilde{D}S(u_h)\right]\Big|_{V_h}$. Note that we will verify the existence of the inverse operator $\left[I-\widetilde{D}S(u_h)\right]_h^{-1}$ from the nonsingularity of the matrix corresponding to $\left[I-\widetilde{D}S(u_h)\right]\Big|_{V_h}$ in actual calculations.

Next we define the operator T:V→V as follows:

$$T(u) \equiv N_h(u) + (I - P_{K_h})S(u). \tag{19}$$

Then T is considered as the Newton-like operator for the former part of (18), but as the simple iterative operator for the latter part. T becomes a compact and continuous map on V by properties of S. Using some techniques, for a given problem we can not only define the Newton-like operator, but also devise a Newton-like Method. Furthermore, we obtain the following proposition and theorem.

Proposition 3. *Given the assumption that* $N_h(u) \in K_h$,

$$u = S(u) \Leftrightarrow u = T(u). \tag{20}$$

Theorem 4. *If there exists a nonempty, bounded, convex, and closed subset* U⊂K *such that* T(U)={T(u)|u∈U}⊂U, *then by the Schauder fixed point theorem, there exists a solution* u∈U *of* u=S(u).

When we decompose the set U as $U=U_h \oplus U_\perp$ in Theorem 8.1, where $U_h \subset K_h$ and $U_\perp \subset K_h^*$, the verification condition can be written by

$$\begin{cases} N_h(U) \subset U_h, \\ (I - P_{K_h})S(U) \subset U_\perp. \end{cases} \tag{21}$$

Here, U_h is represented as the linear combination of the base functions of V_h with interval coefficients, whereas $U_\perp$ is the intersection of K_h^* with a ball in V. That is,

$$U_h = \left\{ \varphi_h \in K_h : \varphi_h = \sum_{j=1}^{M} A_j \phi_j \ \text{ with } \ a_j \in \left[\underline{A_j}, \overline{A_j}\right] \right\},$$

$$U_\perp = \{\varphi \in K_h^* : \|\varphi\|_V \le \alpha\},$$

respectively.

Note that $N_h(U)$ can be directly computed from U_h and $U_\perp$ with additional information on the a priori error estimates. On the other hand, $(I - P_{K_h})S(U)$ is evaluated using (14), by the following constructive error estimates for the finite approximate solution of variational inequality (8):

$$\left\|(I - P_{K_h})S(U)\right\|_V \le Ch \sup_{u \in U} \|f(u)\|_{L^2(\Omega)}.$$

Therefore, the former condition in (21) is validated as the inclusion relations of corresponding coefficient intervals; the latter part can be checked by comparing two nonnegative real numbers.

Next we show a computer algorithm to construct the set U which satisfies the verification condition (21). In order to realize it, we use the iteration method described in Subsection 2.2. Similarly to that in Subsection 2.2, we now generate the following iteration sequence $\left\{\left(U_h^{(n)},\alpha_n\right)\right\}$ for n=0,1,2,···. For n≥1, the δ-inflation of $\left(U_h^{(n-1)},\alpha_{n-1}\right)$ is denoted by $\left(\widetilde{U}_h^{(n-1)},\widetilde{\alpha}_{n-1}\right)$. Next, for the set $\widetilde{U}^{(n-1)}=\widetilde{U}_h^{(n-1)}\oplus[\widetilde{\alpha}_{n-1}]$, define $\left(U_h^{(n)},\alpha_n\right)$ by

$$\begin{cases} U_h^{(n)}\supset N_h\left(\widetilde{U}^{(n-1)}\right), \\ \alpha_n=Ch\sup\limits_{u\in\widetilde{U}^{(n-1)}}\parallel f(u)\parallel_{L^2(\Omega)}. \end{cases} \tag{22}$$

Finally, the verification condition in a computer is given by the following theorem. The proof of Theorem 4 will be given here for the sake of completeness; it is based on Proposition 3 and Schauder's fixed point theorem.

Theorem 5. *For an integer* NN, *if two relationships*

$$U_h^{(N)}\subset\widetilde{U}_h^{(N-1)}\quad and\quad \alpha_N<\widetilde{\alpha}_{N-1} \tag{23}$$

hold, then there exists a solution u of (8) in $U_h^{(N)}\oplus[\alpha_N]$. Here, the first term of (21) means the strict inclusion in the sense of each coefficient interval of $U_h^{(N)}$ and $\widetilde{U}_h^{(N-1)}$.

APPLICATIONS

The study for the numerical verification method for elliptic variational inequalities has been still made less progress than for the differential equation case. The author's method in the present chapter can be also applied, in principal, to the verification of solutions of the practical problems. Namely, in Section 3.1, we first give, a slightly detailed descroption of the basic principle and formulation of our numerical verification method for the solution of obstacle problems with a homogeneous condition. This should be an appropriate introduction to another applications of our idea. The basic approach of the method consisits of the fixed point formulation of the problems and construction of the function set, in a computer, satisying the validation condition of a certain infinite dimensional fixed point theorem.

We also mention that it is possible to extend the method to more general problems with non-homogenerous obstacles. Moreover, in order to apply our method to the problem whose associated operator is not retractive in a neighborhood of the solution, a Newton-like method is introduced. Next, in Section 3.2, we apply our method to another type of free boundary problem with appears in the elasto-plastic deformation theory. This problem causes some properties of non-smoothness in tha associated finite dimensional equations. But, we can also overcome such a difficulty by appling the solution method for non-smooth problems developed by [29, 32, 33]. In the Section 3.3, we briefly remark that our enclosure method can also be applied to the so-called simplified Signorini problem which is a simplified version of a problem accurring in the elasticity theory [43]. Finally, in Section 3.4, we show the way to apply our approach to elliptic variational inequalities of the second kind appearing in the flow problems of a viscos-plastic fulied in a pipe.

Obstacle Problems

We introduce the verification method for solutions of the obstacle problem which is known as a free boundary problem to cahracterize the contacted zone by an obstacle ψ in an elastic membrane region.

Homogeneous case

Here, 'homogeneous' stands for the case that obstacle $\psi \equiv 0$ in the whole domain.

Basic formulation of verification

Though the basic idea of verification is given in other places [26, 27, 28], in order to keep the paper as self-contained as possible, we describe rather detailed formulation and verification procedure for the present case.

Let Ω be a bounded convex domain in $\mathbb{R}^n$, $1 \leq n \leq 2$, with piecewise smooth boundary $\partial\Omega$. We set $V \equiv H_0^1(\Omega) = \{v \in H^1(\Omega) : v|_{\partial\Omega} = 0\}$ and
$a(u, v) = (\nabla u, \nabla v)$

which is adopted as the inner product on V, where $(\cdot,\cdot)$·· stands for the inner product on $L^2(\Omega)$. We define $K := \{v \in V : v \geq 0 \text{a.e.on} \Omega\}$.

First, we note that, by well-known result [49], for any $g \in L^2(\Omega)$, the problem:

$$a\left(u,v-u\right)\geq\left(g,v-u\right),\quad \forall v\in K,\quad u\in K, \tag{24}$$

has a unique solution $u\in V\cap H^2(\Omega)$, and the estimate

$$\left|u\right|_{H^2(\Omega)}\leq\left\|g\right\|_{L^2(\Omega)} \tag{25}$$

holds [49], where $|w|_{H2}$ implies the semi-norm of w in $H^2(\Omega)$ defined by

$$\left|w\right|_{H^2(\Omega)}^2\equiv\sum_{i,j=1}^{n}\left\|\frac{\partial^2 w}{\partial x_i\partial x_j}\right\|_{L^2(\Omega)}^2.$$

Now consider the following elliptic variational inequalities with nonlinear right-hand side;

$$\begin{cases}\text{Find } w\in K \text{ such that}\\ a\left(w,v-w\right)\geq\left(f\left(w\right),v-w\right),\quad \forall v\in K.\end{cases} \tag{26}$$

We take an appropriate finite dimensional subspace V_h of V for $0<h<1$. Usually, V_h is taken to be a finite element subspace with mesh size h. We then define K_h, an approximation of K, by

$$\begin{cases}\text{Find } w\in K \text{ such that}\\ a\left(w,v-w\right)\geq\left(f\left(w\right),v-w\right),\quad \forall v\in K.\end{cases}$$

We also define the projection P_K from V onto K. That is, $v=P_K(w)$, the projection of $w\in V$ into K, is defined as the unique solution of the following problem:

$$v\in K:\quad a\left(v,\zeta-v\right)\geq a\left(w,\zeta-v\right),\quad \forall\zeta\in K. \tag{27}$$

And define the projection P_{Kh} from V onto K_h. That is, $vh=P_{Kh}(w)$, the projection of w into K_h, is defined as follows:

$$v_h\in K_h:\quad a\left(v_h,\zeta-v_h\right)\geq a\left(w,\zeta-v_h\right),\quad \forall\zeta\in K_h. \tag{28}$$

Now, as one of the approximation properties of K_h, assume that.

For each $w\in K\cap H^2\left(\Omega\right)$, there exists a positive constant C_1, independent of h, such that

$$\left\|w-P_{K_h}w\right\|_V\leq C_1 h\left|w\right|_{H^2(\Omega)}. \tag{29}$$

Here, C_1 has to be numerically determined. For example, it is known that we may take $C_1=\frac{\sqrt{5}}{\pi}$ for the linear element in one dimensional case [27]. Furthermore, it will be readily seen that the same constant can be taken for the two dimensional bilinear element from the consideration on the proof of Theorem 5.1 in [27]. To verify the existence of a solution of (26) in a computer, we use the fixed point formulation.

First, note that, for each $w\in V$, there exists a unique $F(w)\in V$ such that

$$\left(\nabla F\left(w\right),\nabla v\right)=\left(f\left(w\right),v\right),\quad \forall v\in V, \tag{30}$$

which also implies that

$$\begin{cases} -\Delta F(w) = f(w) & \text{in } \Omega, \\ F(w) = 0 & \text{on } \partial\Omega. \end{cases} \quad (31)$$

Then the map $F:V \to V$ is compact. By (30), the problem (26) is equivalent to finding $w \in V$ such that

$$a(w, v-w) \geq a(F(w), v-w), \qquad \forall v \in K. \quad (32)$$

Using the definition (27) and (32), we now have the following fixed point problem for the compact operator $P_K F$.

$$\text{Find } \exists w \in V \text{ such that } w = P_K F(w). \quad (33)$$

Verification condition

We introduce two concepts, rounding and rounding error, which enable us to deal with the infinite dimensional problem by finite procedures, that is, in a computer.

Now we define the dual cone of K_h by

$$K_h^* = \{w \in V : a(w,v) \leq 0, \quad \forall v \in K_h\},$$

and note that K_h^* is also closed convex cone in V with vertex at 0 which is the only point common to K_h and K_h^*. From (28) it follows that K_h^* is the set of points whose projections into K_h is 00.

Lemma 6. *Any* $w \in V$ *can be uniquely decomposed into the sum of two orthogonal elements. That is,*

$$w = P_{K_h} w \oplus (I - P_{K_h})w = P_{K_h} w \oplus P_{K_h^*} w.$$

Here, $\oplus$ *denotes the sum of two orthogonal elements in the sense of* V.

For any $w \in V$, we now define the rounding $R(P_K F(w)) \in K_h$ by the solution of the following problem:

$$a(R(P_K F(w)), v_h - R(P_K F(w))) \geq (f(w), v_h - R(P_K F(w))), \quad \forall v_h \in K_h.$$

Next, for any subset $W \subset V$, we define the rounding $R(P_K FW) \subset K_h$ by

$$R(P_K FW) = \{w_h \in K_h : w_h = R(P_K F(w)), \quad w \in W\}.$$

Usually, $R(P_K FW)$ is enclosed and represented as a linear conbination of the base functions in V_h with interval coefficients.

Moreover, for $W \subset V$, we define $RE(P_K FW)$, the rounding error of $P_K FW$, as a subset of K_h^*, that is,

$$RE(P_K FW) = \{v \in K_h^* : \| v \|_V \leq C_0 h \| f(W) \|_{L^2}\}, \tag{34}$$

where

$$\| f(W) \|_{L^2} \equiv \sup_{w \in W} \| f(w) \|_{L^2}.$$

Here, $C_0 \equiv C_1 C_2$, where C_1 is the same positive constant as in (29), and C_2 is determined by the following regularity estimate for the solution to (24) of the form

$$|u|_{H^2} \leq C_2 \|g\|_{L^2}. \tag{35}$$

Thus we may take as $C_2=1$ for the present case from (25). Then, we have

$$P_K F(w) - R(P_K F(w)) \in RE(P_K F(w)), \quad \forall w \in W.$$

Therefore, the following verification condition is obtained by Schauder's fixed point theorem.

Lemma 7.*If there exists a nonempty, bounded, convex, and closed subset* $W \subset K$ *such that*

$$R(P_K FW) \oplus RE(P_K FW) \subset W, \tag{36}$$

then there exists a solution of $w=P_K F(w)$ in W.

We sometimes refer the above set W as *a candidate set*, which we generate in computer so that it satisfies the condition (36).

Verification procedures

We describe the method to find a set W satisfying (36) in the below.

Consider the following approximate solution $w_h \in K_h$ of (24):

$$a(w_h, v_h - w_h) \geq (g, v_h - w_h), \quad \forall v_h \in K_h, \quad w_h \in K_h. \tag{37}$$

Since the bilinear form a(·,·)a·· is symmetric, (37) is reduced to the quadratic programming problem:

$$\min_{v \in K_h} \left[\frac{1}{2} a(v, v) - (g, v)\right. . \tag{38}$$

Let $\{\phi_j\}_{j=1 \cdots M}$ be a basis of V_h with usual linear functions such that $\phi_j(x) \geq 0, \quad \forall x \in \Omega$ and satisfying

$$\phi_j(x_i) = \begin{cases} 1, & i = j, \\ 0, & i \neq j, \end{cases}$$

where x_i is an interior node of the finite element mesh. Then (38) reduces to the following vector form:

$$\min_{w\geq 0}\left[\frac{1}{2}w'Dw - P'w\right], \tag{39}$$

where w≥0 means the componentwise relation. Here, $D := (d_{ij})_{1\leq i,j\leq M}$ with $d_{ij} = (\nabla\phi_i, \nabla\phi_j)$, and w is the coefficient vector with $\{\phi_j\}$ of the function v in (38). Also, $P:=((g,\phi_j))_{1\leq j\leq M}$.

Furthermore, we define for any $\alpha \in R^+$, nonnegative real number, we set $[\alpha] \equiv \{\phi \in K_h^*; \quad \|\phi\|_V \leq \alpha\}$.

Then, for a given candidate set $W=W_h\oplus[\alpha]$ with $W_h \subset K_h$, the computation of the rounding $R(P_K FW)$ reduces to enclose an interval vector $Z=(Z_j)$ and $Y=(Y_j)$ satisfying the following nonlinear system of equations [27]:

$$\begin{cases} Y - DZ = -(f(W), \phi_j), & 1 \leq j \leq M, \\ Y_j Z_j = 0, & 1 \leq j \leq M. \end{cases} \tag{40}$$

Here, $(f(W),\phi_j)$ is evaluated as an interval B_j such that $\{(f(w),\phi_j)|w\in W\}\subset Bj$.. In order to solve (40) with guaranteed accuracy, we use some interval approaches for nonlinear system of equations [19, 20]. Thus, using the solution of (40), we can enclose the set $R(P_K FW)$ in (36). Combining this with (34), we can successfully compute the left-hand side of (36) for any candidate set $W=W_h\oplus[\alpha]$.

Thus we can present a computational verification condition. In the actual computation, we use an iterative procedure with δ-inflationinflation technique to find the set W satisfying (36). Several numerical examples for verification are presented in [27] for one dimensional problem using linear finite element.

Non-homogeneous case

In this subsection, we consider the two-dimensional case. In order to verify solutions numerically, it is necessary to determine some constants that appear in the a priori error estimates. For the non-homogeneous case, we define $K:=\{v\in V: v\geq\psi \text{ a.e. on } \Omega\}$, where ψ is a given $H^2(\Omega)$ function such that ψ≤0 on ∂Ω and is not identically equal to 0. Let Ω be a square with side 1 and let $\mathscr{T}_h$ be the uniform triangulation of Ω. We introduce $\Sigma_h = \{p; \quad p\in\overline{\Omega}, p \text{ is a vertex of } T\in\mathscr{T}_h\}$ and define the approximate V_h of $H_0^1(\Omega)$ by $V_h = \{v_h; \quad v_h \in H_0^1(\Omega)\cap C^0(\overline{\Omega}), v_h|_T \in P_1,$

$\forall T \in \mathscr{T}_h\}$. Here, $v_h|_T$ denotes the restriction of v_h to T and P_1 representing the space of polynomials in two variables of degree ≤1. It is then quite natural to approximate K by

$$K_h = \{v_h \in V_h; v_h(p) \geq \psi(p), \forall p \in \Sigma_h\}.$$

Note that, in general, $K_h \neq V_h \cap K$. Then, P_K and P_{K_h} are similarly defined as before, and we also have the constructive error estimates of the form, $\forall v_h \in K_h$ and $\forall v \in K$,

$$\|u_h - u\|_{H_0^1(\Omega)} \leq C(g, \psi, h), \tag{41}$$

where,

$$C(g,\psi,h) \leq \sup_{g \in L^2(\Omega)} \sqrt{(0.494)^2 h^2 |u|_{H^2}^2 + 2(\|g\|_{L^2} + \|Au\|_{L^2})\left((0.494)^2 h^2 |u|_{H^2} + 6h^2 |\psi|_{H^2}\right)}.$$

We provide a numerical example of verification in the two-dimensional case according to the procedures described in the previous section. Let Ω=(0,1)×(0,1). We consider the case $f(u) = Ku + \sin \pi x \sin 2\pi y$ and $\psi = \sin \pi x \sin \pi y$. For simplicity, we only consider the uniform mesh here. First, we divide the domain into small triangles with a uniform mesh size h and choose the basis of V_h as the pyramid functions.

The execution conditions are as follows (Figures 1–3):

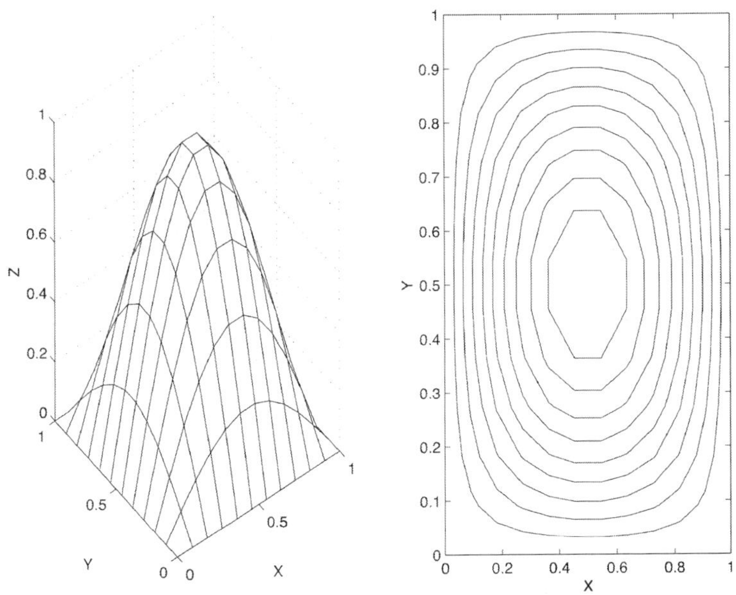

Figure 1: Obstacle function ψ.

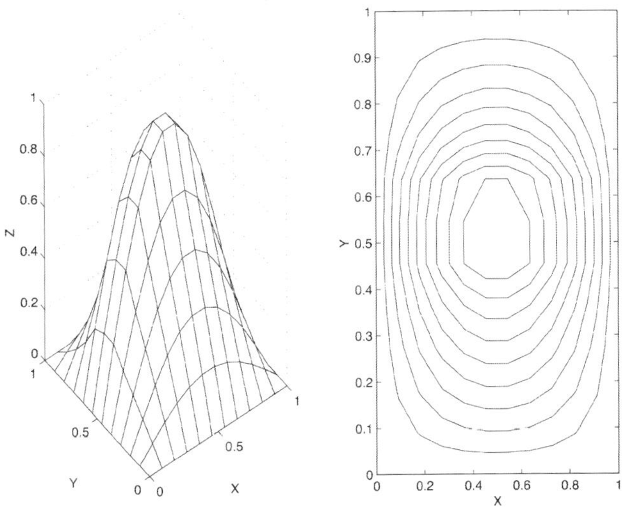

Figure 2: Approximate solution $u_h^{(0)}$.

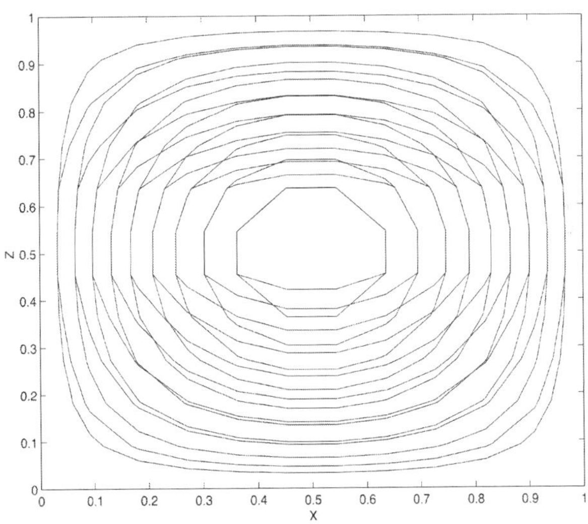

Figure 3: Illustration of the contact zone.

K=0.1, dimV_h=10

Obstacle function $\psi = \sin \pi\, x \sin \pi\, y$

The outline of ψ is shown in Figure1..

Initial $u_h^{(0)} = \text{Galerkin approximation}, \alpha_0 = 0$

The outline of $u_h^{(0)}$ is shown in Figure2.

Illustration of contact zone between obstacle and approximate solution is shown in Figure3.

Extension parameters:δ=10^{-5}.

Results are as follows:

Iteration numbers for verification : 2

$H_0^1(\Omega)$ – error bound : 0.15437

Maximum width of coefficient intervals in $\left\{A_j^{(N)}\right\} = 0.00001$.

Detailed arguments and with numerical examples are presented in [42].

A Newton-type verification method

The idea of the enclosure method for solutions of obstacle problems is based upon simply sequential iterations for the original fixed point operator $P_K F$. Therefore, it is difficult to apply the method to the problem of which associated operator is not retractive in a neighborhood of the solution. In order to overcome such a difficulty, we introduce an another formulation using a Newton-like operator. The essential point is the way to devise the Newton-like operator for a kind of non-differentiable map which defines the original problem.

To formulate a Newton-type verification condition, we need a Fréchet derivative of the operator $P_K F$. However, $P_K F$ is not Fréchet differentiable at all. Therefore, we define the approximate Fréchet-like derivative $\widetilde{D}_K F(u_h)$ on V_h for some $u_h \in K_h$ instead of the Fréchet derivative. Assume that $\{\phi_j\}_{j=1\cdots M}$ is a basis of V_h, where $M=\dim V_h$, such that $\phi_j(x) \geq 0$ on Ω and satisfying

$$\phi_j(x_i) = \begin{cases} 1, & i = j, \\ 0, & i \neq j, \end{cases}$$

where x_i is an interior node of the finite element mesh.

And, for $v_h \in V_h$, we represent it such as

$v_h = \sum_{j=1}^{M} v_{hj}\phi_j.$

Here, $(v_{hj})_{j=1,\cdots,M}$ is called as the coefficient vector of v_h. Now we take a fixed subset $N_0 \subset \{1,2,\cdots,M\}$, define V_h,N_0, the closed subspace of V_h, by
$V_{h,N_0} = \{v_h | v_h \in V_h, v_{hj} = 0 \ \ for \ \ j \notin N_0\}.$

And let P_{h,N_0} be a H_0^1-projection from V onto V_h,N_0 defined by
$a(u - P_{h,N_0}u, v) = 0, \quad \forall v \in V_{h,N_0}, P_{h,N_0}u \in V_{h,N_0}.$

In order to define $\widetilde{D}_K F(u_h) : V_h \to V_{h,N_0}$, we differentiate the first equation of (40) in W at $W=u_h$ to get, for arbitrary $\delta \in V_h$,

$$\partial Y^* - D\partial Z^* = -\{(f'(u_h)\delta, \phi_j)\}_{1 \le j \le M}. \tag{42}$$

Here, $\partial Y^* = \left(\widetilde{Y}_j^*\right)_{1\le j\le M}$ and $\partial Z^* = \left(\widetilde{Z}_j^*\right)_{1\le j\le M}$, where $\widetilde{Y}_j^* = 0$ for $j \in N_0$ and $\widetilde{Z}_j^* = 0$ for $j \notin N_0$, respectively.

Then we define the approximate Fréchet-like derivative of $P_K F(u)$ at $u=u_h$, as the linear map $\widetilde{D}_K F(u_h) : V_h \to V_{h,N_0}$ such that, for each $\delta \in V_h$,
$\widetilde{D}_K F(u_h)(\delta) := \sum_{j=1}^{M} \widetilde{Z}_j^{\,*} \phi_j.$

We now assume that.

A4. The restriction to V_h,N_0 of the operator $P_{h,N_0}\left[I - \widetilde{D}_K F(u_h)\right] : V_h \to V_{h,N_0}$ has the inverse operator

$$\left[P_{h,N_0} - \widetilde{D}_K F(u_h)\right]_h^{-1} : V_{h,N_0} \to V_{h,N_0}.$$

Here, II means the identity map on V_h.

By using the above approximate Fréchet-like derivative, we define the Newton-like operator N_h:V→V_h by

$$N_h(w) \equiv P_{K_h}w - \left[P_{h,N_0} - \widetilde{D}_K F(u_h)\right]_h^{-1} P_{h,N_0}(P_{K_h} - P_{K_h}P_K F)(w)\Big).$$

Next we define the operator T:V→V as follows:
$T(w) \equiv N_h(w) + (I - P_{K_h})P_K F(w).$

Then T becomes a compact map on VV and it follows the fixed point problem $w=P_K Fw$ is equivalent to $w=T(w)$. Detailed arguments and with numerical examples are presented in [35].

Elasto-plastic Torsion Problems

In this subsection, we consider an enclosure metnod of solutions for elasto-plastic torsion problems governed by an elliptic variational inequalities [25, 32, 33]. The nonlinear elasto-plastic torsion problem is defined as the same type elliptic variational inequalities as (26) with

$$K := \{v \in H_0^1(\Omega); |\nabla v| \leq 1 \quad \text{a.e. on} \quad \Omega\}. \tag{43}$$

As is well known [56, 58], two sub-domains Ω_p and Ω_e defined by

$$\Omega_p = \{x; x \in \Omega, |\nabla u| = 1\},$$

and

$$\Omega_e = \Omega \backslash \Omega_p = \{x; x \in \Omega, |\nabla u| < 1\}$$

correspond to the plastic and elastic regions, respectively. The elastic region Ω_e and the plastic region Ω_p are not known beforehand and should be determined, therefore $\partial\Omega_e \cap \partial\Omega_p$ is actually the free boundary of the problem (26). The problem (26) has been formulated as the problem of finding u satisfying

$$\begin{cases} -\Delta u = f(u) & \text{in} \quad \Omega_e, \\ |\nabla u| = 1 & \text{in} \quad \Omega_p, \\ u = 0 & \text{on} \quad \partial\Omega. \end{cases} \tag{44}$$

The finite dimensional convex subset K_h is also defined similarly as before:

$$K_h := V_h \cap K = \{v_h | \; v_h \in V_h, \; ||\nabla v_h| \leq 1 \text{ a.e. on } \Omega\}. \tag{45}$$

In order to formulate the verification procedure, we need a verified computational method for solving the finite dimensional part (rounding) and a constructive estimates for infinite dimensional part (rounding error) as in the previous subsection.

Following [49, 56], we define the Lagrangian functional $\mathscr{L}$ associated with (1) by

$$\mathscr{L}(v,\mu) = \frac{1}{2}\int_\Omega |\nabla v|^2 dx - (g,v) + \frac{1}{2}\int_\Omega \mu\left(|\nabla v|^2 - 1\right) dx.$$

It follows, from [49, 56], that if $\mathscr{L}$ has a saddle point $\{u,\lambda\} \in H_0^1(\Omega) \times L_+^\infty(\Omega)$, then u is a solution of (1), where $L_+^\infty(\Omega) = \{q \in L^\infty(\Omega); q \geq 0 \;\; a.e. \;\; in \;\; \Omega\}$. We use the Uzawa algorithm to solve (1). Thus we can claculate the rounding $R(P_K F(W))$, for a candidate set W, by solving the following problem with guaranteed error bounds:

$$\begin{cases} \text{Find } \{u_h, \lambda_h\} \in K_h \times \Lambda_h \quad \text{such that} \\ \lambda_h = \max\left[\lambda_h + \rho\left(|\nabla u_h|^2 - 1\right), 0\right] \quad \text{with } \rho > 0. \\ \int_\Omega (1+\lambda_h)\nabla u_h \cdot \nabla v_h dx = (f(W), v_h), \forall v_h \in V_h, \quad u_h \in V_h, \end{cases} \tag{46}$$

The problem (46) can be formulated as a system of nonlinear and nonsmooth (nondifferentiable) equations. A verification method for nonsmooth equations by a generalized Krawczyk operator is studied in [1, 55]. We briefly describe the method presented by [55] in the below.

We consider the following equivalent system of nonlinear(and nondifferentiable) equation to (46) for a fixed w∈W

$$H(x) = 0. \tag{47}$$

Here, we assume that $H:R^n \rightarrow R^n$ is locally Lipschitz continuous. The equivalence means that x^* solves (46) if and only if x^* solves (47). The method is based on the mean value theorem for local Lipschitz functions of the form

$$H(x) - H(y) \in co\,\partial H([x])(x-y), \quad \text{for all} \quad x, y \in [x],$$

where [x] stands for an interval vector, "co" denotes the convex hull, and ∂H the generalized Jacobian in Clarke's sense [57], which is also considered as a slope function, and

$$co\,\partial H([x]) := co\{V \in \partial H(x); \quad x \in [x]\}.$$

Let $[L_{[x]}]$ be an interval matrix such that $co\,\partial H([x]) \subseteq [L_{[x]}]$. Then for any $x, y \in [x] \subseteq R^n$ it holds that $H(x) - H(y) \in [L_{[x]}](x-y)$.

Then an interval operator for nonsmooth equations is defined by

$$G(x, A, [x]) := x - A^{-1}H(x) + \left(I - A^{-1}[L_{[x]}]\right)([x] - x). \tag{48}$$

The mapping G(x,A,[x]) is called a generalized Krawczyk operator. Therefore, the verification condition of solutions for (46) in [x] is given by

$$G(x, A, [x]) \subseteq [x] \subset D.$$

Thus, we can compute the solution of (46) with guaranteed accuracy. That is, we can enclose the rounding $R(P_K F(U))$. On the other hand, in order for the calculation of the rounding error $RE(P_K F(U))$, the similar arguements can also be applied for one dimensional problem. Actually, we can prove that the same constant $C_0 = \frac{\sqrt{5}}{\pi}$ is also valid for the present problem in one dimensinal case, which implies that we can give a verification procedure besed on the same principle as before [25, 32, 33]. In [33], we extended the approach to the numerical proof of existence of solutions for elasto-plastic torsion problems as well as gave a numerical example for one dimensional case. The verification method in [33] is based on the generalized Krawczyk operator for solving a system of nonsmooth (nondifferentiable) equations. In order to use the generalized Krawczyk operator, we need to calculate the Jacobian. In that case, we need some complicated techniques. However, in many cases, calculating the generalized Jacobian is very difficult. To overcome such difficulties, we proposed a numerical verification method without using the generalized Krawczyk operator. This method is attractive, since calculating the generalized Jacobian is not required in the computational performance. Furthermore, up to know, our verification methods are mainly based on the enclosure of solutions in the sense of L^2 or H^1 norms. We considered a numerical verification method with guaranteed L^∞ error bounds for the solution of elasto-plastic torsion problem.

Simplified Signorini Problems

A simplified Signorini problem is also given by the elliptic variational inequalities of the form (26) with

$$K := \{v \in H_0^1(\Omega); \mid v \geq 0 \quad \text{on} \quad \partial\Omega\} \tag{49}$$

and

$$a(u, v) = \int_\Omega \nabla u \cdot \nabla v dx + \int_\Omega uv dx. \tag{50}$$

where

$$\nabla u \cdot \nabla v = \frac{\partial u}{\partial x_1}\frac{\partial v}{\partial x_1} + \frac{\partial u}{\partial x_2}\frac{\partial v}{\partial x_2}.$$

As well known, the solution u of this elliptic variational inequalities can be characterized as a solution of the following free boundary problem finding u and two subsets Γ_0 and Γ_+ such that $\Gamma_0 \cup \Gamma_+ = \partial\Omega$ and $\Gamma_0 \cap \Gamma_+ = \emptyset$

$$\begin{cases} -\Delta u + u = f(u) \quad \text{in} \quad \Omega, \\ u = 0 \quad \text{on} \quad \Gamma_0, \frac{\partial u}{\partial n} \geq 0 \quad \text{on} \quad \Gamma_0, \\ u > 0 \quad \text{on} \quad \Gamma_+, \frac{\partial u}{\partial n} = 0 \quad \text{on} \quad \Gamma_+, \end{cases} \tag{51}$$

where $\frac{\partial}{\partial n}$ the outer normal derivative on $\partial\Omega$. In the present case, the approximation subspace K_h is taken as

$$K_h := V_h \cap K = \{v_h | \quad v_h \in V_h, \quad v_h \geq 0 \quad \text{on} \quad \partial\Omega\}. \tag{52}$$

For a candidate set W, the computation of rounding $R(P_K F(W))$ is also reduced to the quadratic programming problem as in the Section 3.1 [56].

Since the constant C_2 in (25) is easily estimated as $C_2=1$, the standard approximation property of the interpolation by K_h gives a constructive error estimates to compute the rounding error $RE(P_K F(W))$. For a simplified Signorini problem [43], we constructed a computing algorithm which automatically encloses the solution within guaranteed error bounds. In particular, the method proposed in [43] enables us to verify the free boundary of a simplified Signorini problem, which has been impossible so far. Concerning the numerical verification of solutions for elliptic variational inequalities, we would like to mention that the inclusion method described in this article can be applied to the solution of the elliptic variational inequalities on large space domains.

Some Other Problems

In this subsection, we show that our idea of verification method can also be applied to the elliptic variational inequalities of the second kind.

Now, we define the functional $j(v)=\int\Omega|\nabla v|dx$.. We consider the following problem of the flow of a viscous plastic fluid in a pipe:

$$\begin{cases} \text{Find } u \in H_0^1(\Omega) & \text{such that} \\ a(u, v-u) + j(v) - j(u) \geq (f(u), v-u), & \forall v \in H_0^1(\Omega). \end{cases} \tag{53}$$

As in the previous section, we consider the following auxiliary problem associated with (53) for a given $g \in L^2(\Omega)$:

$$a(u, v-u) + j(v) - j(u) \geq (g, v-u), \forall v \in H_0^1(\Omega), u \in H_0^1(\Omega). \tag{54}$$

By the well known result, we have the following lemma.

Lemma 8. There exists a unique solution $u \in H_0^1(\Omega) \cap H^2(\Omega)$ of (54) for any $g \in L^2$, such that

$$\|u\|_{H^2(\Omega)} \leq \hat{C} \|g\|_{L^2(\Omega)}.$$

When we denote the solution u of (54) by u=Ag and define the composite map F on $H_0^1(\Omega)$ by F(u)≡Af(u), which is a little bit of different from the previously appeared symbol F in Section 2, we have.

Theorem 9. F is compact on $H_0^1(\Omega)$ and the problem (53) is equivalent to the fixed point problem

u=F(u).

Proof. First, for a bounded subset U⊂$L^2(\Omega)$, we show that AU⊂$H_0^1(\Omega)$ is relatively compact. Secondly, prove that A:$L^2(\Omega)\rightarrow H_0^1(\Omega)$ is continuous. By Lemma 3, AU⊂$H^2(\Omega)\cap H_0^1(\Omega)$ and AU is bounded in $H^2(\Omega)$. Since UU is bounded in $L^2(\Omega)$, by the Sobolev imbedding theorem, we have AU is relatively compact in $H_0^1(\Omega)$. Next, for arbitrary $f_1,f_2\in L^2(\Omega)$, setting $u_1=Af_1$ and $u_2=Af_2$, by using (54), we obtain

$$a(u_1, u_2 - u_1) + j(u_2) - j(u_1) \geq (f_1, u_2 - u_1),$$
$$a(u_2, u_1 - u_2) + j(u_1) - j(u_2) \geq (f_2, u_1 - u_2).$$

With the above inequalities, we obtain $a(u_2 - u_1, u_2 - u_1) = -a(u_1, u_2 - u_1) + a(u_2, u_2 - u_1) \leq j(u_2) - \mathscr{T}_h$

Hence, by the Poincaré inequality, we have

$$\| u_2 - u_1 \|^2_{H_0^1(\Omega)} \leq \| f_2 - f_1 \|_{L^2(\Omega)} \| u_2 - u_1 \|_{L^2(\Omega)} \leq \overline{C} \| f_2 - f_1 \|_{L^2} \| u_2 - u_1 \|_{H_0^1(\Omega)}.$$

Therefore, we obtain

$$\| u_2 - u_1 \|_{H_0^1(\Omega)} \leq \overline{C} \| f_2 - f_1 \|_{L^2(\Omega)}.$$

That is, A is Lipschitz continuous as a map $L^2(\Omega) \rightarrow H_0^1(\Omega)$. Hence A is compact. The latter half in the theorem is straightforward from the definition of F.

We now define the approximate problem corresponding to (54) as

$$a(u_h, v_h - u_h) + j(v_h) - j(u_h) \geq (g, v_h - u_h), \forall v_h \in V_h, u_h \in V_h. \tag{55}$$

In order to apply our verification method to enclose the solutions of (53), we need a guaranteed computation of the exact solution of the problem (55), *a rounding procedure*, as well as the constructive error estimates between the solution of (54) and (55), *rounding error estimates*.

A major difficulty in solving the problem (55) numerically is the processing of the nondifferentiable term $j(u) = \int_\Omega |\nabla u| \, dx$. One approach is the method of Lagrange multiplier on that term, whose continuous version is as follows [56].

Let us define $\Lambda = \{q \mid q \in L^2(\Omega) \times L^2(\Omega), |q(x)| \leq 1 \quad a.e. \quad x \in \Omega\}$ with $|q(x)| = \sqrt{q_1(x)^2 + q_2(x)^2}$. Then the solution u of (54) is equivalent to the existence of q satisfying

$$\begin{cases} a(u,v) + \int_\Omega q \cdot \nabla v = (g,v), \forall v \in H_0^1(\Omega), u \in H_0^1(\Omega), \\ q \cdot \nabla u = |\nabla u| \quad a.e. \quad , q \in \Lambda. \end{cases} \tag{56}$$

Moreover, it is known that (56) is equivalent to the following problem:

$$\begin{cases} a(u,v) + \int_\Omega q \cdot \nabla v = (g,v), \forall v \in H_0^1(\Omega), u \in H_0^1(\Omega), \\ q = \frac{q+\rho\nabla u}{\sup(1,|q+\rho\nabla u|)}. \end{cases} \tag{57}$$

Here ρ is a positive constant. Let $\mathscr{T}_h$ be a triangulation of Ω, and let define L_h and Λh (approximation of $L^\infty(\Omega) \times L^\infty(\Omega)$ and Λ, respectively) by

$$L_h = \left\{ q_h \middle| q_h = \sum_{T \in \mathscr{T}_h} q_T \chi_T, q_T \in R^2 \right\} \text{ and } \Lambda_h = \Lambda \cap L_h ,$$

where χ_τ is the characteristic function of τ.

Then our first purpose, computing the rounding RF(U), is to enclose the solution of the following approximation problem of (57):

$$\begin{cases} a(u_h, v_h) + \int_\Omega q_h \cdot \nabla v_h = (g, v_h), \forall v_h \in V_h, u_h \in V_h, \\ q_h = \frac{q_h + \rho\nabla u_h}{\sup(1,|q_h+\rho\nabla u_h|)}. \end{cases} \tag{58}$$

The Eq. (58) leads to a kind of finite dimensional, nonlinear but nondifferentiable problem. We use a slope function method proposed by Rump [18, 19, 20] to enclose the solutions of (58) with g=f(W) for a candidate set W. On the other hand, the rounding error RE(F(U)) can be computed by using the following constructive error estimates:

Theorem 10. Let u and u_h be solutions of (54) and (55), respectively. If $g \in L^2(\Omega)$, then there exists a constant C(h) such that

$$\|u_h - u\|_{H_0^1(\Omega)} \leq C(h) \|g\|_{L^2(\Omega)}.$$

Here, we may take $C(h) = \frac{\sqrt{5}}{\pi} h$ for the linear element in one dimensional case, and C is also numerically estimated such that $C(h) \approx O\left(h^{\frac{1}{2}}\right)$ for the two dimensional linear element. A proof of this theorem is described in Ryoo and Nakao [34]. Thus we can also implement the verification algorithm for the solution of (53) as in the previous section. For details on this subsection, please refer to Ref. [47].

CONCLUSIONS

We have surveyed numerical verification methods for differential equations, especially around partial differential equations, variational inequalities and the author's works. But the period of this research is shorter than the history of the numerical methods for differential equations by computer and we can say it is still in the stage of case studies. Indeed, recently, this kind of studies have been referred little by little for practical applications in PDEs and variational inequalities but there are many open problems to be resolve. Therefore, we can make no safe prediction that these approaches will grow into really useful methods for various kinds of equations and variational inequalities in mathematical analysis. Also, since the program description of the verification algorithm is very complicated in general, there is another problem like software technology associated with assurance for the correctness of the verification program itself. Actually, some of the mathematician would not give credit the computer assisted proof in analysis as correct as they believe the theoretical proof, which might cause a kind of seriously emotional problem in the methodology of mathematical sciences. And there is another difficulty from the huge scale of numerical computations which often exceed the capacity of the concurrent computing facilities.

However, in the twenty-first century, the computing environment would make more and more rapid progress, which should be beyond conception in the present state. In any case, a realistic study for partial differential equations and variational inequalities should be the future subject of the numerical computations with guaranteed accuracy. The authors believe that numerical methods with guaranteed accuracy for differential equations and variational inequalities would highly improve the reliability in the numerical simulation of the complicated phenomena in both mathematical and engineering sciences.

ACKNOWLEDGMENTS

This work was supported by the National Research Foundation of Korea (NRF) grant funded by the Korea government (MEST) (No. 2017R1A2B4006092).

REFERENCES

1. Chen X, Yamamoto T. On the convergence of some quasi-Newton methods for nonlinear equations with nondifferentiable operators. Computing. 1992;48:87-94
2. Minamoto T, Nakao MT. Numerical verifications of solutions for nonlinear parabolic equations in one-space dimensional case. Reliable Computing. 1997;3:137-147
3. Minamoto T. Numerical verifications of solutions for nonlinear hyperbolic equations in one-space dimensional case. Applied Mathematics Letters. 1997;10:91-96
4. Minamoto T, Yamamoto N, Nakao MT. Numerical verification method for solutions of the perturbed Gelfand equation. Methods and Applications of Analysis. 1998;7(200):251-262
5. Nagatou K, Yamamoto N, Nakao. An approach to the numerical verification of solutions for nonlinear elliptic problems with local uniqueness. Numerical Functional Analysis and Optimization. 1999;20:543-565
6. Nakao MT. A numerical approach to the proof of existence of solutions for elliptic problems. Japan Journal of Industrial and Applied Mathematics. 1988;5:313-332
7. Nakao MT. A numerical verification method for the existence of weak solutions for nonlinear boundary value problems. Journal of Mathematical Analysis and Applications. 1992;164:489-507
8. Nakao MT, Yamamoto N. Numerical6verifications of solutions for elliptic equations with strong nonlinearity. Numerical Functional Analysis and Optimization. 1991;12:535-543
9. Nakao MT, Watanabe Y. On computational proofs of the existence of solutions to nonlinear parabolic problems. Journal of Computational and Applied Mathematics. 1994;50:401-410
10. Nakao MT, Lee SH, Ryoo CS. Numerical verification of solutions for elasto-plastic torsion problems. Computers & Mathematics with Applications. 2000;39(200):195-204
11. Nakao MT, Yamamoto N, Kimura S. On the best constant in the error bound for the H10H01-projection into piecewise polynomial space. Journal of Approximation Theory. 1998;93:491-500

12. Nakao MT, Yamamoto N, Watanabe Y. A posteriori and constructive a priori error bounds for finite element solutions of Stokes equations. Journal of Computational and Applied Mathematics. 1998;91:137-158
13. Nakao MT, Ryoo CS. Numerical verifications of solutions for variational inequalities using Newton-like method. Information. 1999;2:27-35
14. Oishi S. Numerical verification of existence and inclusion of solutions for nonlinear operator equations. Journal of Computational and Applied Mathematics. 1995;60:171-185
15. Plum M. Computer-assisted existence proofs for two-point boundary value problems. Computing. 1991;46:19-34
16. Plum M. Numerical existence proofs and explicit bounds for solutions of nonlinear elliptic boundary value problems. Computing. 1992;49:25-44
17. Plum M. Computer assisted enclosure methods for elliptic differential equations. Journal of Linear Algebra and Its Applications. 2001;324:147-187
18. Rump SM. INTLAB-INTerval LABoratory. In: Csendes T, editor. Developments in Reliable Computing. Amsterdam: Kluwer Academic Publishers; 1999. pp. 77-104
19. Rump SM. In: Kulish UW, Miranker WL, editors. Solving Algebraic Problems with High Accuracy, A New Approach to Scientific Computation. New York: Academic Press; 1983. pp. 51-120
20. Rump SM. In: Herzberger J, editor. Verification Methods for Dense and Sparse Systems of Equations, Topics in Validated Computations. Amsterdam: North-Holland; 1994. pp. 63-136
21. Tsuchiya T, Nakao MT. Numerical verification of solutions of para me trized nonlinear boundary value problems with turning points. Japan Journal of Industrial and Applied Mathematics. 1997;14:357-372
22. Watanabe Y, Nakao MT, Yamamoto N. Verified computation of solutions for nondifferentiable elliptic equations related to MHD equilibra. Nonlinear Analysis. 1997;28:577-587
23. Yamamoto N. A numerical verification method for solutions of boundary value problems with local uniqueness by Banach's fixed point theorem. SIAM Journal on Numerical Analysis. 1998;35:2004-2013
24. Agarwal RP, Ryoo CS. Numerical verifications of solutions for obstacle problems. Computing Supplementum. 2001;15:9-19

25. Lee SH, Ryoo CS. A numerical verification of the approximate solutions for elasto-plastic torsion problem. Kyungpook Mathematical Journal. 1999;39:149-157
26. Ryoo CS. A computational verification method of solution with uniqueness for obstacle problems. Bulletin of Informatics and Cybernetics. 1998;30:133-144
27. Ryoo CS, Nakao MT. Numerical verification of solutions for variational inequalities. Numerische Mathematik. 1998;81:305-320
28. Ryoo CS. A priori estimates for the finite element approximation of an obstacle problem. Journal of Applied Mathematics and Computing. 2000;7:175-181
29. Ryoo CS. Numerical verification of solutions for obstacle problems using Newton-like method. Computers and Mathematics with Applications. 2000;39:185-194
30. Ryoo CS. Numerical verification of solutions for some unilateral problems. Applied Mathematics Letters. 2000;13:13-17
31. Ryoo CS. A computational verification method of exiatence of solution for elastoplastic torsion problem with uniqueness. Applied Mathematics Letters. 2000;13:25-29
32. Ryoo CS. Computational proofs of the existence of solutions to elasto-plastic torsion problems. Mathematical and Computer Modelling. 2000;32:289-298
33. Ryoo CS. Verified computation of solutions for obstacle problems with guaranteed L∞L∞ error bound. Applied Mathematics Letters. 2001;14:769-773
34. Ryoo CS, Nakao MT. Numerical verification of solutions for variational inequalities of the second kind. Computers and Mathematics with Applications. 2002;43:1371-1380
35. Ryoo CS, Agarwal RP. Numerical inclusion methods of solutions for variational inequalities. International Journal for Numerical Methods in Engineering. 2002;54:1535-1556
36. Ryoo CS, Song H, Kim SD. Numerical verification of solutions for some unilateral boundary value problems. Computers and Mathematics with Applications. 2002;44:787-797
37. Ryoo CS. Solving obstacle problems with guaranteed accuracy. Computers and Mathematics with Applications. 2003;45:823-834

38. Ryoo CS, Nakao MT. Numerical verification of solutions for obstacle problems. Journal of Computational and Applied Mathematics. 2003;161:405-416
39. Ryoo CS. A numerical verification of solutions of free boundary problems. Computers and Mathematics with Applications. 2004;48:429-435
40. Ryoo CS. Verification for existence of solutions for some generalized obstacle problems. Applied Mathematics Letters. 2005;18:635-640
41. Ryoo CS. Numerical enclosure methods to verify the existence of solutions to variational inequalities. Proceeding of the Jangjeon Mathematical Society. 2005;8:219-228
42. Ryoo CS. An approach to the numerical verification of solutions for obstacle problems. Computers and Mathematics with Applications. 2007;53:842-850
43. Ryoo CS. Numerical verification of solutions for Signorini problems using Newton-like method. International Journal for Numerical Methods in Engineering. 2008;73:1181-1196
44. Ryoo CS. A numerical approach to the proof of existence of solutions for some generalized obstacle problems. Applied Mathematics and Computation. 2010;216:3365-3369
45. Ryoo CS. An approach to the numerical verification of solutions for variational inequalities using Schauder fixed point theory. Boundary Value Problems. 2014;2014:235
46. Ryoo CS. Verified computations of solutions for some unilateral boundary value problems for second order equations, Journal of. Applied Mathematics and Informatics. 2021;39:295-302
47. Bazaraa MS, Shetty CM. Nonlinear Programming. New York: John Wiley; 1979
48. Falk RS. Error estimates for the approximation of a class of variational inequalities. Mathematics of Computation. 1974;28:963-971
49. Glowinski R. Numerical Methods for Nonlinear Variational Problems. New York: Springer; 1984
50. Kinderlehrer D, Stampacchia G. An Introduction to Variational Inequalities and Their Applications. New York: Academic Press; 1980
51. Mosco U. Convergence of convex sets and of solutions of variational inequalities. Advances in Mathematics. 1969;3:510-585

52. Mosco U, Strang G. One sided approximation and variational inequalities. Bulletin of the American Mathematical Society. 1974;80:308-312
53. Adams RA. Sobolev Spaces. New York: Academic Press; 1975
54. Adams E, Kulisch U. Scientific Computing with Automatic Result Verification. San Diego: Academic Press; 1993
55. Chen X. A verification method for solutions of nonsmooth equations. Computing. 1997;58:281-294
56. Ciarlet PG. The Finite Element Method for Elliptic Problems. Amsterdam: Noth–Holland; 1978
57. Clarke FH. Optimization and Nonsmooth Analysis. New York: John Wiley; 1983
58. Collatz L. The Numerical Treatment of Differential Equations. Berlin-Heidelberg: Springer; 1960
59. Zeidler E. Applied Functional Analysis: Applications to Mathematical Physics. Vol. 108. Berlin, Germany: Springer-Verlag; 1995
60. Zeidler E. Applied Functional Analysis: Main Principles and Their Applications. Vol. 109. Berlin, Germany: Springer-Verlag; 1995
61. Zeidler E. Nonlinear Functional Analysis and its Applications I. Berlin, Germany: Springer-Verlag; 1986

Chapter

ON TRIGONOMETRIC NUMERICAL INTEGRATOR FOR SOLVING FIRST ORDER ORDINARY DIFFERENTIAL EQUATION

4

A. A. Obayomi, S. O. Ayinde, O. M. Ogunmiloro

Department of Mathematics, Faculty of Science, Ekiti State University, Ado-Ekiti, Nigeria

ABSTRACT

In this paper, we used an interpolation function with strong trigonometric components to derive a numerical integrator that can be used for solving first order initial value problems in ordinary differential equation. This numerical integrator has been tested for desirable qualities like stability, convergence and consistency. The discrete models have been used for a numerical experiment which makes us conclude that the schemes are suitable for the solution of first order ordinary differential equation.

Citation: Obayomi, A. , Ayinde, S. and Ogunmiloro, O. (2019), "On Trigonometric Numerical Integrator for Solving First Order Ordinary Differential Equation". Journal of Applied Mathematics and Physics, 7, 2564-2578. doi: 10.4236/jamp.2019.711175.

Keywords: Numerical Integrator, Ordinary Differential Equation, Initial Value Problems, Stability Analysis, Nonstandard Methods, Interpolation Methods

INTRODUCTION

Formulation of the Interpolating Function

Finite difference schemes have been in the forefront of the methods of using discrete models to approximate the solution of ordinary differential equations. Among the techniques used in building finite difference, scheme is the use of interpolation which requires the design of a basis function that is adequately differentiable in the domain of the numerical integration. Such basis function is then used to create a discrete version of the differential equation involved. This method has been used in the works of [1] - [6], etc. Notable among latest works on interpolating with trigonometric function includes [7], who constructed a new piecewise rational quadratic trigonometric spline with four local positive shape parameters in each subinterval is to visualize some given planar data. The order of approximation of the developed interpolating function was found to $0(h^3)$. [8] developed a new method for smooth rational cubic trigonometric interpolation based on values of function which is being interpolated. This rational cubic trigonometric spline is used to constrain the shape of the interpolant in such a way that the uniqueness of the interpolating function for the given data would be replaced by uniqueness of the interpolating curve for the given data. [9] constructed a new quadratic trigonometric B-spline with control parameters to address the problems related to two dimensional digital image interpolation. The newly constructed spline is then used to design an image interpolation scheme. Most of these works concentrated on application of trigonometric splines to digital imaging. We are interested in a general use of trigonometric interpolation functions for creating discrete models for the solution of ordinary differential Equation.

Nonstandard Modeling Techniques

The need for the nonstandard method came up due to some shortcomings of the standard method, in which the qualitative properties of the exact solutions are not usually transferred to the numerical solution. These shortcomings may create a lot of problems, which may affect the stability properties of the standard approach [10].

The concept of numerical instability and its proof by [10] [11], has led to the establishment of five major modeling rules proposed for the construction of difference schemes that will exhibit numerical stability. He also used these techniques to derive numerical models that are exact schemes for some classes of ordinary differential equations.

A finite difference scheme is called nonstandard finite difference method, if at least one of the following conditions is met [12] :

a) In the discrete derivative, the traditional denominator is replaced by a non-negative function φ such that, $\varphi(h)=h+0(h^2)$ as $h\to 0$;

b) Non-linear terms that occur in the differential equation are approximated in a non-local way, i.e. by a suitable function of several points of the mesh. The concept of nonstandard finite difference schemes was proposed by [10] as a solution to the numerical instability that exist in the use of finite difference schemes.

Since the discovery of this method, researchers like [5] [12] among others have suggested ways of building numerically reliable schemes using the nonstandard modeling rules. In this work we will create a scheme using the both of these two techniques above.

DERIVATION OF THE SCHEME

Let assume that a solution of a differential equation can be represented by a function

$$F(x) = A\cos x + Be^{-\propto x} - Qx \tag{1}$$

where $-\propto$ and Q are simulation parameters and A and B are arbitrary constants.

Let a first order ordinary differential equation possess a real valued solution and be differentiable in its domain several times, then from (1) we can write:

$$y = A\cos x + Be^{-\propto x} - Qx$$

$$y' = -A\sin x - \propto Be^{-\propto x} - Q \tag{2}$$

$$y'' = -A\cos x + \propto^2 Be^{-\propto x} \tag{3}$$

$$y''' = +A\sin x - \propto^3 Be^{-\propto x} \tag{4}$$

$$y^{iv} = A\cos x + \propto^4 Be^{-\propto x} \tag{5}$$

From (3) and (5) we have

$$y'' + y^{iv} = B\left(\propto^4 + \propto^2\right)e^{-\propto x}$$

$$B = \frac{\left(y'' + y^{iv}\right)}{\left(\propto^4 + \propto^2\right)e^{-\propto x}} \tag{6}$$

From (2) and (4) we have

$$y' + y''' = -B\left(\propto^3 + \propto\right)e^{-\propto x}$$

$$y''' - y' = 2A\sin x + B\left(\propto - \propto^3\right)e^{-\propto x}$$

$$A = \frac{\left(y''' - y'\right) - B\left(\propto - \propto^3\right)e^{-\propto x}}{2\sin x} \tag{7}$$

Also from (3) and (5) we have

$$y^{iv} - y'' = 2A\cos x + \left(\propto^4 - \propto^2\right)Be^{-\propto x}$$

$$A = \frac{\left(y^{iv} - y''\right) - B\left(\propto^4 - \propto^2\right)e^{-\propto x}}{2\cos x} \tag{8}$$

For the discrete representation

$$y = A\cos x + Be^{-\propto x} - Qx$$

$$y(x) = A\cos x + Be^{-\propto x} - Qx$$

$$y(x_n) = A\cos x_n + Be^{-\propto x_n} - Qx_n \tag{9}$$

$$y(x_{n+1}) = A\cos x_{n+1} + Be^{-\propto x_{n+1}} - Qx_{n+1} \tag{10}$$

$$y(x_{n+1}) \equiv y_{n+1} \text{ and } y(x_n) \equiv y_n$$

$$y(x_{n+1}) - y(x_n) \equiv y_{n+1} - y_n$$

$$e^{\propto x_{n+1}} - e^{\propto x_n} = e^{\propto x_n}\left(e^{\propto h} - 1\right)$$

$$M = \left(e^{\propto h} - 1\right)$$

$$y_{n+1} = y_n - 2A\sin\left(x_n + \frac{h}{2}\right)\sin\frac{h}{2} + BMe^{-\propto x_n} - Qh \tag{11}$$

From (7), (8) and (11)

$$y_{n+1} = y_n - 2\left\{\frac{\left(y^{iv} - y''\right) - B\left(\propto^4 - \propto^2\right)e^{-\propto x_n}}{2\cos x}\right\}\sin\left(x_n + \frac{h}{2}\right)\sin\left(\frac{h}{2}\right)$$
$$+\left\{\frac{\left(y'' + y^{iv}\right)}{\left(\propto^4 + \propto^2\right)e^{-\propto x_n}}\right\}\left(e^{\propto h} - 1\right)e^{-\propto x_n} - Qh \tag{12}$$

Or

$$y_{n+1} = y_n - 2\left\{\frac{\left(y''' - y'\right) - B\left(\propto - \propto^3\right)e^{-\propto x}}{2\sin x_n}\right\}\sin\left(x_n + \frac{h}{2}\right)\sin\left(\frac{h}{2}\right)$$
$$+\left\{\frac{\left(y'' + y^{iv}\right)}{\left(\propto^4 + \propto^2\right)e^{-\propto x_n}}\right\}\left(e^{\propto h} - 1\right)e^{-\propto x_n} - Qh \tag{13}$$

Let $f_n = y'$, $f_n^1 - y''$, $f_n^2 = y'''$, $f_n^3 = y^{iv}$

Then

$$y_{n+1} = y_n - 2\left\{\frac{\sin\left(x_n + \frac{h}{2}\right)\sin\left(\frac{h}{2}\right)}{2\cos x_n}\left(f_n^3 - f_n^1\right) - \left\{\frac{\propto^4 - \propto^2}{\propto^4 + \propto^2}\right\}\left(f_n^3 + f_n^1\right)\right\}$$
$$+\left\{\frac{\left(e^{\propto h} - 1\right)\left(f_n^3 + f_n^1\right)}{\left(\propto^4 + \propto^2\right)}\right\} - Qh \tag{14}$$

Or

$$y_{n+1} = y_n - 2\left\{\frac{\sin\left(x_n + \frac{h}{2}\right)\sin\left(\frac{h}{2}\right)}{2\sin x_n}\left(f_n^2 - f_n\right) - \left\{\frac{\propto - \propto^3}{\propto^4 + \propto^2}\right\}\left(f_n^3 + f_n^1\right)\right\}$$
$$+\left\{\frac{\left(e^{\propto h} - 1\right)\left(f_n^3 + f_n^1\right)}{\left(\propto^4 + \propto^2\right)}\right\} - Qh \tag{15}$$

$$E = \frac{\sin\left(x_n + \frac{h}{2}\right)\sin\left(\frac{h}{2}\right)}{2\cos x_n}, \quad F = \frac{\propto^4 - \propto^2}{\propto^4 + \propto^2},$$

$$G = \frac{e^{\propto h} - 1}{\propto^4 + \propto^2}, \quad H = \frac{\propto - \propto^3}{\propto^4 + \propto^2} \text{ and } I = \frac{\sin\left(x_n + \frac{h}{2}\right)\sin\left(\frac{h}{2}\right)}{2\sin x_n} \tag{16}$$

$$y_{n+1} = y_n - 2\left\{E\left(f_n^3 - f_n^1\right) - F\left(f_n^3 + f_n^1\right)\right\} + \left\{G\left(f_n^3 + f_n^1\right)\right\} - Qh, \quad \cos x_n \neq 0$$

$$y_{n+1} = y_n - 2\left\{I\left(f_n^2 - f_n\right) - H\left(f_n^3 + f_n^1\right)\right\} + \left\{G\left(f_n^3 + f_n^1\right)\right\} - Qh, \quad \sin x_n \neq 0$$

$$y_{n+1} = y_n + (2F - 2E + G)f_n^3 + (2E + 2F + G)f_n^1 - Qh, \quad \cos x_n \neq 0 \tag{17}$$

$$y_{n+1} = y_n + (2H + G)f_n^3 - (2I)f_n^2 + (2H + G)f_n^1 + (2I)f_n - Qh, \quad \sin x_n \neq 0 \tag{18}$$

Let

$$R = (2F - 2E + G), \quad S = (2E + 2F + G),$$
$$U = (2H + G), \quad T = (-2I), \quad M = (2I) \tag{19}$$

Substitute (19) into (17) and (18), we have the integrator in the form (20) and (21) respectively

$$y_{n+1} = y_n + Rf_n^3 + Sf_n^1 - Qh \tag{20}$$

$$y_{n+1} = y_n + Uf_n^3 + Tf_n^2 + Uf_n^1 + Mf_n - Qh \tag{21}$$

PROPERTIES OF THE INTEGRATION METHOD

Definition [13]

Any algorithm for solving a differential equation in which the approximation y_{n+1} to the solution at x_{n+1} can be calculated iff x_n, y_n and h are known is called a one step method. It is a common practice to write the functional dependence y_{n+1} on the quantities x_n, y_n and h in the form

$$y_{n+1} = y_n + \phi(x_n, y_n, h)$$

where $\phi(x_n, y_n, h)$ is the incremental function.

Definition [1]

A numerical scheme with an incremental $\phi(x_n,y_n,h)$ is said to be consistent with the initial value problem $y'=f(x,y),y(x_0)=y_0$ if the incremental function is identically zero at t_0 when h=0.

Theorem [13]

Let the incremental function of the scheme defined in the one step scheme above be continuous and jointly as a function of its arguments in the region defined by

$x\in[a,b]$ and $y\in(-\infty,\infty)$, $0\le h\le h_0$

where $h_0>0$ and let there exists a constant L such that $\phi(x_n,y_n,h)-\phi(x_n,y_n^*,h)\le L|y_n-y_n^*|$ for all (x_n,y_n,h) and (x_n,y_n^*,h) in the region just defined then the relation $(x_n,y_n,0)=(x_n,y_n^*)$ is a necessary condition for the convergence of the new scheme.

Theorem [1]

Let $y_n=y(x_n)$ and $p_n=p(x_n)$ denote two different numerical solution of the differential equation with the initial condition specified a $y_0=y(x_0)=\xi$ and $p_0=p(x_0)=\xi^*$ respectively such that $|\xi-\xi^*|<\varepsilon,\ \varepsilon>0$.

If the two numerical estimates are generated by the integration scheme, we have

$$y_{n+1}=y_n+h\phi(x_n,y_n,h)$$

$$p_{n+1}=p_n+h\phi(x_n,p_n,h)$$

The condition that $|y_{n+1}-p_{n+1}|\le K|\xi-\xi^*|$ is the necessary and sufficient condition for the stability and convergence of the schemes.

Proof of Convergence of the Integration Method

The increment function $\emptyset(x_n,y_n;h)$ can be written in the form

$$\emptyset(x_n,y_n;h)=\{Mf(x_n,y_n)+Uf^{(1)}(x_n,y_n)+Tf^{(2)}(x_n,y_n)+Uf^{(3)}(x_n,y_n)-Qh\} \quad (22)$$

Consider Equation (19), we can also write

$$\varnothing\left(x_n,y_n^*;h\right)=\left\{Mf\left(x_n,y_n^*\right)+Lf^{(1)}\left(x_n,y_n^*\right)+Tf^{(2)}\left(x_n,y_n^*\right)+Uf^{(3)}\left(x_n,y_n^*\right)-Qh\right\}$$

$$\begin{aligned}&\varnothing\left(x_n,y_n^*;h\right)-\varnothing\left(x_n,y_n;h\right)\\&=M\left[f\left(x_n,y_n^*\right)-f\left(x_n,y_n\right)\right]+U\left[f^{(1)}\left(x_n,y_n^*\right)-f^{(1)}\left(x_n,y_n\right)\right]\\&\quad+T\left[f^{(2)}\left(x_n,y_n^*\right)-f^{(2)}\left(x_n,y_n\right)\right]+U\left[f^{(3)}\left(x_n,y_n^*\right)-f^{(3)}\left(x_n,y_n\right)\right]-Qh+Qh\end{aligned} \tag{23}$$

Let $\bar{y}$ be defined as a point in the interior of the interval whose points are y and y*, applying mean value theorem, we have

$$f\left(x_n,y_n^*\right)-f\left(x_n,y_n\right)=\frac{\partial f\left(x_n,\bar{y}\right)}{\partial y_n}\left(y_n^*-y_n\right)$$

$$f^{(1)}\left(x_n,y_n^*\right)-f^{(1)}\left(x_n,y_n\right)=\frac{\partial f^{(1)}\left(x_n,\bar{y}\right)}{\partial y_n}\left(y_n^*-y_n\right)$$

$$f^{(2)}\left(x_n,y_n^*\right)-f^{(2)}\left(x_n,y_n\right)=\frac{\partial f^{(2)}\left(x_n,\bar{y}\right)}{\partial y_n}\left(y_n^*-y_n\right)$$

$$f^{(3)}\left(x_n,y_n^*\right)-f^{(3)}\left(x_n,y_n\right)=\frac{\partial f^{(3)}\left(x_n,\bar{y}\right)}{\partial y_n}\left(y_n^*-y_n\right)$$

We define

$$L=\sup_{(x_n,y_n)\in D}\frac{\partial f\left(x_n,y_n\right)}{\partial y_n},\quad L_1=\sup_{(x_n,y_n)\in D}\frac{\partial f^{(1)}\left(x_n,y_n\right)}{\partial y_n}$$

$$L_2=\sup_{(x_n,y_n)\in D}\frac{\partial f^{(2)}\left(x_n,y_n\right)}{\partial y_n},\quad L_3=\sup_{(x_n,y_n)\in D}\frac{\partial f^{0}\left(x_n,y_n\right)}{\partial y_n}$$

Therefore

$$\begin{aligned}&\varnothing\left(x_n,y_n^*;h\right)-\varnothing\left(x_n,y_n;h\right)\\&=M\left[f\left(x_n,y_n^*\right)-f\left(x_n,y_n\right)\right]+U\left[f^{(1)}\left(x_n,y_n^*\right)-f^{(1)}\left(x_n,y_n\right)\right]\\&\quad+T\left[f^{(2)}\left(x_n,y_n^*\right)-f^{(2)}\left(x_n,y_n\right)\right]+U\left[f^{(3)}\left(x_n,y_n^*\right)-f^{(3)}\left(x_n,y_n\right)\right]\\&=ML\left(y_n^*-y_n\right)+UL_1\left(y_n^*-y_n\right)+TL_2\left(y_n^*-y_n\right)+UL_3\left(y_n^*-y_n\right)\end{aligned} \tag{24}$$

Taking the absolute value of both sides

$$\begin{aligned}&\left|\varnothing\left(x_n,y_n^*;h\right)-\varnothing\left(x_n,y_n;h\right)\right|\\&\leq\left|ML\left(y_n^*-y_n\right)+UL_1\left(y_n^*-y_n\right)+TL_2\left(y_n^*-y_n\right)+UL_3\left(y_n^*-y_n\right)\right|\\&\leq\left|ML+UL_1+TL_2+UL_3\right|\left|y^*-y\right|\end{aligned} \tag{25}$$

If we let K=|ML+UL$_1$+TL$_2$+UL$_3$|
then our Equation (23) turns to

$$\left|\varnothing\left(x_n, y_n^*; h\right) - \varnothing\left(x_n, y_n; h\right)\right| \leq K\left|y^* - y\right| \tag{26}$$

which is the condition for convergence.

Consistence of the Integration Method

Consider an initial value problem of the form

$$y' = f(x, y), \; y(x_o) = y_o \tag{27}$$

Having an integrator of the form

$$y_{n+1} = y_n + \phi(x_n, y_n, h)$$

which can be obtained using (17) and (18) above and applying the rule in Section 1.2 a).

The renormalized nonstandard form of the schemes will be

$$y_{n+1} = y_n + \varphi\left\{Rf_n^3 + Sf_n^1 - Qh\right\}, \quad \cos x_n \neq 0 \tag{28}$$

$$y_{n+1} = y_n + \varphi\left\{Uf_n^3 + Tf_n^2 + Uf_n^1 + Mf_n - Qh\right\}, \quad \sin x_n \neq 0 \tag{29}$$

where $y_{n+1} = y_n + \varphi(x_n, y_n; h), \; \varphi = \sin(\propto h)$ then

If $h=0$, $E=0$, $F=\frac{\propto^4 - \propto^2}{\propto^4 + \propto^2}$, $G=0$ and $H=\frac{\propto - \propto^3}{\propto^4 + \propto^2}$, $I=0$ and φ=0 then (28) and (29) reduced to $y_{n+1}=y_n$

$$\Rightarrow \varphi(x_n, y_n; 0) = 0 \tag{30}$$

It is a known fact that a consistent method has order of at least one. Therefore, the new numerical integrator is consistent since Equations (28) and (29) can be reduced to (30) when h=0.

Stability Analysis of the Integration Method

We shall establish the stability analysis of the integrator by considering the theorem established by [14].

Let $y_n = y(x_n)$ and $P_n = P(x_n)$ denote two different numerical solutions of initial value problem of ordinary differential Equation (25) with the initial conditions specified as $y(x_o)=\eta$ and $p(x_o)=\eta^*$ respectively, such that $\left|\eta-\eta^*\right|<\varepsilon,\ \varepsilon>0$. If the two numerical estimates are generated by the integrator (19). From the increment function (26), we have

$$y_{n+1}=y_n+\varphi\varnothing(x_n,y_n;h) \tag{31}$$

$$P_{n+1}=P_n+\varphi\varnothing(x_n,p_n;h) \tag{32}$$

The condition that

$$\left|y_{n+1}-P_{n+1}\right|\le K\left|\eta-\eta^*\right| \tag{33}$$

is the necessary and sufficient condition that our new method (19) be stable and convergent.

Proof

From (29) we have

$$y_{n+1}=y_n+\varphi(2H+G)f_n^3-\varphi(2I)f_n^2+\varphi(2H+G)f_n^1+\varphi(2I)f_n-\varphi Qh$$

$$y_{n+1}=y_n+\varphi\{Uf_n^3+Tf_n^2+Uf_n^1+Mf_n-Qh\} \tag{34}$$

Then let

$$y_{n+1}=y_n+\varphi\{Mf(x_n,y_n)+Uf^{(1)}(x_n,y_n)+Tf^{(2)}(x_n,y_n)+Uf^{(3)}(x_n,y_n)-Qh\} \tag{35}$$

and

$$p_{n+1}=p_n+\varphi\{Mf(x_n,p_n)+Uf^{(1)}(x_n,p_n)+Tf^{(2)}(x_n,p_n)+Uf^{(3)}(x_n,p_n)-Qh\} \tag{36}$$

Therefore,

$$\begin{aligned} &y_{n+1}-p_{n+1}\\ &=y_n-p_n+\varphi\{M[f(x_n,y_n)-f(x_n,p_n)]+U[f^1(x_n,y_n)-f^1(x_n,p_n)]\\ &\quad+T[f^2(x_n,y_n)-f^2(x_n,p_n)]+U[f^3(x_n,y_n)-f^3(x_n,p_n)]\} \end{aligned} \tag{37}$$

Applying the mean value theorem as before, we have

$$\begin{aligned} y_{n+1}-p_{n+1}&=y_n-p_n+\varphi\left\{M\left[\frac{\partial f(x_n,p_n)}{\partial y_n}(y_n-p_n)\right]\right\}\\ &\quad+\varphi\left\{U\left[\frac{\partial f^{(1)}(x_n,p_n)}{\partial y_n}(y_n-p_n)\right]+T\left[\frac{\partial f^{(2)}(x_n,p_n)}{\partial y_n}(y_n-p_n)\right]\right.\\ &\quad\left.+U\left[\frac{\partial f^{(3)}(x_n,p_n)}{\partial y_n}(y_n-p_n)\right]\right\} \end{aligned} \tag{38}$$

We define

$$L = \sup_{(x_n, y_n)\in D} \frac{\partial f(x_n, \bar{y}_n)}{\partial y_n}, \quad L_1 = \sup_{(x_n, y_n)\in D} \frac{\partial f^{(1)}(x_n, \bar{y})}{\partial y_n}$$

$$L_2 = \sup_{(x_n, y_n)\in D} \frac{\partial f^{(2)}(x_n, \bar{y})}{\partial y_n}, \quad L_3 = \sup_{(x_n, y_n)\in D} \frac{\partial f^{3}(x_n, \bar{y})}{\partial y_n}$$

Therefore

$$\begin{aligned} y_{n+1} - p_{n+1} &= \varphi ML(y_n - p_n) + \varphi UL_1(y_n - p_n) \\ &+ \varphi TL_2(y_n - p_n) + \varphi UL_3(y_n - p_n) \end{aligned} \tag{39}$$

Taking the absolute value of both sides

$$\begin{aligned} &\left|\varnothing(x_n, y_n^*; h) - \varnothing(x_n, y_n; h)\right| \\ &\le \left|\varphi ML(y_n - p_n) + \varphi UL_1(y_n - p_n) + \varphi TL_2(y_n - p_n) + \varphi UL_3(y_n - p_n)\right| \\ &\le \left|\varphi ML + \varphi UL_1 + \varphi TL_2 + \varphi UL_3\right|\left|y_n - p_n\right| \end{aligned} \tag{40}$$

If we let $K = \left|\varphi ML + \varphi UL_1 + \varphi TL_2 + \varphi UL_3\right|$

then our Equation (34) turns to

$$\left|\varnothing(x_n, y_n; h) - \varnothing(x_n, p_n; h)\right| \le K\left|y_n - p_n\right| \tag{41}$$

which is the condition for convergence.

and $y(x_o) = \eta$, $P(x_o) = \eta^*$, given ε>0, then

$$\left|y_{n+1} - p_{n+1}\right| \le N\left|y_n - p_n\right| \tag{42}$$

and

$$\left|y_{n+1} - p_{n+1}\right| \le N\left|\eta - \eta^*\right| < \varepsilon, \text{ for every } \varepsilon > 0 \tag{43}$$

Then we conclude that our method (29) is stable and convergent.

Note: Similar arguments can be used to proof the stability, convergence and consistency of the scheme $y_{n+1} = y_n + \varphi\left\{Rf_n^{'3} + Sf_n^{'1} - Qh\right\}$.

THE IMPLEMENTATION OF THE INTEGRATION METHOD

Application of the Finite Difference Schemes to a Deferential Equation I

We derive a scheme for the first equation thus

$$y' = 25 + y^2,\ \ y(0) = 0,\ \ y = 5\tan(5x) \tag{44}$$

$$y'' = 2yy' = 2y\left(25 + y^2\right) = 50y + 2y^3 \tag{45}$$

$$y''' = 100 + 50y^2 + 150y^2 + 6y^4 = 100 + 200y^2 + 6y^4 \tag{46}$$

$$y^{iv} = 5000 + 800y^3 + 24y^5 \tag{47}$$

Since we are integrating from zero and $\cos x_n \neq 0$

$$y_{n+1} = y_n + \varphi\left\{Rf_n^3 + Sf_n^1 - Qh\right\}$$

where $f_n = y'$, $f_n^1 - y''$, $f_n^2 = y'''$, $f_n^3 = y^{iv}$.

Application of the Finite Difference Schemes to Logistic Model

We now derive a scheme for the first model

$$y' = y - y^2,\ \ y = \frac{y_0}{y_0 + (1 - y_0)e^t},\ \ y(0) = 0.5 \tag{48}$$

$$y'' = y' - 2yy'$$

$$y'' = y - 3y^2 + 2y^3 \tag{49}$$

$$y''' = y - 6y^2 + 6y^3 - y^2 + 6y^3 - 6y^4 \tag{50}$$

$$y^{iv} = y - 15y^2 + 50y^3 - 60y^4 + 24y^5 \tag{51}$$

Since we are integrating from zero we will use

$$y_{n+1} = y_n + \varphi\left\{Rf_n^3 + Sf_n^1 - Qh\right\},$$

where $f_n = y'$, $f_n^1 = y''$, $f_n^2 = y'''$, $f_n^3 = y^{iv}$.

The Hybrid Nonstandard Schemes

The new scheme will be obtained by substituting the derivatives and applying it to our scheme

$$y_{n+1} = y_n + \varphi\{Ry'' + Sy'' - Qh\} \tag{52}$$

The new standard scheme will be named (New SCH STD) with

$$\varphi = 1, \quad h = h \tag{53}$$

For each of the examples the hybrid schemes will be obtained by applying the renormalization techniques.

The hybrid scheme (NEW SCH h) is obtained by using φ=sin(∝h), step size

$$h = h \tag{54}$$

The hybrid scheme (NEW SCH SIN) by using φ=sin(∝h), step size

$$h = \sin(rh), \quad r \in R \tag{55}$$

The hybrid scheme (NEW SCH EXP) by using φ=sin(∝h), step size

$$h = \frac{\left(e^{\lambda h} - 1\right)}{\lambda}, \quad \lambda \in R \tag{56}$$

The choice of this denominator function as step size is informed by the works of [5] [12].

Experimentation and Result

The following are the 3D graphs obtained from the schemes when applied to the two models (Figures 1-14). We have used same parameters, step size, denominator functions and simulation parameters ∝ and Q to test the two differential equation models.

The Graph for the Method of Differential Equation Model I

Graph of the Schemes with parameters: h = 0.001, ∝ = (4.25 - 14), Q = 36, λ = 0.26.

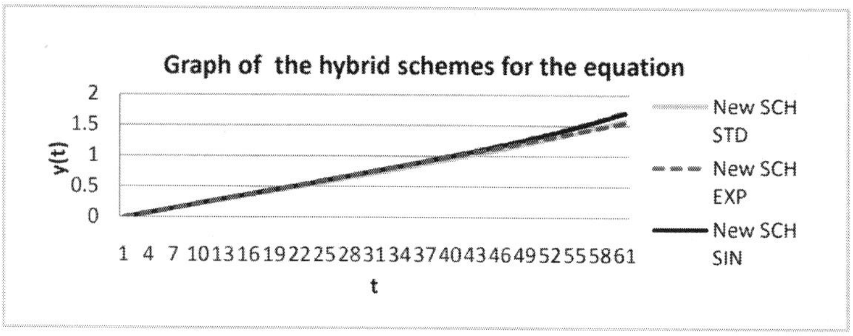

Figure 1: Solution curves for the schemes of Model I.

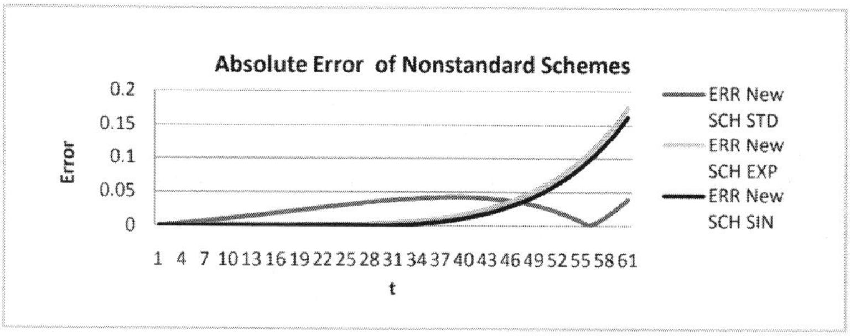

Figure 2: Graph of absolute error for the schemes in Figure 1.

Graph of the Schemes with parameters: h = 0.0001, ∝ = (4.25 - 14), Q = 36, λ = 0.26.

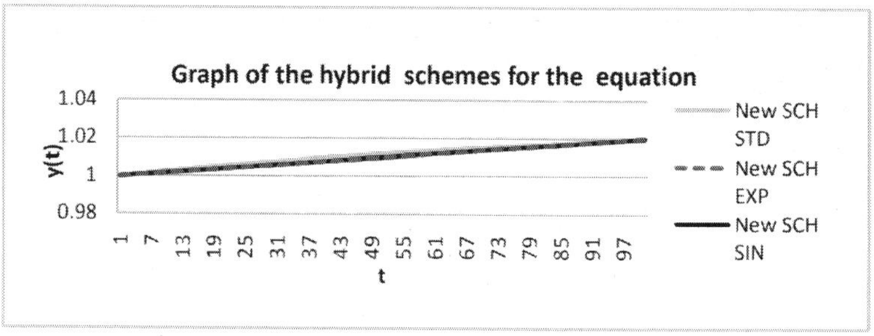

Figure 3: Solution curves for the schemes of model II.

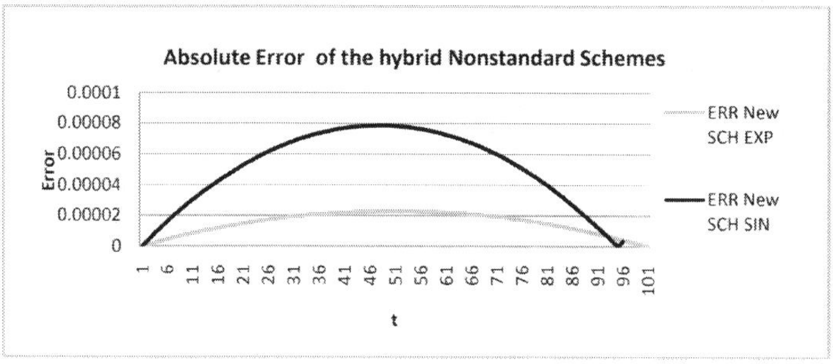

Figure 4: Graph of absolute Error for schemes in Figure 3.

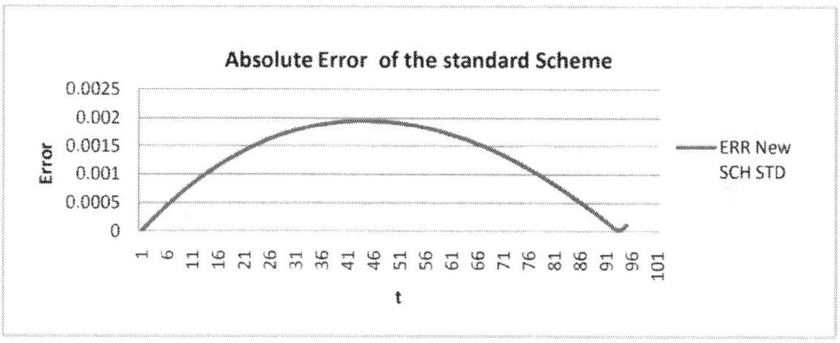

Figure 5: Graph of absolute Error for schemes in Figure 3.

The Graph for the Schemes of the Logistic Model II

Graph of the Schemes with parameters: h = 0.001, ∝ = (6.39 - 7.3), Q = 36, λ = 0.0026.

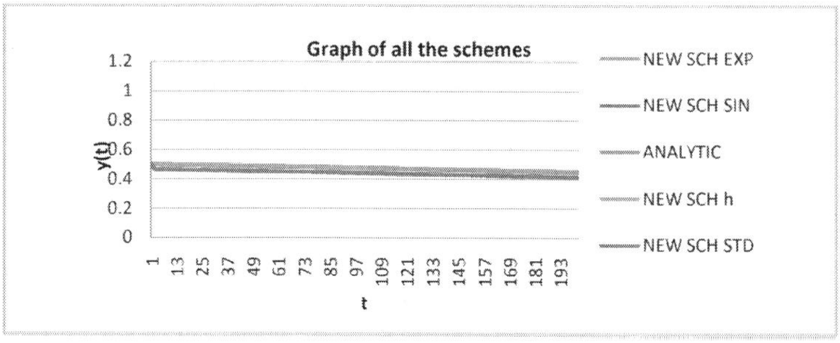

Figure 6: Solution curves for the schemes of Model II.

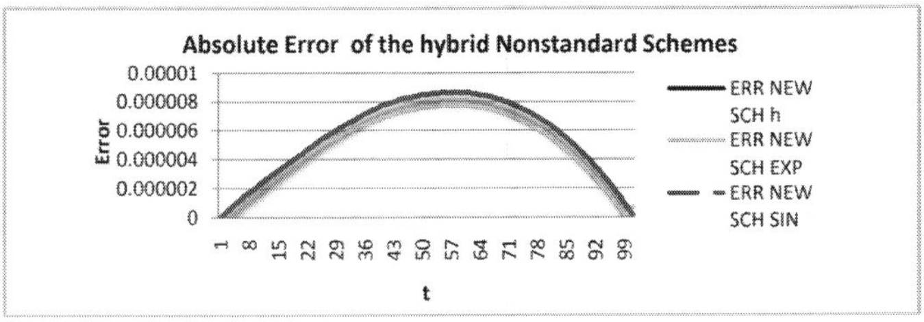

Figure 7: Graph of absolute error for the schemes in Figure 6.

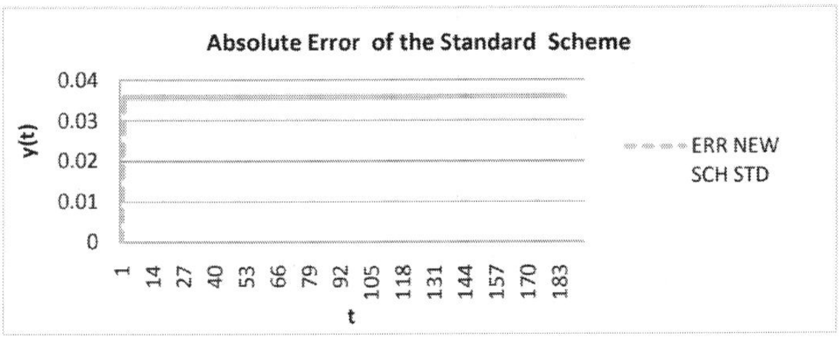

Figure 8: Graph of absolute error for the standard scheme in Figure 6.

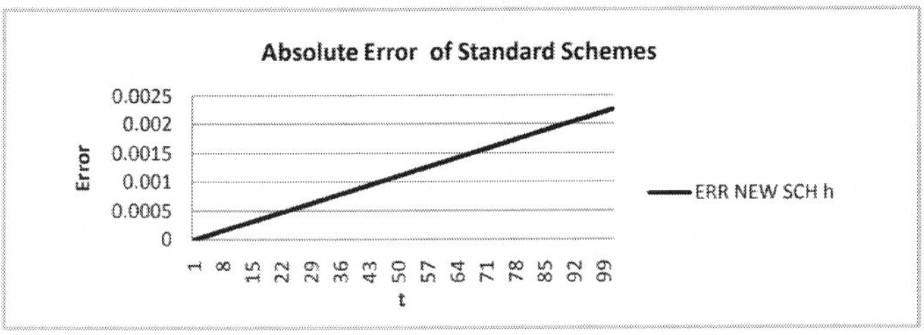

Figure 9: Error curve for the standard scheme in Figure 6.

Graph of the Schemes of model II with parameters: h = 0.0001, ∝ = (6.39 - 7.3), Q = 36, λ = 0.0016.

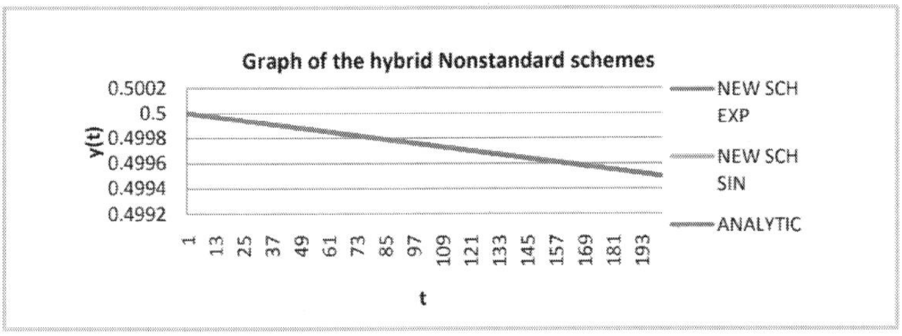

Figure 10: Solution curves for the schemes of model II.

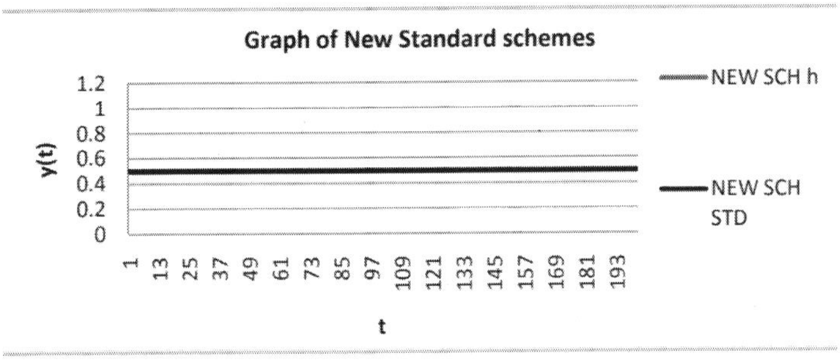

Figure 11: Solution curves for the standard schemes of model II.

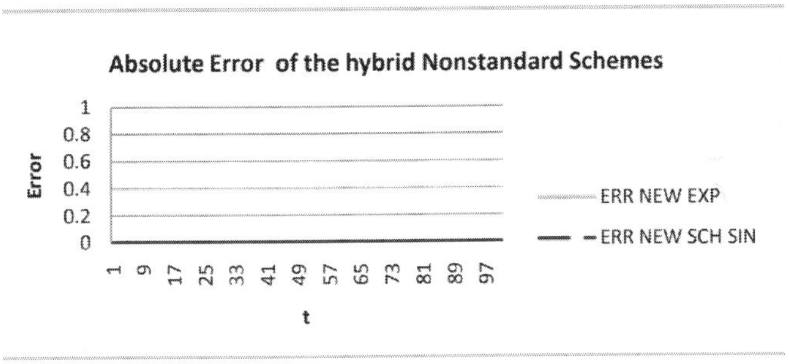

Figure 12: Graph of absolute Error for the hybrid Nonstandard schemes in Figure 10.

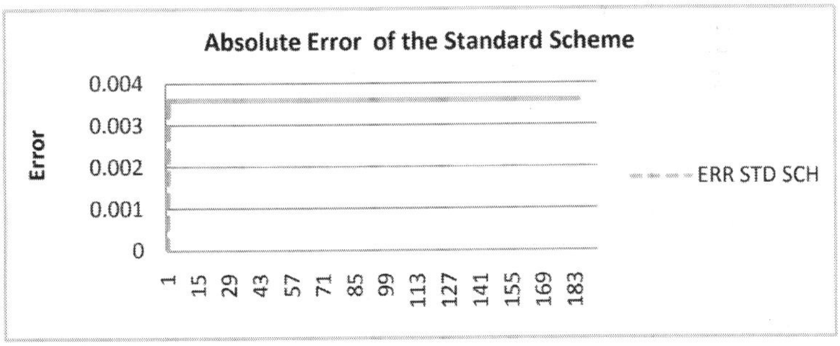

Figure 13: Graph of absolute Error for the Standard schemes in Figure 11.

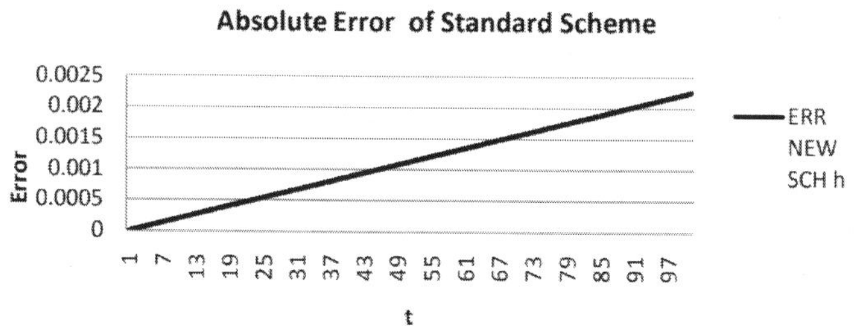

Figure 14: Graph of absolute Error for the Standard schemes in Figure 11.

DISCUSSION AND CONCLUSION

The derived simulation models have been tested with the control parameters ∝ and Q. We also applied the Nonstandard method by modifying denominator function φ, which also provides for parameters λ and r that can be chosen to obtain iteratively assigned step size as denominator. The discrete model worked for the tested differential equation. The solution curves of the hybrid schemes follow the analytical solutions of the respective equations monotonically are shown in Figure 1, Figure 3, Figure 5, Figure 10 and Figure 12. The numerical properties of the schemes like linear stability, convergence and consistency have been proved analytically. During the course of simulating the equations, we varied these control parameters to obtain family of curves that are very close to the analytic solution. The scheme NEW SCH h and NEW SCH STD are the standard schemes because we maintain the h=h as step size in both cases. The result showed that the two schemes do not possess smooth monotonicity properties (see Figure 6 and Figure 11) and diverge if φ=1 (see Figure 14). These two schemes possess the highest absolute error of deviation from the analytic solution (see Figure 8, Figure 9, Figure 13 and Figure 14). These results confirm the good qualities of Nonstandard modeling techniques [15]. The choice of appropriate values for variables λ and r can be determined using the conditions set by [5] [12]. The graph of Absolute error in Figure 4, Figure 7 and Figure 12 has demonstrated these qualities. The Nonstandard schemes of Model II produced absolute errors that are very close to zero. It is also observed that these two hybrid schemes are exact schemes for the logistic equation when h≤0.0001 because the result of the schemes is the same with the analytic solution (see Figure 12). This work can be extended by add higher degree polynomial to the basis function for a possible improved performance [14] [16]. We can conclude that the discrete model is suitable for the solution of first order ordinary differential equation as proposed.

REFERENCES

1. Fatunla, S.O. (1988) Numerical Methods for Initial Values Problems on Ordinary Differential Equations. Academic Press, New York.
2. Ibijola, E.A. (2000) Some Methods of Numerical Solutions of Ordinary Differential Equations. Far East Journal of Applied Mathematics, 4, 371-390.
3. Partidar, K.C. (2005) On the Use of Nonstandard Finite Difference Methods. Journal of Difference Equations and Applications, 11, 735-758. https://doi.org/10.1080/10236190500127471
4. Ibijola, E.A. and Obayomi, A.A. (2012) A New Numerical Scheme for the Solution of the Logistic Equation. Advances in Differential Equations and Control Processes, 10, 1-18.
5. Obayomi, A.A. and Oke, M.O. (2016) Development of New Nonstandard Denominator Function for Finite Difference Schemes. Journal of Nigerian Association of Mathematical Physics, 33, 50-60.
6. Obayomi, A.A. (2018) Development of a Discrete Model for the Tsunami Tidal Waves. Journal of Mathematics and Computer Science, 8, 98-113.
7. Liu, S.J., Chen, Z.L. and Zhu, Y.P. (2015) Rational Quadratic Trigonometric Interpolation Spline for Data Visualization. Mathematical Problems in Engineering, 2015, Article ID: 983120. https://doi.org/10.1155/2015/983120
8. Rana, S.S., Dube, M. and Tiwari, P. (2013) Rational Trigonometric Interpolation and Constrained Control of the Interpolant Curves. International Journal of Computer Applications, 67, 40-44. https://doi.org/10.5120/11387-6671
9. Hussain, M.Z., Abbas, S. and Irshad, M, (2017) Quadratic Trigonometric B-Spline for Image Interpolation Using GA. PLoS ONE, 12, e0179721. https://doi.org/10.1371/journal.pone.0179721
10. Mickens, R.E. (1994) Nonstandard Finite Difference Models of Differential Equations. World Scientific, 115, 144-162. https://doi.org/10.1142/2081
11. Mickens, R.E. (1981) Nonlinear Oscillations. Cambridge University Press, New York.

12. Anguelov, R. and Lubuma, J.M.S. (2003) Nonstandard Finite Difference Method by Nonlocal Approximation. Mathematics and Computers in Simulation, 61, 465-475. https://doi.org/10.1016/S0378-4754(02)00106-4
13. Henrici, P. (1962) Discrete Variable Methods in ODE. John Willey & Sons, New York.
14. Zill, D.G. and Cullen, R.M. (2005) Differential Equations with Boundary Value Problems. 6th Edition, Cole Thompson Learning Academic Resource Center, 90-101.
15. Lambert, J.D. (1973) Introductory Mathematics for Scientists and Engineers. John Wiley & Sons, New York.
16. Dreyer, T.P. (2005) Modeling with Ordinary Differential Equations. CRC Press, Boca Raton, FL.

Chapter 5

AUGMENTED LAGRANGIAN METHODS FOR NUMERICAL SOLUTIONS TO HIGHER ORDER DIFFERENTIAL EQUATIONS

Xuefeng Li

Department of Mathematical Sciences, Loyola University, New Orleans, LA, USA

ABSTRACT

A large number of problems in engineering can be formulated as the optimization of certain functionals. In this paper, we present an algorithm that uses the augmented Lagrangian methods for finding numerical solutions to engineering problems. These engineering problems are described by differential equations with boundary values and are formulated as optimization of some functionals. The algorithm achieves its simplicity and versatility by choosing linear equality relations recursively for the augmented Lagrangian associated with an optimization problem. We demonstrate

Citation: Li, X. (2017), "Augmented Lagrangian Methods for Numerical Solutions to Higher Order Differential Equations". Journal of Applied Mathematics and Physics, 5, 239-251. doi: 10.4236/jamp.2017.52021.

the formulation of an optimization functional for a 4th order nonlinear differential equation with boundary values. We also derive the associated augmented Lagrangian for this 4th order differential equation. Numerical test results are included that match up with well-established experimental outcomes. These numerical results indicate that the new algorithm is fully capable of producing accurate and stable solutions to differential equations.

Keywords: Augmented Lagrangian Methods, Method of Multipliers, Finite Element Solutions, Differential Equations

INTRODUCTION

Many problems in engineering are described by boundary value problems of differential equations of order 2k in the form of

$$g\left(x, y, y', y'', \cdots, y^{(2k)}\right) = 0,\ x \in (a,b), \tag{1}$$

subject to certain boundary conditions, where x is the independent variable in interval $(a,b) \subset \mathcal{R}$, k is a positive integer, $g \in \mathcal{C}^1\left(\mathcal{R}^{2k+2}, \mathcal{R}\right)$ and $y = y(x) \in \mathcal{C}^{2k}(a,b)$ is a solution to (1). Such a solution is also referred to as a strong solution because of the requirement that $y \in \mathcal{C}^{2k}(a,b)$.

However, a strong solution to (1) may not exist for some problems. Even when a strong solution to (1) exists, such a solution may be too costly to calculate numerically in practice.

Therefore, weak formulations of (1) are preferred. Assume that (1) admits a weak formulation in the following form.

$$\begin{cases} a(y,\eta) \equiv \int_a^b G\left(x, y, y', \cdots, y^{(k)}, \eta, \eta', \cdots, \eta^{(k)}\right) \mathrm{d}x, \\ \exists y \in \mathcal{V},\ a(y,\eta) = 0,\ \forall \eta \in \mathcal{V}_0, \end{cases} \tag{2}$$

where $\mathcal{V}$ is usually a subset of the Hilbert space $\mathcal{H}^k(a,b)$ subject to some boundary conditions, $\mathcal{V}_0$ is a subspace of $\mathcal{H}^k(a,b)$ satisfying some homogeneous boundary conditions, and $G \in \mathcal{C}^1\left(\mathcal{R}^{2k+3}, \mathcal{R}\right)$. In other words, a solution to (1) also satisfies (2).

A solution to (2) is referred to as a weak solution to (1) because of $y \in \mathcal{H}^k(a,b)$ instead of $y \in \mathcal{C}^{2k}(a,b)$.

We also assume that for $y \in \mathcal{V}$, there exists a certain functional

$$J(y) = \int_a^b \mathcal{G}\left(x, y, y', \cdots, y^{(k)}\right) \mathrm{d}x, \tag{3}$$

where $\mathcal{G} \in \mathcal{H}^1\left(\mathcal{R}^{k+2}, \mathcal{R}\right)$ such that a stationary point of functional $J(y)$ satisfies (2). A stationary point y of (3) satisfies

$$\frac{\mathrm{d}}{\mathrm{d}t} J(y + t\eta)\Big|_{t=0} = 0, \forall \eta \in \mathcal{V}_0, \tag{4}$$

where η is referred to as a test function. Equivalently, a stationary point y of (3) satisfies

$$\begin{cases} \forall \eta \in \mathcal{V}_0, \\ \int_a^b \left[\frac{\partial \mathcal{G}}{\partial y} \eta + \frac{\partial \mathcal{G}}{\partial y'} \eta' + \cdots + \frac{\partial \mathcal{G}}{\partial y^{(k)}} \eta^{(k)} \right] \mathrm{d}x = 0. \end{cases} \tag{5}$$

Refer to [1] for more details.

We see that a weak solution of (1) corresponds to an optimization problem of functional $J(y)$ defined in (3). Such a weak solution is simply a stationary point of (3).

As pointed out in [1], a boundary value problem of inhomogeneous boundary conditions can be effectively treated as a problem with homogeneous boundary conditions by introducing a special function satisfying the inhomogeneous boundary conditions. For simplicity, we will assume homogeneous boundary conditions for the rest of this paper. Under this assumption,

$$\mathcal{V} \equiv \mathcal{V}_0. \tag{6}$$

Hestenes [2] and Powell [3] introduced the augmented Lagrangian methods or the method of multipliers in 1969 for the study of optimization problems. Comprehensive applications of augmented Lagrangian methods for optimization and boundary value problems were investigated by Bertsekas [4], Fortin and Glowinski [5].

Even though most research on the augmented Lagrangian methods have been focusing on their direct applications in optimization, there have been consistent interests in using the augmented Lagrangian methods in other fields over the years. Recent applications of the augmented Lagrangian methods include image processing and computer vision [6], numerical solutions to Laplace equation with various boundary values [7], mechanics [7] [8] [9], medical imaging and computational biology [10], geophysics [11], ontology regarding big data representation and storage [12], and elastica theory [13], to name just a few.

In this paper, we'll investigate the applications of the augmented Lagrangian methods in boundary value problems of differential equations associated with problems from engineering. These are usually higher order differential equations (4th or higher). The feature of higher order of these problems is exploited to allow the development of simple and stable methods for their numerical solutions using finite elements.

The rest of the paper is organized in the following way. In section 2, we briefly review the augmented Lagrangian methods for optimization problems, and the finite element methods for solving weak formulation (2). In section 3, we'll introduce a special formulation for the construction of an augmented Lagrangian for finding numerical solutions to higher order differential equations using finite elements. We show a sample boundary value problem of a higher order nonlinear differential equation, its weak formulation, and the associated optimization problem in section 4. We'll present numerical tests for this sample differential equation in section 5. The numerical results demonstrate that this new version of the augmented Lagrangian methods is accurate, stable, versatile, and easy to implement.

AUGMENTED LAGRANGIAN METHODS AND FINITE ELEMENT SOLUTIONS

Generally speaking, an analytic solution to an optimization problem is not available. We therefore focus on numerical solutions using augmented Lagrangian methods and finite elements.

Augmented Lagrangian Methods

For a constrained optimization problem

$$\begin{cases} \min_{y \in W} J(y) \\ \text{subject to} \\ c_j(y) = 0, \ \forall j \in K, \end{cases} \tag{7}$$

where K is a subset of the set of all natural numbers, and $\mathcal{W}$ is some functional space, an augmented Lagrangian method for (7) introduces an associated augmented Lagrangian

$$\mathcal{L}_{\{r_j\}}\left(y,\{\mu_j\}\right) \equiv J(y)+\sum_{j\in K}\mu_j c_j(y)+\frac{1}{2}\sum_{j\in K} r_j c_j^2(y), \tag{8}$$

and the associated unconstrained optimization problem

$$\min_{y\in\mathcal{W}} \mathcal{L}_{\{r_j\}}\left(y,\{\mu_j\}\right), \tag{9}$$

where for each $j \in K$, μ_j is a Lagrangian multiplier, and r_j is a pre-chosen constant serving as a penalty when $c_j(y)$ is not sufficiently small.

Notice that the function spaces in (7) and (9) are identical. In particular, both (7) and (9) impose the same regularity requirements on y through $\mathcal{W}$.

When an optimization problem is associated with a boundary value problem of a differential equation, we have the leeway for constructing the associated augmented Lagrangian that imposes weaker regularity requirements than those of the original optimization problem. We'll present the formulation later in a separate section.

Finite Element Methods

Assume that $\mathcal{U}_N$ is a N-dimensional subspace that approximates $\mathcal{V}\subset\mathcal{H}^k(a,b)$ in (2). Assume also that $\{\phi_j\}_1^N$ is a basis for $\mathcal{U}_N$. The Galerkin method of finite elements amounts to finding a function $y_N \in \mathcal{U}_N$ such that

$$a(y_N,\eta)=0, \forall \eta\in\mathcal{U}_N, \tag{10}$$

where $a(\cdot,\cdot)$ is defined in (2).

Because $\mathcal{U}_N$ is finite dimensional, there exist unique coefficients $\{\alpha_j\}_1^N$ such that

$$y_N=\sum_{j=1}^{N}\alpha_j\phi_j. \tag{11}$$

Because (10) is true for all $\eta\in\mathcal{U}_N$, it must be true for every $\phi_j, 1\le j\le N$.

$$a\left(\sum_{j=1}^{N}\alpha_j\phi_j,\phi_j\right)=0, 1\le j\le N. \tag{12}$$

In fact, (12) forms a system of N equations for N unknown coefficients $\{\alpha_j\}_1^N$. These equations are linear or nonlinear depending on $a(\cdot,\cdot)$ being linear or nonlinear. Finding an approximate solution to (2) corresponds to solving the system of Equations (12) for coefficients $\{\alpha_j\}_1^N$.

In particular, because $\mathcal{U}_N \subset \mathcal{V} \subset \mathcal{H}^k(a,b)$, it is required that $\phi_j \in \mathcal{H}^k(a,b)$ for $1 \le j \le N$. Such regularity requirement may become too demanding on the basis for $\mathcal{U}_N$ when K is big.

Using weaker regularity requirements generally results in the following advan- tages.

1) We may use simpler test function η that improves overall numerical efficiency.
2) The condition numbers for linearized matrices of (12) are much smaller [1].

Our goal is to find new approaches that use the least regularity requirement possible, to be introduced next.

AUGMENTED LAGRANGIAN METHODS FOR DIFFERENTIAL EQUATIONS

In this section, we propose a new approach for the construction of augmented Lagrangian associated with boundary value problems of differential equations. This new approach exploits the feature of higher order of the differential equations to arrive at the least regularity requirements on weak solutions to boundary value problems. The approach coincides with those discussed in [5] for differential equations of second order or less.

Recall that to find a weak solution y to (1) means to solve for y in (5). That means all test functions must be from $\mathcal{H}^k(a,b)$. We now propose the following augmented Lagrangian associated with (3).

$$\begin{cases} \mathcal{L}_{\{r_j\}}\left(y,\{p_j\},\{\mu_j\}\right) = \int_a^b \mathcal{G}(x,y,p_1,\ldots,p_k)\,\mathrm{d}x \\ \qquad\qquad + \sum_{j=1}^k \int_a^b \mu_j \left(p'_{j-1} - p_j\right)\mathrm{d}x + \frac{1}{2}\sum_{j=1}^k r_j \int_a^b \left(p'_{j-1} - p_j\right)^2 \mathrm{d}x, \\ y \in \mathcal{W}_0, \\ p_j \in \mathcal{W}_j, 1 \le j < k, \\ p_k, \mu_j \in \mathcal{H}^0(a,b), 1 \le j \le k, \end{cases} \tag{13}$$

where $\mathcal{W}_j, 0 \le j < k$, is a certain subspace of $\mathcal{H}^1(a,b)$ that is associated with some of the homogeneous boundary conditions of (1) (see assumption for (6)), $\{p_j\}_1^k$ are supplementary variables to be linked to y by the recursive linear equality relations

$$p'_{j-1} = p_j, 1 \le j \le k, \tag{14}$$

and $p_0 \equiv y$, $\{\mu_j\}_1^k$ are Lagrangian multipliers, which are also functions of x, and $\{r_j\}_1^k$ are pre-chosen constants serving as penalty in case any of the linear equality relations in (14) is not met satisfactorily.

We immediately recognize the major difference between functional J(y) of (3) and the augmented Lagrangian of (13). Whereas the regularity requirement for J(y) of (3) is

$$y \in \mathcal{V} \subset \mathcal{H}^k(a,b), \tag{15}$$

the regularity requirements for augmented Lagrangian of (13) are

$$p_j \in \mathcal{W}_j \subset \mathcal{H}^1(a,b), 0 \le j \le k-1,$$
$$p_k, \mu_j \in \mathcal{H}^0(a,b), 1 \le j \le k, \tag{16}$$

independent of k, the order of the differential equation.

The following is the major conclusion of this paper.

Theorem 3.1 If $(y,\{p_j\},\{\mu_j\})$ is a stationary point of (13) where $p_0 \equiv y$, then $\{p_j\}$ satisfies (14), and $y \in \mathcal{V} \subset \mathcal{H}^k(a,b)$ is a weak solution to J(y) of (3), which is also a weak solution to (1).

Proof. A stationary point $(y,\{p_j\},\{\mu_j\})$ for the augmented Lagrangian $\mathcal{L}_{\{r_j\}}(y,\{p_j\},\{\mu_j\})$ of (13) satisfies,

$$\frac{\mathrm{d}}{\mathrm{d}t}\mathcal{L}_{\{r_j\}}(y+t\eta_0,\{p_j\},\{\mu_j\})\Big|_{t=0} = 0, \tag{17}$$

$$\frac{\mathrm{d}}{\mathrm{d}t}\mathcal{L}_{\{r_j\}}(y,\ldots,p_i+t\eta_{1,i},\ldots,\{\mu_j\})\Big|_{t=0} = 0, 1 \le i < k, \tag{18}$$

$$\frac{\mathrm{d}}{\mathrm{d}t}\mathcal{L}_{\{r_j\}}(y,\ldots,p_k+t\eta_{1,k},\{\mu_j\})\Big|_{t=0} = 0, \tag{19}$$

$$\frac{\mathrm{d}}{\mathrm{d}t}\mathcal{L}_{\{r_j\}}(y,\{p_j\},\ldots,\mu_i+t\eta_{2,i},\ldots)\Big|_{t=0} = 0, 1 \le i \le k, \tag{20}$$

where $\eta_0 \in \mathcal{W}_0$, $\eta_{1,i} \in \mathcal{W}_i$ for $1 \le i < k$, $\eta_{1,k} \in \mathcal{H}^0(a,b)$, $\eta_{2,i} \in \mathcal{H}^0(a,b)$ for $1 \le i \le k$, are the respective test functions.

The four stationary point relations (17), (18), (19) and (20) expand into

$$\int_a^b \frac{\partial}{\partial y} \mathcal{G}(x, y, p_1, \ldots, p_k)\eta_0 \mathrm{d}x + \int_a^b \mu_1 \eta_0' \mathrm{d}x + r_1 \int_a^b (y' - p_1)\eta_0' \mathrm{d}x = 0, \tag{21}$$

$$\int_a^b \frac{\partial}{\partial p_i} \mathcal{G}(x, y, \ldots, p_i, \ldots)\eta_{1,i} \mathrm{d}x - \int_a^b \mu_i \eta_{1,i} \mathrm{d}x + \int_a^b \mu_{i+1} \eta_{1,i}' \mathrm{d}x$$
$$- r_i \int_a^b (p_{i-1}' - p_i)\eta_{1,i} \mathrm{d}x + r_{i+1} \int_a^b (p_i' - p_{i+1})\eta_{1,i}' \mathrm{d}x = 0, \quad 1 \le i < k, \tag{22}$$

$$\int_a^b \frac{\partial}{\partial p_k} \mathcal{G}(x, y, \ldots, p_k)\eta_{1,k} \mathrm{d}x - \int_a^b \mu_k \eta_{1,k} \mathrm{d}x - r_k \int_a^b (p_{k-1}' - p_k)\eta_{1,k} \mathrm{d}x = 0, \tag{23}$$

$$\int_a^b (p_{i-1}' - p_i)\eta_{2,i} \mathrm{d}x = 0, \quad 1 \le i \le k. \tag{24}$$

Because (24) is true for any $\eta_{2,i} \in \mathcal{H}^0(a,b)$ for $1 \le i \le k$, it proves that $\{p_j\}$ satisfies the recursive linear equality relations (14). That is,

$$p_0 = y, p_1 = p_0' = y', p_2 = p_1' = y'', \ldots, p_k = p_{k-1}' = y^{(k)} \in \mathcal{H}^0(a,b), \tag{25}$$

in weak sense, implying that $y \in \mathcal{H}^k(a,b)$.

Consequently, (21), (22) and (23) reduce to

$$\int_a^b \left[\frac{\partial}{\partial y} \mathcal{G}(x, y, p_1, \ldots, p_k)\eta_0 + \mu_1 \eta_0' \right] \mathrm{d}x = 0, \tag{26}$$

$$\int_a^b \left[\frac{\partial}{\partial p_i} \mathcal{G}(x, y, \ldots, p_i, \ldots)\eta_{1,i} - \mu_i \eta_{1,i} + \mu_{i+1} \eta_{1,i}' \right] \mathrm{d}x = 0, \quad 1 \le i < k, \tag{27}$$

$$\int_a^b \left[\frac{\partial}{\partial p_k} \mathcal{G}(x, y, \ldots, p_k)\eta_{1,k} - \mu_k \eta_{1,k} \right] \mathrm{d}x, \tag{28}$$

for any test functions $\eta_0 \in \mathcal{W}_0$, $\eta_{1,i} \in \mathcal{W}_i$ for $1 \le i < k$, $\eta_{1,k} \in \mathcal{H}^0(a,b)$. $\forall \eta \in \mathcal{V}_0 \subset \mathcal{H}^k(a,b)$, if we choose functions η_0 and $\eta_{1,i}$ for $1 \le i \le k$ in the following ways,

$$\eta_0 = \eta,$$

$$\eta_{1,i} = \eta^{(i)}, 1 \le i \le k,$$

then (26), (27) and (28) become

$$\int_a^b \left[\frac{\partial}{\partial y} \mathcal{G}(x, y, p_1, \ldots, p_k)\eta + \mu_1 \eta' \right] \mathrm{d}x = 0, \tag{29}$$

$$\int_a^b \left[\frac{\partial}{\partial p_i} \mathcal{G}(x, y, \ldots, p_i, \ldots) \eta^{(i)} - \mu_i \eta^{(i)} + \mu_{i+1} \eta^{(i+1)} \right] \mathrm{d}x = 0, \ 1 \le i < k, \tag{30}$$

$$\int_a^b \left[\frac{\partial}{\partial p_k} \mathcal{G}(x, y, \ldots, p_k) \eta^{(k)} - \mu_k \eta^{(k)} \right] \mathrm{d}x. \tag{31}$$

After adding up all equations in (29), (30) and (31), we arrive at

$$\int_a^b \left[\frac{\partial}{\partial y} \mathcal{G}(x, y, p_1, \ldots, p_k) \eta + \sum_{i=1}^{k} \frac{\partial}{\partial p_i} \mathcal{G}(x, y, \ldots, p_i, \ldots) \eta^{(i)} \right] \mathrm{d}x = 0, \tag{32}$$

Using results in (25) along with (32), we have proved that $y \in \mathcal{V} \subset \mathcal{H}^k(a,b)$ satisfies (5) for any test function $\eta \in \mathcal{V}_0 \subset \mathcal{H}^k(a,b)$. That completes the proof.

We therefore have three approaches for finding approximate solutions to (1).

1) Solve (1) for an approximate strong solution in $C^{2k}(a,b)$ directly, e.g., by a certain finite difference method. This approach demands the highest regularity requirement where functions involved must belong to $C^{2k}(a,b)$.

2) Solve (1) for an approximate weak solution in $\mathcal{H}^k(a,b)$ via (2), (10) and (12). This approach demands that functions involved belong to $\mathcal{H}^k(a,b)$. This has been the standard approach found in most work related to finite element methods.

3) Solve (1) for an approximate weak solution in $\mathcal{H}^1(a,b)$ by finding stationary point of the augmented Lagrangian (13). This approach demands that functions involved belong to $\mathcal{H}^1(a,b)$, regardless of the order 2K of differential Equation (1). This approach achieves the minimum regularity requirement possible for any differential equation of order second or higher.

The method that uses the augmented Lagrangian (13) has the least regularity requirement among the three approaches.

Regardless of the order of differential Equation (1), finding approximate solution using the augmented Lagrangian (13) with finite elements is simple and standard as illustrated below.

- Because of (16), we can always use Hermite cubic polynomials to approximate $p_j, 0 \le j < k$, and use piecewise linear functions to approximate p_k and $\mu_j, 1 \le j \le k$.

- The bases for the corresponding finite dimensional spaces for finite elements are the Hermite cubic shape functions and linear shape functions, respectively.
- The approximate solutions are obtained from (12) by substituting Hermite cubic shape functions and linear shape functions for $\{\phi_j\}$, respectively.

We'll implement the augmented Lagrangian methods associated with Lagrangian (13) for finding approximate solution to a boundary value problem of a 4th order nonlinear differential equation in the next section.

A SAMPLE DIFFERENTIAL EQUATION

Many of the problems in engineering are described by boundary value problems of differential equations of orders two or higher. An example is the following nonlinear Euler-Bernoulli beam equation [14] [15] [16],

$$(EIv'')'' - (T_r + T_a)v'' - f(v) = 0,\ 0 < x < L, \quad (33)$$

with various boundary conditions (natural boundary conditions are not enforced by standard arguments), such as:

$$v(0) = v'(0) = 0, \quad (34)$$

for a rectangular beam of length L, width w and thickness t. Here, vis the deflection of the beam, E Young's modulus, I moment of inertia of cross-section of the beam, T_r represents residual force which is independent of v,

$$T_a = \frac{Ewt}{2L}\int_0^L |v'|^2\, dx, \quad (35)$$

represents the axial force in the beam, and f(v) is the intensity of external force exerted on the beam which is assumed to be continuous in v. We see that (33) along with (34) is just a special case of (1) for k=2. A strong solution v of (33) requires that $v \in C^4(0, L)$.

Subspace $\mathcal{V}$ associated with (33) becomes

$$\mathcal{V} = \{v | v \in \mathcal{H}^2(0, L),\ v \text{ satisfies } (34)\}. \quad (36)$$

Because boundary conditions in (34) are themselves homogeneous, the subspace $\mathcal{V}_0$ of all test functions is identical to $\mathcal{V}$, as has been assumed back in (6).

When multiplying any test function $\eta \in \mathcal{V}_0$ to (33), integrating the result over $[0, L]$, and applying integration by parts, we arrive at the following.

$$\int_0^L \left[EIv''\eta'' + (T_r + T_a)v'\eta' - f(v)\eta \right] \mathrm{d}x = 0, \forall \eta \in \mathcal{V}_0. \tag{37}$$

In other words, a solution to (33) also satisfies (37).

To find a solution to (33) becomes to find a function $v \in \mathcal{V}$ such that (37) is true. A solution to (37) is therefore called a weak solution to (33) because the regularity requirement has become

$$v \in \mathcal{V} \subset \mathcal{H}^2(0, L). \tag{38}$$

We also call (37) the weak formulation of (33).

Furthermore, for any $v \in \mathcal{V}$, if we define

$$J(v) = \int_0^L \left[\frac{EI}{2}|v''|^2 + \frac{2T_r + T_a}{4}|v'|^2 - F(v) \right] \mathrm{d}x, \tag{39}$$

where $F'(v) = f(v)$, we can verify with some algebraic manipulations that a stationary point of J(v) in (39) is a weak solution to (33). Details are omitted here for brevity.

We've shown that (33) indeed admits a weak formulation (37). A weak solution of (33) corresponds to a stationary point of (39). Such a stationary point is a solution to an optimization problem of functional J(v) of (39).

Based on functional J(v) in (39), we can define the associated augmented Lagrangian as shown below.

$$\begin{aligned} \mathcal{L}_{r_1, r_2}(v, p_1, p_2, \mu_1, \mu_2) = &\int_0^L \left[\frac{EI}{2} p_2^2 + \frac{2T_r + T_a}{4} p_1^2 - F(v) \right] \mathrm{d}x \\ &+ \int_0^L \mu_1 (v' - p_1) \mathrm{d}x + \int_0^L \mu_2 (p_1' - p_2) \mathrm{d}x \\ &+ \frac{1}{2} r_1 \int_0^L (v' - p_1)^2 \mathrm{d}x + \frac{1}{2} r_2 \int_0^L (p_1' - p_2)^2 \mathrm{d}x, \end{aligned} \tag{40}$$

Where $(v, p_1, p_2, \mu_1, \mu_2) \in \mathcal{W}_0 \times \mathcal{W}_1 \times \mathcal{H}^0 \times \mathcal{H}^0 \times \mathcal{H}^0$,

$$\begin{aligned} \mathcal{W}_0 &= \{ v | v \in \mathcal{H}^1(0, L), v \text{ satisfies } (34) \}, \\ \mathcal{W}_1 &= \{ p_1 | p_1 \in \mathcal{H}^1(0, L), p_1(0) = 0 \}, \end{aligned} \tag{41}$$

and

$$T_a \equiv T_a(p_1) = \frac{Ewt}{2L} \int_0^L p_1^2 \mathrm{d}x. \tag{42}$$

A stationary point of (40) hence satisfies the following five equations.

$$\begin{cases} -\int_0^L f(v)\eta_0 \mathrm{d}x + \int_0^L \mu_1 \eta_0' \mathrm{d}x + r_1 \int_0^L (v' - p_1)\eta_0' \mathrm{d}x = 0, \\ \int_0^L \frac{2T_r + T_a}{2} p_1 \eta_{1,1} \mathrm{d}x + \int_0^L \frac{T_a'(p_1, \eta_{1,1})}{4} p_1^2 \mathrm{d}x - \int_0^L \mu_1 \eta_{1,1} \mathrm{d}x + \int_0^L \mu_2 \eta_{1,1}' \mathrm{d}x \\ \quad - r_1 \int_0^L (v' - p_1)\eta_{1,1} \mathrm{d}x + r_2 \int_0^L (p_1' - p_2)\eta_{1,1}' \mathrm{d}x = 0, \\ \int_0^L EIp_2 \eta_{1,2} \mathrm{d}x - \int_0^L \mu_2 \eta_{1,2} \mathrm{d}x - r_2 \int_0^L (p_1' - p_2)\eta_{1,2} \mathrm{d}x = 0, \\ \int_0^L (v' - p_1)\eta_{2,1} \mathrm{d}x = 0, \\ \int_0^L (p_1' - p_2)\eta_{2,2} \mathrm{d}x = 0, \end{cases} \tag{43}$$

for their respective test functions

$$\eta_0 \in \mathcal{W}_0, \eta_{1,1} \in \mathcal{W}_1, \eta_{1,2} \in \mathcal{H}^0(0,L), \eta_{2,1} \in \mathcal{H}^0(0,L), \eta_{2,2} \in \mathcal{H}^0(0,L), \tag{44}$$

and

$$T_a'(p_1, \eta_{1,1}) = \frac{Ewt}{L} \int_0^L p_1 \eta_{1,1} dx. \tag{45}$$

Notice that boundary conditions (34) become

$$v(0) = v'(0) = 0, \tag{46}$$

and

$$p_1(0) = 0. \tag{47}$$

Finding an exact solution to a stationary point from (43) is highly unlikely, if not impossible. Fortin and Glowinski [5] suggested an iterative algorithm (called ALG1) for finding approximate solution to (43) as shown below.

Steps of ALG1.

1) Choose arbitrary initial guesses $\mu_1^{(0)}$ and $\mu_2^{(0)}$.

2) For $j \geq 0$, calculate $\left(v^{(j)}, p_1^{(j)}, p_{32}^{(j)}\right) \in \mathcal{W}_0 \times \mathcal{W}_1 \times \mathcal{H}^0$ by solving

$$\begin{cases} -\int_0^L f\left(v^{(j)}\right)\eta_0 \mathrm{d}x + \int_0^L \mu_1^{(j)} \eta_0' \mathrm{d}x + r_1 \int_0^L \left(\left(v^{(j)}\right)' - p_1^{(j)}\right)\eta_0' \mathrm{d}x = 0, \\ \int_0^L \frac{2T_r + T_a\left(p_1^{(j)}\right)}{2} p_1^{(j)} \eta_{1,1} \mathrm{d}x + \int_0^L \frac{T_a'\left(p_1^{(j)}, \eta_{1,1}\right)}{4} \left(p_1^{(j)}\right)^2 \mathrm{d}x - \int_0^L \mu_1^{(j)} \eta_{1,1} \mathrm{d}x \\ \quad + \int_0^L \mu_2^{(j)} \eta_{1,1}' \mathrm{d}x - r_1 \int_0^L \left(\left(v^{(j)}\right)' - p_1^{(j)}\right)\eta_{1,1} \mathrm{d}x \\ \quad + r_2 \int_0^L \left(\left(p_1^{(j)}\right)' - p_2^{(j)}\right)\eta_{1,1}' \mathrm{d}x = 0, \\ \int_0^L EIp_2^{(j)} \eta_{1,2} \mathrm{d}x - \int_0^L \mu_2^{(j)} \eta_{1,2} \mathrm{d}x - r_2 \int_0^L \left(\left(p_1^{(j)}\right)' - p_2^{(j)}\right)\eta_{1,2} \mathrm{d}x = 0, \end{cases} \tag{48}$$

where test functions η_0, $\eta_{1,1}$ and $\eta_{1,2}$ satisfy (44), then update $\mu_1^{(j+1)}, \mu_2^{(j+1)} \in \mathcal{H}^0$ by

$$\begin{cases} \mu_1^{(j+1)} = \mu_1^{(j)} + \rho_1\left(\left(v^{(j)}\right)' - p_1^{(j)}\right), \\ \mu_2^{(j+1)} = \mu_2^{(j)} + \rho_2\left(\left(p_1^{(j)}\right)' - p_2^{(j)}\right), \end{cases} \tag{49}$$

where ρ_1 and ρ_2 are two pre-chosen positive constants.

3) Check for convergence. Repeat step 2 when needed. Convergence is reached when the following conditions are met,

$$\begin{cases} \left\|v^{(j+1)} - v^{(j)}\right\| \le \varepsilon_r \left\|v^{(j)}\right\|, \\ \left\|p_1^{(j+1)} - p_1^{(j)}\right\| \le \varepsilon_r \left\|p_1^{(j)}\right\|, \\ \left\|p_2^{(j+1)} - p_2^{(j)}\right\| \le \varepsilon_r \left\|p_2^{(j)}\right\|, \\ \left\|\mu_1^{(j+1)} - \mu_1^{(j)}\right\| \le \varepsilon_r \left\|\mu_1^{(j)}\right\|, \\ \left\|\mu_2^{(j+1)} - \mu_2^{(j)}\right\| \le \varepsilon_r \left\|\mu_2^{(j)}\right\|, \end{cases} \tag{50}$$

for a pre-chosen relative tolerance ε_r.

It seems that (48) is just as difficult to solve as (43) is. Fortin and Glowinski [5] suggested yet another iterative algorithm (called ALG2) for finding approximate solution to (48) as shown below.

Steps of ALG2.

1) For a fixed j, let $p_1^{(j,0)} = p_1^{(j-1)}$, $p_2^{(j,0)} = p_2^{(j-1)}$.

2) For $m \ge 1$,

a) $\forall \eta_0 \in \mathcal{W}_0$, solve (51) for $v^{(j,m)}$,

$$-\int_0^L f\left(v^{(j,m)}\right)\eta_0 \mathrm{d}x + \int_0^L \mu_1^{(j)}\eta_0' \mathrm{d}x + r_1\int_0^L \left(\left(v^{(j,m)}\right)' - p_1^{(j,m-1)}\right)\eta_0' \mathrm{d}x = 0, \tag{51}$$

subject to boundary conditions (46).

b) $\forall \eta_{1,1} \in \mathcal{W}_1$, $\forall \eta_{1,2} \in \mathcal{H}^0$, solve system of Equations (52) for $p_1^{(j,m)}$ and $p_2^{(j,m)}$,

$$\begin{cases} \int_0^L \frac{2T_r + T_a\left(p_1^{(j,m)}\right)}{2} p_1^{(j,m)} \eta_{1,1} \mathrm{d}x + \int_0^L \frac{T_a'\left(p_1^{(j,m)}, \eta_{1,1}\right)}{4} \left(p_1^{(j,m)}\right)^2 \mathrm{d}x \\ - \int_0^L \mu_1^{(j)} \eta_{1,1} \mathrm{d}x + \int_0^L \mu_2^{(j)} \eta_{1,1}' \mathrm{d}x - r_1 \int_0^L \left(\left(v^{(j,m)}\right)' - p_1^{(j,m)} \right) \eta_{1,1} \mathrm{d}x \\ + r_2 \int_0^L \left(\left(p_1^{(j,m)}\right)' - p_2^{(j,m)} \right) \eta_{1,1}' \mathrm{d}x = 0, \\ \int_0^L EIp_2^{(j,m)} \eta_{1,2} \mathrm{d}x - \int_0^L \mu_2^{(j)} \eta_{1,2} \mathrm{d}x - r_2 \int_0^L \left(\left(p_1^{(j,m)}\right)' - p_2^{(j,m)} \right) \eta_{1,2} \mathrm{d}x = 0, \end{cases} \tag{52}$$

subject to boundary conditions (47).

3) Repeat step 2 for a certain number of times, or until convergence. Then

$$\begin{cases} v^{(j)} = \lim_{m \geq 1} v^{(j,m)}, \\ p_1^{(j)} = \lim_{m \geq 1} p_1^{(j,m)}, \\ p_2^{(j)} = \lim_{m \geq 1} p_2^{(j,m)}. \end{cases} \tag{53}$$

In particular, we solve (51) and (52) using the standard finite element approach where we approximate $v^{(j,m)}$ and $p_1^{(j,m)}$ by the Hermit cubic polynomials, and approximate $p_2^{(j,m)}$, $\mu_1^{(j)}$ and $\mu_2^{(j)}$ by piecewise linear functions.

We'll present numerical results from solving (33) using ALG1 and ALG2 in the next section.

NUMERICAL TESTS

In this section, we study a microbeam switch that is electrostatically actuated by an applied voltage. Hu, Chang and Huang [17] first studied such a problem in 2004. The structure [17] can be schematically described as a thin metal beam hanging over a substrate separated by some insulator, where one end of the microbeam is fixed and the other is free (fixed-free beam). The beam is pulled (deflected) towards the substrate when a voltage is applied between the beam and the substrate. The action of such a structure is modelled using (33) subject to boundary conditions (34). We are interested in the gap between the free end of the microbeam and the substrate subject to different applied voltages.

The specifics of the microbeam structure are listed below.

- Beam length, width and thickness are 20 mm, 5 mm and 57 μm, respectively.
- Initial gap between microbeam and substrate is 92 μm.
- Young's modulus is 1.558×10^{11} Pa.
- Permittivity of vacuum is 8.85×10^{-12} F/m.
- Poisson's ratio of 0.06 is used because the microbeam is considered wide, i.e., width is significantly greater than thickness.

We present in Table 1 numerical results from solving (33) subject to boundary conditions (34). Our numerical results indicated that the newly introduced augmented Lagrangian methods (ALM) are fully capable of producing quality solutions that match up well with experiment data. The number of finite elements used is denoted by ne in Table 1.

The relative errors of results obtained from ALM are comparable to those in [17] and [16], for all cases of finite element approximations. In fact, it appears that the augmented Lagrangian methods have already converged with relatively small number (i.e., 12) of finite elements because increasing the number of finite elements used does not cause much of a change in accuracy.

Table 1: Comparison of results, $\rho_1 = \rho_2 = 0.0208$, $r = s = 330$, and $\varepsilon_r = 0.0001$

Voltage (V)	End gaps (μm) by [17]	End gaps (μm) by ALM and corresponding relative errors							
		ne = 12		ne = 24		ne = 48		ne = 96	
20	90.5	90.3	−0.22%	90.3	−0.22%	90.3	−0.22%	90.3	−0.22%
40	84.6	84.6	−0.01%	84.6	−0.01%	84.6	−0.01%	84.6	−0.01%
60	70.0	71.2	1.7%	71.2	1.70%	71.2	1.69	71.2	1.70%
65	64.0	64.1	0.13%	64.1	0.13%	64.1	0.13%	64.1	0.13%
67	59.0	59.3	0.49%	59.3	0.49%	59.3	0.49%	59.3	0.49%

CONCLUSION

An algorithm is developed based on the augmented Lagrangian methods and the finite element, exploiting the order of the differential equation it solves. Independent of the order of the differential equation, we are always able to use only Hermite cubic and linear finite elements to approximate variables involved. As a result, this algorithm is easy to implement, and is capable of producing accurate and stable solutions to engineering problems that admit weak formulations associated with optimization of some functionals. Extensions of this algorithm for solving engineering problems described by higher order partial differential equations are being investigated by the author. Results will be submitted for publication in the near future.

REFERENCES

1. Axelsson, O. and Barker, V.A. (2001) Finite Element Solution of Boundary Value Problems. Theory and Computation. SIAM. https://doi.org/10.1137/1.9780898719253
2. Hestenes, M.R. (1969) Multiplier and Gradient Methods. Journal of Optimization Theory and Applications, 4, 303-320. https://doi.org/10.1007/BF00927673
3. Powell, M.J.D. (1969) A Method for Nonlinear Constraints in Minimization Problems. In: Fletcher, R., Ed., Optimization, Academic Press, New York, NY, 283-298.
4. Bertsekas, D.P. (1996) Constrained Optimization and Lagrange Multiplier Methods. Athena Scientific (First Published 1982).
5. Fortin, M. and Glowinski, R. (1983) Augmented Lagrangian Methods: Applications to the Numerical Solution of Boundary-Value Problems. North-Holland.
6. Liu, X. (2016) Augmented Lagrangian Method for Total Generalized Variation Based Poissonian Image Restoration. Comput. Math. Appl., 71, 1694-1705. https://doi.org/10.1016/j.camwa.2016.03.005
7. Zhang, S. and Li, X. (2016) Boundary Augmented Lagrangian Method for the Signorini Problem. Appl. Math., 61, 215-231. https://doi.org/10.1007/s10492-016-0129-7
8. Bielski, W.R., Galka, A. and Telega, J.J. (2002) Augmented Lagrangian Methods for a Class of Nonconvex Contact Problems in Structural Mechanics. Contact Mechanics (Praia da Consolação, 2001), Solid Mech. Appl., Kluwer Acad. Publ., Dordrecht, 261-268.
9. Zhang, S. and Li, X. (2015) Boundary Augmented Lagrangian Method for Contact Problems in Linear Elasticity. Eng. Anal. Bound. Elem., 61, 127-133. https://doi.org/10.1016/j.enganabound.2015.07.006
10. Kim, S. and Lee, C. (2015) Accurate Surface Reconstruction in 3D Using Two- Dimensional Parallel Cross Sections. J. Math. Imaging Vision, 53, 182-195. https://doi.org/10.1007/s10851-015-0566-z
11. Ionescu, I.R., Mangeney, A., Bouchut, F. and Roche, O. (2015) Viscoplastic Modeling of Granular Column Collapse with Pressure-Dependent Rheology. J. Non-Newton. Fluid Mech., 219, 1-18.
12. Gao, W., Zhu, L. and Wang, K. (2015) Ontology Sparse Vector Learning Algorithm for Ontology Similarity Measuring and Ontology Mapping via ADAL Technology. International Journal of Bifurcation and

Chaos, 25, 1540034. http://www.worldscientific.com/doi/abs/10.1142/S0218127415400349

13. Duan, Y., Wang, Y. and Hahn, J. (2013) A Fast Augmented Lagrangian Method for Euler's Elastica Models. Numer. Math. Theory Methods Appl., 6, 47-71.
14. Gere, J.M. (2004) Mechanics of Materials. Thomson Learning, 6th Edition.
15. Zhang, L.X. and Zhao, Y.P. (2003) Electromechanical Model of RF MEMS Switches. Microsyst. Technol., 9, 420-426. https://doi.org/10.1007/s00542-002-0250-2
16. Sadeghian, H., Rezazadeh, G. and Osterberg, P.M. (2007) Application of the Generalized Differential Quadrature Method to the Study of Pull-In Phenomena of MEMS Switches. Journal of Microelectromechanical Systems, 16, 1334-1340. https://doi.org/10.1109/JMEMS.2007.909237
17. Hu, Y.C., Chang, C.M. and Huang, S.C. (2004) Some Design Considerations on the Electrostatically Actuated Microstructures. Sensors and Actuators A, 112, 155-161. https://doi.org/10.1016/j.sna.2003.12.012

SECTION 2: SYMBOLIC COMPUTING METHODS

Chapter 6

LEARNING NETWORK ASSISTED BY MEANS OF SYMBOLIC COMPUTATION

J.J. Miralles, J.A. Martinez, I. Marin and M.M. Artigao

University of Castilla-La Mancha Spain

INTRODUCTION

Background of this work is summarized in (Miralles et al., 2004) and (Marin et al., 2008), where Network Learning (NL) and Learning Network (LN) are described. Their differences are also introduced in such work. However, despite of the need of a learning model from a pedagogical point of view, this pays more attention in the point of view of Computational Science. This is due to the main knowledge area where the work is framed. In (Downes, 2004) are described learning elements, which are lacked in cited work. That paper contains the basic principles and useful elements to develop a model

Citation: J.J. Miralles, J.A. Martinez, I. Marin and M.M. Artigao, "Learning Network Assisted by Means of Symbolic Computation", IntechOpen Technology, 2009, DOI: 10.5772/7283.

of network learning. In this chapter, we propose complete the open area in previous works.

Under these assumptions, the principal goal of this chapter is to introduce a model for the teaching assisted by network in the framework of connectivism as a learning theory in the digital age. Our model uses a Learning Network-Assisted by using symbolic computation through the web technology. We also present some guidelines in order to develop these web applications. Moreover, a set of paradigmatic examples are implemented by means of this model. Particularly, we show some selected examples that they require that teachers of different subjects to interact amongst them. These applications were based on examples specifically choose in order to illustrate how to increase the transversal character amongst different teaching subjects in Industrial Engineering in the context of ECTS. These examples described here emphasize the web technologies as a well-suited for the learning of scientific and technological subjects. The key idea of these activities, in the context of connective knowledge paradigm, is the use of symbolic computation like a link both among teachers and between students and different matters, and not only as a computational tool.

This chapter is organized as follows. First of all, Section 2 presents a needed criticism of analogical learning theories. Complex networks and connectivism are described in Section 3. Network Learning (NL) and Learning Network (LN) are also described. In Section 4, the AALN is presented and Computer Algebra Systems are introduced. webMathematica technology is discussed in Section 5. This is a useful web technology for support a LN based on symbolic computation. An example of LN based on symbolic computation is described in Section 6. Finally, the conclusions are outlined in Section 7.

THE NEED FOR CRITICISM IN THE CONTEXT OF THE THEORIES OF ANALOGICAL LEARNING

This section will present the arguments of the developers of the connectivism's theory, compared to classical theories of learning in the context of the digital age. Learning theories have two chief values according to (Hill, 2002). One is in providing us with vocabulary and a conceptual framework for interpreting the examples of learning that we observe. The other is in suggesting where to look for solutions to practical problems. The theories do not give us solutions, but they do direct our attention to those variables that are crucial in finding solutions.

There are three main categories or philosophical frameworks under which learning theories fall: behaviorism, cognitivism, and constructivism. Behaviorism focuses only on the objectively observable aspects of learning. Cognitive theories look beyond behaviour to explain brain-based learning. And constructivism views learning as a process in which the learner actively constructs or builds new ideas or concepts. These theories make up the different educational environments, which we usually have a certain familiarity. We would classify these theories as analogics, because they deal about constructions of learning developed in an era where teaching and learning processes does not interact with the technological development of the digital society, and where, therefore, any methodology learning from either of these epistemologies, was implemented under an analogic support. Over the last twenty years, technology has reorganized how we live, how we communicate, and how we learn. Learning needs and theories that describe learning principles and processes, should be reflective of underlying social environments (Vaill, 1996) emphasizes that: "..learning must be a way of being - an ongoing set of attitudes and actions by individuals and groups that they employ to try to keep abreast of the surprising, novel, messy, obtrusive, recurring events.."

One might consider that the most relevant results from the interaction between teaching and learning processes and the revolution of the digital information, is that the half-life of knowledge has been shortened by at least an order of magnitude, only in the last twenty years. Knowledge grows exponentially, life in many fields of knowledge are now measured in months and years. As a result of the interaction-learning digital revolution, we can highlight the following current trends in learning:

- Many learners will move into a variety of different, possibly unrelated fields over the course of their lifetime.
- Informal learning is a significant aspect of our learning experience. Formal education no longer comprises the majority of our learning. Learning now occurs in a variety of ways - through communities of practice, personal networks, and through completion of work-related tasks.
- Learning is a continual process, lasting for a lifetime. Learning and work related activities are no longer separate. In many situations, they are the same.

- Technology is altering (rewiring) our brains. The tools we use define and shape our thinking.

The organization and the individual are both learning organisms. Increased attention to knowledge management highlights the need for a theory that attempts to explain the link between individual and organizational learning.

- Many of the processes previously handled by learning theories (especially in cognitive information processing) can now be off-loaded to, or supported by, technology.
- Know-how and know-what is being supplemented with know-where (the understanding of where to find knowledge needed).
- As theories of learning share many attributes and new ones build progressively on previous ones, any consideration of learning requires a review of existing theories. (Driscoll, 2000) categorizes learning into three broad epistemological frameworks:
- Objectivism states that reality is external and objective, and that knowledge is gained through experiences.
- Pragmatism states that reality is provisional, and knowledge is negotiated through experience and thinking.
- Interpretivism states that reality is internal, and knowledge is constructed. Analogic theories of learning consider that learning occurs inside a person. Even social constructivist approaches, which argue that learning is a social process, promote the role of the individual (and physical presence, i.e. brain-based) in learning. These theories do not address learning that occurs outside of people (i.e. learning that is stored and manipulated by technology). They also fail to describe how learning happens within organizations. When the existing learning theories are seen through the prism of technology, many important questions:
- How are learning theories impacted when knowledge is no longer acquired in the linear manner?
- What adjustments need to made with learning theories when technology performs many of the cognitive operations previously performed by learners (information storage and retrieval).
- How can we continue to stay current in a rapidly evolving information ecology?

- How do learning theories address moments where performance is needed in the absence of complete understanding?
- What is the impact of networks and complexity theories on learning?
- What is the impact of chaos as a complex pattern recognition process on learning?
- With increased recognition of interconnections in differing fields of knowledge, how are systems and ecology theories perceived in light of learning tasks?

COMPLEX NETWORKS AND CONNECTIVISM

Complex Networks

The twenty-first century has begun intellectually under the paradigm of complex networks under the sign of the connection. In particular, the 1998 Nature paper (Watts, & Strogatz, 1998), entitled "*Collective dynamics of small-world networks*, is widely regarded as a seminal contribution to the new interdisciplinary field of "complex networks," whose applications reach from graph theory and statistical physics to sociology, business, epidemiology, and neuroscience. As one measure of this paper's impact, it is the most highly cited article about networks in the past decade, according to the ISI Web of Science. It was also the sixth most highly cited paper in physics, with 2700 citations, between January 1, 1998 and August 31, 2008.

The science of complex systems is a new multidisciplinary field aiming at understanding the complex real world around us. Complex networks allow the study of model complex phenomena in different contexts, and many scientific areas can be considered to fall into the realm of complex systems and can be studied by using nonlinear mathematical models, statistical methods and computer modelling approaches (Watts & Strogatz, 1998), (Watts, 1999), (Strogatz, 2001), (Watts, 2003), (Barabasi, 2002), (Albert & Barabasi, 2002), (Wang, 2002), (Newman, 2003), (Bornholdt & Schuster , 2003), (Boccalettia et al., 2006). The first models of networks were used by mathematicians to model real objects described by the theory of probabilities; these models are called random networks or random graphs. The random graph theory was founded by Paul Erdos and Alfred Renyi. A detailed review of the item is in the book (Bollobas, 1985). Moreover, physicists since the early twentieth century have used networks to describe various physical ordered structures. The paradigm of complex networks arose in the late twentieth century and early twenty-first century, when technology has allowed measurements in networks of various kinds and it has been empirically discovered that real-world networks are in an intermediate state between the networks ordered from the physical networks and random mathematical networks. Although many quantities and measures of complex networks have been proposed and investigated in the past few years, three quantities: the average path length, clustering coefficient, and degree distribution play a key role in the recent development of complex network theory and modelling.

Complex network has been a hotspot in study of abroad fields and achieves many meaningful results in mathematics, physics, biology, communication, economics and sociology. In fact any complex system in nature can be modelled as a network, where vertices are the elements of the system and edges represent the interactions between them. The study and characterization of the statistical properties of complex networks has received much attention in the last few years. It is shown by Barabasi, Newman, Watts and many other researchers that a great variety of real networks exhibit a small world property and scale-free character.

The popularization of the World Wide Web (Bernes Lee et al., 1994), as a medium for commerce, communication, information sharing, and education has raised the profile of networks as a means of human organization. Research from fields as diverse as sociology, physics, information and knowledge (Benkler, 2006), and organizational effectiveness (Stephenson, 2002) suggest that networks fundamentally alter the hierarchical structure found in many traditional institutions. Academic journals reflect exponential growth in focus on complex networks in sociology (Borgatti & Foster, 2003), as well as mathematics, physics, chemistry, and other fields (Scharnhorst, 2003). The growth of interest in, and research on, networks as organizational models for all aspects of society is significant. The essential information that the complex networks provide us, is that the vast majority of natural and artificial systems, which could be modelled as networks, the measures undertaken to determine the nature of its topology, show us that these networks have a structure that is an interpolation between absolute order and complete randomness, the topology of these networks inhabits the "edge of chaos".

In a pedagogical perspective, the complex thinking is focused on learning to learn (E. Morin, 1999, 2001), (Clergue, 1997), (Mason, 2008). The digital age has led to significant changes in how knowledge circulates. Would have been a sort of off shoring on knowledge, to the extent that the same began to escape from the places, where the society had the legitimacy for learning. At present, there is a multiplicity of knowledge without a proper place or spaces not traditionally accepted for the production of knowledge. What differentiates the university from other institutions is the search for better ways of making explicit the extra- scientific conditions, and deconstructed by controlling their influence.

Networks, as models of organizing education, are part of a larger general shift, beginning in the second half of the 20th century, away from individualist, essentialist, and atomistic explanations to more relational, contextual, and systemic understandings (Borgatti & Foster, 2003), (Baumeister, 2005) status that networking is having an impact on all aspects of university life. Networks, while generally associated with the development of the Internet, have long served a vital role in the management of and functioning in complex information environments (Wright, 2007). Viewing networks as structural models for education and learning is the new context and this is the context and the new challenge, in which the university training is immersed.

A network can simply be defined as connections between entities a graph from a mathematical point of view. Computer networks, power grids, and social networks all function on the simple principle that people, groups, systems, nodes, entities can be connected to create an integrated whole. Alterations within the network have ripple effects on the whole. From a mathematical point of view a network can be defined as a graph with additional information on vertices and for edges. A network $N = (V, E, T, W)$ consists of : a graph $G= (V, E)$, where V is the set of vertices or nodes and E is the set of pairs (i, j), with $i, j, c\ V$, called edges (or lines, links, ties, arcs) that denote nodes i and j are connected. Undirected lines are called edges, and directed lines are arcs. We will denote by n and m the cardinalities of the sets V and E, $n = cad\ (V)$, $m = card\ (E)$. T are vertex value functions: t: $V \rightarrow A$, and W are edge value functions: $w: E \rightarrow B$. Normally one imposes that $i \neq j$ so that self- connections are avoided. Along with this definition, we can also consider that the elements in E are ordered pairs $(i,j) \neq (j,i)$. In this case we will talk of a directed network (or diagraph). It is also very common to assign weights (numbers) to the edges so that we have a weighted (or valued) network. The cardinality of V and E can tell us about the nature of the graph.

Taking into account that the maximal cardinality of *E* is $\binom{n}{2}$ we will talk about a sparse network when *m* << *n2* and a dense one when m $\approx$ n^2. The elements of the sets (V, E) receive different names depending on the context or scientific field in which we move. The following Table 1 shows some of them:

Table 1: Networks are everywhere

Scientifc Field	*v*	*E*
Mathematics	Vertex	Edge
Physics	Site	Bond
Computer Science	Node	Link
Sociology	Actor	Tie

Overview of Connectivism

The Connectivism is a theoretical framework for understanding learning developed mainly by Stephen Downes and George Siemens, (Downes, 2005, 2006, 2007), (Siemens, 2005, 2006, 2008). In connectivism, the starting point for learning occurs when knowledge is actuated through the process of a learner connecting to and feeding information into a learning community. Siemens (2005) states, "*Connectivism is the integration of principles explored by chaos, network, and complexity and self-organization theories. Learning is a process that occurs within nebulous environments of shifting core elements - not entirely under the control of the individual. Learning (defined as actionable knowledge) can reside outside of ourselves (within an organization or a database), is focused on connecting specialized information sets, and the connections that enable us to learn more are more important than our current state of knowing.*"

In the connectivist model, a learning community is described as a node, which is always part of a larger network. Nodes arise out of the connection points that are found on a network. A network is comprised of two or more nodes linked in order to share resources. Nodes may be of varying size and strength, depending on the concentration of information and the number of individuals who are navigating through a particular node (Downes, 2006). In this framework a community is the clustering of similar areas of interest that allows for interaction, sharing, dialoguing, and thinking together. According to connectivism, knowledge is distributed across an information network and can be stored in a variety of digital formats. Since information is constantly changing, its validity and accuracy may change over time, depending on the discovery of new contributions pertaining to a subject. By extension, one's understanding of a subject, one's ability to learn about the subject in question, will also change over time. Connectivism stresses that two important skills that contribute to learning are the ability to seek out current information, and the ability to filter secondary and extraneous information. The learning process is cyclical, in that learners will connect to a network to share and find new information, will modify their beliefs on the basis of new learning, and will then connect to a network to share these realizations and find new information once more. Learning is considered a knowledge creation process and not only knowledge consumption. One's personal learning network is formed on the basis of how one's connections to learning communities are organized by a learner.

The view of knowledge as composed of connections and networked entities, and the concept of emergent, connected, and adaptive knowledge provides the epistemological framework for connectivism (Siemens, 2005) as a learning theory. Connectivism posits that knowledge is distributed across networks and the act of learning is largely one of forming a diverse network of connections and recognizing attendant patterns (Siemens, 2006).

Principles of connectivism:

- Learning and knowledge rests in diversity of opinions.
- Learning is a process of connecting specialized nodes or information sources.
- Learning may reside in non-human appliances.
- Capacity to know more is more critical than what is currently known
- Nurturing and maintaining connections is needed to facilitate continual learning.
- Ability to see connections between fields, ideas, and concepts is a core skill.
- Currency (accurate, up-to-date knowledge) is the intent of all connectivist learning activities.
- Decision-making is itself a learning process. Choosing what to learn and the meaning of incoming information is seen through the lens of a shifting reality.

While there is a right answer now, it may be wrong tomorrow due to alterations in the information climate affecting the decision.

The next table indicates how different theories of learning relate based on Ermer's and Newby's five questions (Mergel, 1998).

Table 2: Learning Theories (Siemens, 2008)

Property	***Behaviouris m***	***Cognitivism***	***Constructivism***	***Connectivism***
How learning occurs	Black box- observable behaviour main focus	Structure, computatio nal	Social, meaning created by each learner (personal)	Distributed within a network, social, technologically enhanced, recognizing and interpreting patterns
Influencing factors	Nature of reward, punishment, stimuli	Existing schema, previous experiences	Engagement, participation, social, cultural	Diversity of network, strength of ties
Role of memory	Memory is the hardwiring of Repeated experiences - where reward and punishment are most influential	Encoding, storage, retrieval	Prior knowledge remixed to current context	Adaptive patterns, representative of current state, existing in networks
How transfer occurs	Stimulus, response	Duplicating knowledge constructs of "knower"	Socialization	Connecting to (adding) nodes
Types of learning best explained	Task-based learning	Reasoning, clear objectives, problem solving	Social, vague ("ill defined")	Complex learning, rapid changing core, diverse knowledge sources

Network Learning versus Learning Network

At this point, it is necessary to differentiate between Network Learning (NL) and Learning Network (LN). The first one is the logic-didactic-expert layer that determines the implementation of a learning network, while the second one is the set of software, hardware, institutions and even human factors that enable network learning. This is a subset of connectivism (Siemens, 2005). Downes provides connective knowledge as the epistemological foundation of connectivism (Downes, 2005).

"A property of one entity must lead to or become a property of another entity in order for them to be considered connected; the knowledge that results from such connections is connective knowledge". The main goal of a LN is to change the traditional model of knowledge transmission (represented by a one-way teaching and learning) towards a new teacher-student relationship. In this case, the role of teacher has not the full knowledge, but this is distributed by peer groups. This forces to change the traditional role of the teacher in the classroom, and therefore gives a certain role outside them. A key idea is Learning Network strengthens the role of learning guides performed by teachers respect to the traditional class, which strengthens the role of teaching carried out by ones.

Why the Learning Networks are relevant in the implantation of the European Credit Transfer System (ECTS)? Currently, there is a fundamental change for the role of educational networks in Spanish University. What is the reason for so important change respect to the use of educational networks? The answer is ECTS, and therefore the educational university must be renewed. ECTS computes the student time inside classroom and laboratory, as well as study hours, time spent on exams and the one spent in the performance of guided works. This measure unit changes the university activity approach, so the teacher time giving a lecture must be decreased, while the student time "doing things" must be increased. The role of LN are going to perform in this new system is obvious. Teachers design the networks according to their knowledge areas, and turn into mentors for their use by students. The traditional class must be strictly designed to be employed for "everything" outside of LN. Students must to spend time of doing, to do in the LN.

COMPUTER ALGEBRA SYSTEMS AND ADA AUGUSTA LEARNING NETWORK

The symbolic calculus, formal calculus or computer algebra is used to perform mathematical calculations by using numbers, symbols, expressions and formulas in an accurate way, while the numerical calculus works only with floating point numbers (and therefore approximations).

The numerical and symbolic systems have evolved quickly, so quite powerful and versatile scientific computing systems have been released. These ones have many functions and tasks that facilitate solving mathematical problems. For this reason, systems of symbolic computation are very useful tools in science, technology, research, and particularly in the learning of scientific subjects. These systems have changed the use of computers in the learning of physical and mathematical areas.

You might think that the idea that computers running calculations with symbols, and as a result give us symbols, is a contemporary idea. The reality is that this idea was introduced in science, at the origin of computers, in the nineteenth century by Charles Babbage and Augusta Ada, (Larcombe, 1999), (Babbage, 1836), (Menabrea, 1961).

Lady Ada Augusta (1815-1852), Countess of Lovelace, which are daughter of Lord Bairon and, friend and patron of Charles Babbage (1791-1871) wrote [11]:

"There are many ways in which it may be desired in special cases to distribute and keep separate the numerical values of different parts of an algebraic formula; and the power of effecting such distributions any extent is essential to the algebraic character of the Analytical Engine. Many persons who are not conversant with mathematical studies imagine that because the business of the engine is to give its results in numerical notation, the nature of its process must consequently be arithmetical rather than algebraic and analytical. This is an error. The engine can arrange and combine its numerical quantities exactly as if they were letters or any other general symbols; and, in fact, it might bring out its results in algebraic notation were provisions made accordingly. It might develop three sets of results simultaneously, viz, symbolic results.; numerical results (its chief and primary object); and algebraic results in literal notation."

The Ada Augusta Learning Network (AALN) was formed taking the collaborative work made by members of some departments and knowledge areas from the University of Castilla la Mancha, Spain. Their beginning are two end of degree projects carried out in cited University (Miralles et al., 2004). The main goal of Ada Augusta project is to develop a learning network-assisted by using symbolic computation for the learning of scientific- technological subjects at university level.

The AALN is implemented by using WebCT as collaborative system and Mathematica (Wolfram, 1999) as symbolic calculus system by means of web technology. Particularly, *webMathematica* is a new technology allows the generation of dynamic web content with *Mathematica* (Wickham-Jones, 2001 & 2003). *webMathematica* adds interactive calculations and visualization to a web site by integrating *Mathematica* with the latest web server technology. This technology enables you to create websites that allow users to compute and visualize results directly from a web browser. You can find more details about this technology in Section 5.

The AALN has been consolidated through educational innovation projects funded by the University of Castilla-La Mancha, in the frame of support for educational networks (2006- 2007 and 2007-2008). Fig. 1 shows the network in 2006. The colours represent the different campus teachers and the geometric shape to their knowledge area. The lines are the links teacher-to-teacher and teacher-to-core computing. The yellow node represents the computational core of the network. Note that student nodes do not appear in this network structure. This is a matter in progress from the connectivism point of view joint to the Social Network Analysis (SNA).

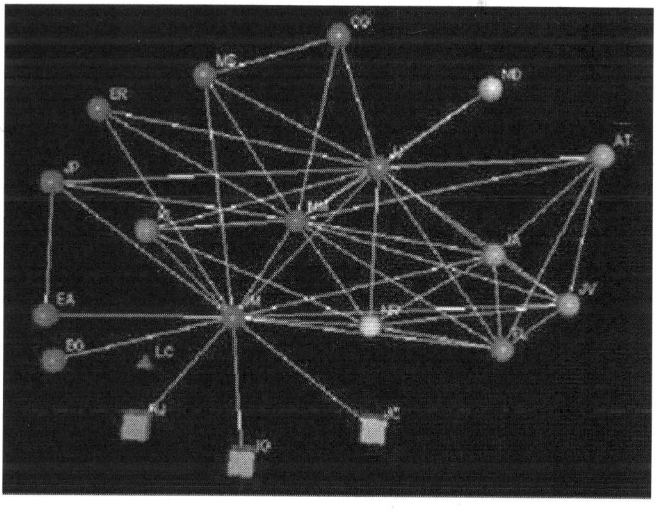

Figure 1: Ada Augusta Learning Network.

WEBMATHEMATICA: A USEFUL WEB TECHNOLOGY FOR LNS BASED ON SYMBOLIC COMPUTATION

webMathematica is a new technology allows the generation of dynamic web content with *Mathematica* (Wickham-Jones, 2001). *webMathematica* adds interactive calculations and visualization to a web site by integrating *Mathematica* with the latest web server technology. This technology enables you to create websites that allow users to compute and visualize results directly from a web browser.

webMathematica and *Mathematica* have the same underlying engine, but they provide fundamentally different user interfaces and are aimed at different types of users. This web technology offers access to specific *Mathematica* applications through a web browser or other web clients. The standard interface provided requires little training to use effectively. In most cases, users neither have to be familiar with *Mathematica* nor need to know they are using *Mathematica*.

There are various important features that *Mathematica* can offer to a web site, including computation, an interactive programming language, connectivity, the *Mathematica* front end, and enhanced support for MathML. Also, there are several areas of use for *webMathematica*. Some of these include web computation, education, publishing, research, and hobbyist calculations.

webMathematica Technology

webMathematica is based on a standard Java technology called servlets (also in JavaServer Pages (JSP) technologies for v2.0). Servlets are special Java programs that run in a Java- enabled web server, which is typically called a servlet container (or sometimes a servlet engine). There are many different types of servlet containers available for several operating systems and architectures. They can also be integrated into other web servers, such as the Apache web Server.

webMathematica allows a site to deliver HTML pages that are enhanced by the addition of *Mathematica* commands. When a request is made for one of these pages, which are called MSP scripts or MSPs, the *Mathematica* commands are evaluated and the computed result is placed in the page. This is done since v2.0 with the standard Java templating mechanism, JavaServer Pages, making use of custom tags.

webMathematica technology uses the requestfresponse standard followed by web servers. A MSP script is processed as part of an HTTP transaction. A client sends a request to the server, which replies with a response. One feature of HTTP requests is that they can send parameters and values to the server. This is essential for any dynamic and interactive behaviour, because parameters are used to select and control the reply. The typical result of a request is an HTML page. However, it could be some other content type, such as a *Mathematica* notebook, or some form of XML. A central component of *webMathematica* technology is the MSP servlet. This is used to process requests and return responses. The MSP servlet deals with each request in a separate thread, allowing more than one request to be processed at the same time. An overview of the workings of a *webMathematica* site is shown in Fig. 2. This is carried out since v2.0 by the kernel manager which calls *Mathematica* in a robust, efficient, and secure manner. The manager maintains a pool of one or more *Mathematica* kernels and, in this way, can process more than one request at a time (Wickham-Jones, 2003).

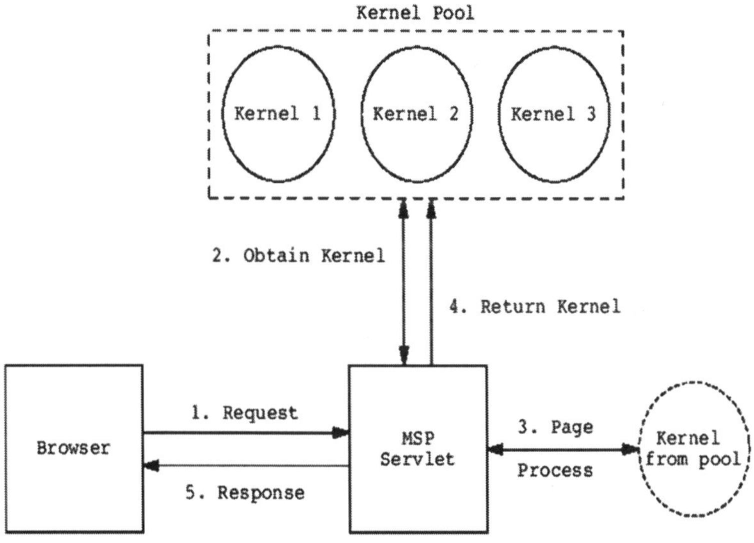

Figure 2: Workings of a *webMathematica* site.

- Browser sends request to *webMathematica* server.
- *webMathematica* server acquires *Mathematica* kernel from the pool.
- *Mathematica* kernel is initialized with input parameters, it carries out calculations,

- and returns result to server.
- *webMathematica* server returns *Mathematica* kernel to the pool.
- *webMathematica* server returns result to Browser.

Mathematica Server Pages

web*Mathematica* v1.0 provides a form of HTML templating based on what are known as MSP scripts. MSPs are standard HTML pages enhanced by the addition of Mathlet tags (<%Mathlet expr %>). When a request is made for one of these pages, *Mathematica* replaces each Mathlet with the result of evaluating and formatting its contents. This usually builds a new HTML document, which is then returned to the client.

MSP scripting technology have been superseded in v2.0 by a new templating mechanism based on JavaServer Pages custom tags. This is a form of JavaServer Pages (JSPs) that use a special library of tags that work with *Mathematica*. JSPs support the embedding of Java into HTML, and are frequently used along with Java Servlets to develop large dynamic web sites. The library of tags is called the MSP Taglib and will work on any compliant servlet engine. One advantage of the use of a tag library is that it can completely hide any use of the Java programming language; this is the case with the MSP Taglib.

The two technologies are very closely related and it is straightforward to convert from one to the other. However, this Section only concerns MSP scripts. More information about *webMathematica* v2.0 can be found in (Wickham-Jones, 2003).

This example (see Fig. 3) evaluates the Date[] function of *Mathematica* each time a user requests the script. The result changes each time the script is requested, demonstrating that this is really a dynamic process.

```
<html>
    ...
    <body>
        ...
        <p>Its current value is:</p>
        <%Mathlet Date[] %>
    </body>
</html>
```

Figure 3: Part of a sample MSP script.

Of course, this example is not interesting, because the user is not doing anything other than retrieving a page. For interactive behaviour, a user needs to send input values with the request.

There are two kinds of variables in *webMathematica*: input variables and page variables. Input variables come with the HTTP request, for example from an input field in an HTML form. These can be identified in *Mathematica* code because they are labelled with a $$ prefix. Page variables are *Mathematica* variables used to hold intermediate values. They are called page variables since they are cleared when the page is finished. Additionally, support for HTTP session variables is provided since v2.0. These are stored in the server and can be used in different pages. This can be useful for saving results from one computation to another. They last for the lifetime of an HTTP session. The next example (see Fig. 4) demonstrates how variables are connected to input values.

This is more elaborated because it contains form and input elements.

```
<html>
    ...
    <body>
        ...
        <form action= "Variables " method= "post "> <!--Form head-->
            Enter something:
            <input type= "text " name= "var " align= "left " size= "10 ">
            <br><%Mathlet $$var %><br>
            <input type= "submit " name= "submitButton " value= "Evaluate ">
        </form>  <!--Form end-->
    </body>
</html>
```

Figure 4: Part of a sample MSP script.

This example has two input tags: the first one allows the user to enter text, and the second one specifies a button that, once pressed, will submit the form. When the form is submitted, it will send information from input elements to the URL specified by the action attribute (in this case, the same MSP). Text entered into the input tag, which uses the name var, will be assigned to the input variable $$var.

The first time the page is accessed there is no value for $$var. When a value is entered in the text field and the Evaluate button pressed, $$var gets a value which is displayed. Note that the value is a *Mathematica* string -if you try and enter a computation such as "5+7", no computation is actually done. If you want the input to be interpreted and evaluated by *Mathematica*, you need to use one of the MSP functions (as MSPBlock or MSPToExpression) that are described in this Section.

webMathematica provides a large library of *Mathematica* commands to handle the many possible ways of working with *Mathematica* computations. Following, it gives a description of cited commands.

a. MSP Functions: Variables

- MSPValue is a function that is useful for extracting the value of variables. Sometimes, it is useful to keep the user input each time the page is used.
- MSPValueQ is a useful function that tests whether variables have values.
- MSPSetDefault is a useful function for setting values of variables.
- MSPToExpression returns the interpreted value of a variable. Also, this function can take a format type for interpretation.
- MSPBlock is one of the key ways to work with variables from the HTTP request.

b. MSP Functions: Showing and formatting results

- MSPExportImage is used to save an image using the *Mathematica* command Export.
- MSPShowAnimation is a convenient way to generate animated GIF images. The argument must be something that evaluates to a list of graphic objects.
- MSPLive3D is a convenient way to work with the LiveGraphics3D graphics applet. This applet displays *Mathematica* three-dimensional graphics and provides support for features such as interactive rotation and resizing.
- MSPFormat is one of the important ways to format results from *Mathematica*. The formatted result can appear in the different format types that *Mathematica* provides for output. In addition the result can be returned as HTML, an image

content, or as MathML.

- MSPShow is the main way to include graphical results from *Mathematica* in an HTML page.

c. MSP Functions: Returning general content

- MSPReturn allows a script to return something different from an HTML result.

- MSPURLStore stores a string of formatted data in the MSP servlet and returns a URL that can be used to retrieve the data.

d. MSP Functions: Other relevant functions

- MSPInclude is a useful function that allows one MSP to include the result of processing another MSP.
- MSPPageOptions sets up global options concerning the current script.
- Several MSP functions throw an MSPException when some error situation appears. These are caught by the script processing code, but it would be permissible for a script author to catch them and process them in some intermediate step.

Also, new functions have been added in *webMathematica* v2.0.

Requeriments and Other Technical Issues

The aim of *webMathematica* technology is to reduce the amount of extra knowledge required for developing a site to a minimum. In practice, this means knowing something about HTML and *Mathematica*. *webMathematica* also aims to automate the management of the site to make running, maintenance, and configuration as convenient as possible.

The minimum technical components for *webMathematical* are:

- A Servlet container supporting the 2.0 API (or higher)
- A Java Runtime Environment (JRE) 1.1 (or higher), Java 2 Version 1.3 (or higher) is recommended

There are many different combinations of hardware and operating systems that support these components. Most systems that run *Mathematica* will support *webMathematica*. Two relevant files are web.xml and MSP.conf. The configuration file MSP.conf holds various site-specific parameters and may need modification for your site. The deployment descriptor file web.xml contains various settings that govern the operation of the web application.

The location for an MSP script is specified in the configuration file MSP.conf with the setting MSPDirectory. This directory is included on the *Mathematica* $Path, and is thus convenient for holding *Mathematica* packages or applications. Of course, you may not want to keep all MSPs in the same directory, preferring to place them in a variety of different locations. You could then access each script with the url http:ff<your_server>fwebMathematicaf MSPf<path_to_your_script>.

However, the way that web*Mathematica* v2.0 maps URLs onto JSPs is very straightforward. The URL names a JSP which lives directly in the web*Mathematica* web application or in a subdirectory.

For some applications, it is useful to use several pools of Mathematica kernels to serve different requests. You can configure the kernels in each pool differently, perhaps with different timeout parameters or different initialization files.

Running a general computation system like *Mathematica* inside of a web site is something that has many potential security hazards for the server. A major danger to any *Mathematica* site is that someone will try to send commands to *Mathematica* that may breach the security of the server. These commands can be sent as the value of input variables passed in from the server. However, *webMathematica* has a fully configurable security system, which is designed to prevent commands that may pose a security risk. You can set your own security model by placing a file in your MSPDirectory called SecurityConfiguration.m. However, the mechanism for locating the security configuration file has changed since v2.0.

The kernel monitor is a servlet that collects information on the running of your site. Upon access, the monitor brings up a page showing the current status of web*Mathematica*, describing various parameters of the site, and giving status information for each kernel. You can find more information about *webMathematica* in (Wickham-Jones, 2001 & 2003).

BUILDING A LEARNING NETWORK BASED ON SYMBOLIC COMPUTATION

Following are described pedagogical elements for making real a Learning Network based on Symbolic Computation. Some resulting applications are also cited. They use previous web technology for generating practical and useful materials, which makes up a Virtual Laboratory. However, this Section is aimed on pedagogical purposes and not on technological ones.

Virtual Laboratory

The Virtual Laboratory is an interactive environment for creating and conducting simulated experiments: a playground for experimentation (Mercer et al., 1990). Several and diverses Virtual Labs can be found in research and educational areas. Virtual Laboratory presented here consists of domain-dependent web applications which form a Learning Network2. Particularly, all applications are concerned symbolic computation.

Examples of applications for resolution of algebraic equations systems, Laplace transforms, resolution of electrical networks by means of loops method or Evan's Root Locus are given by AALN.

These applications require a diverse knowledge of *Mathematica* for users. There are applications absolutely transparent for them, who not need to know they are using *Mathematica.* Users are taken through a sequence of web pages in which they select different input parameters and submit data to build up a sequence of results. In opposite, *Mathematica* experience is required for other kind of applications. A user can type and submit a sequence of *Mathematica* expressions and receives a result. Following are cited pedagogical elements for making these applications.

Interactive and Reusable Learning Units

We define Interactive and Reusable Learning Units (IRLU) as any small and programmable tools in which appear questions with a simple difficulty level. The answers of these questions are one or more algebraic expressions that must be written by the student. These answers are not made by means of a comparison between different expressions. This is one step more where the multiple choice tests or numerical results ones are not acceptable. The main advantage in the use of IRLU is the chance of admitting, answers which could be written in different ways but with the same meaning. It is needed that this programmable tool has a high performance for algebraic simplification. Due to this reason, they must be developed by modern Symbolic Calculus platforms. Also, these platforms should be low- cost in order to be widely used.

These tools must be reusable because they could be easily modifiable. In this way, we can get a lot of applications with only a few changes. They are apparently different but with the same structure. Below, we show some examples applied in different technological fields as Physics, Electricity and Systems Theory.

This is thc basis schcmc for our work:

- Formulation of a question
- In a concise way.
- It will be allowed the insertion of figures.
- Not promote the possibility of ambiguities in the nomenclature.

- Specification (not available for the user) of the correct solution.
- It must take the syntactic rules of mathematical package and be the most simple possible.
- It must to distinguish between kind of two cases: one for the homogeneous solution (equal to zero) and others where they can appear anomalous but their solutions are correct.
- To collect the answer of the user.
- A previous planning is needed in order to make the response as short as possible and without ambiguous interpretations.
- In all cases, they must include notes for leading users (some example of response possible and remainders of valid syntax)
- Comparison between the correct solution and provided by user.
- First task is calculating the ratio of available to correct solutions.
- Next task is to make an algebraic simplification of this ratio using the symbolic program.
- The solution given by the user will be right if the result of this simplification be the unit.
- Sometimes it will be needed to take into account situations which look like to be maliciously made, but objectively they are good choices.
- Notification to the user about the result obtained. It can be a simple CORRECTfERROR
- It is possible to make available the correct solution
- Another possibility is to make some learning suggestions depending on the result obtained.
- Including information or help on demand (this is optional)
- It can include a guide (as a summary) of basic concepts with an flexible level.
- Anyway, the premise Reusable always is necessary.

The IRLU are very interesting because of they are easily programmable. It is possible to make a set of templates and then next IRLUs will be more simple to build. They can be associated in small conceptual modules and used in an aleatory way for the evaluation of the students. However, we can emphasize that they are very useful for the self-evaluation of the students, and this is a very important part in the new European Higher Education Area. We show a basic example3 of IRLU, for Electrical network theory. The formulation can be: "Given the circuit in the figure, write resulting equation applying the Kirchhoff's Loop rule to the loop #1, in the form equation = 0"

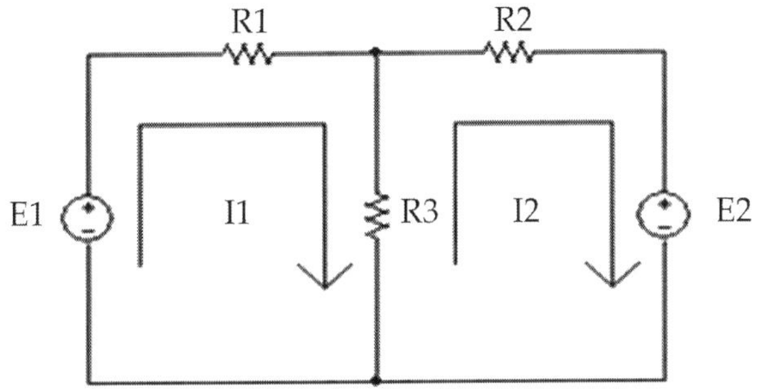

Figure 5: Workings of a webMathematica site.

Notes:

- has to write in function of I1, I2, R1, R2, R3, E1 y E2.
- variables must be separated by some mark (white space is considered as multiplication)
- can be careful with the Mathematica syntax (specially with capital and lower case letters)

A possible solution could be -E1+I1*R1+R3 (I1-I2) = 0

Anyway, other possible solutions are (for example):

- E1-I1*R1+R3(I2-I1) = 0
- -E1+(R1+R3)I1-R2*I2 = 0

An appropriate design of these kind of questions is useful for reinforce the learning in an autonomous and solid way. In this sense, this is fundamental in the use of ECTS.

Applications based on Symbolic Computation for Supervised Works

In this section we describe the incorporation of Symbolic Calculus beyond the evaluation and support tasks to calculus (which are the 90% of cases). This is a different concept (more pretentious) that we can consider as part of guided learning. Our objective is not to replace the presentation lectures, but the student has a help in order to get the pre-established competences, for distance and net education.

Nowadays, there are a lot of application (SCILAB4, Toolboxes de MATLAB5, GNU Octave6, Maple7, Mathematica8, ..) in the field of technology that might be sufficient to achieve our goal. Nevertheless, this is not so. These packages are tools that use both teacher and student, but itself do not give the demanded guide by us. For example:

- There are a subject (Systems Theory) that make use of a methodology (Evans, 1950) that allows to make a diagram which shows the roots variation when varies an adjustable parameter change (normally k) from 0 to +oo. This procedure, known as Root Locus is very important in many applications of different fields. For this reason, the student must know how to construct this geometric site, trough the application of a series of simple rules. Some of the above mentioned computer applications have specific commands ("evans" in SCILAB, "rlocus" in Octave and MATLAB, ..) that allow to construct graphically that figure. However the student must be able to construct this Locus Root by itself and then, to compare their result to provided by other tools. In this case, the applications require more exhaustive programming skills and will use the Symbolic Calculus tools for leading step-by- step to the student to get the correct result.
- The simulation of electric and electronic networks from a numeric point of view is made since 70's. The program SPICE9 is the most known. It is well established that the use of these simulators work in a different way like is used in teaching. The process of analysis of an electric network in the classroom has nothing to do with the management of these programs. However, if we applied appropriately Symbolic Calculus, we could get a similar analysis of an electric network both in the classroom and in the network. This would favour the auto-learning and then to get the desired objectives.

In both cases, we have to add two new components that not appear in the IRLU: to achieve the solution be found by the tool and, on the other side, the followed process for problem solving be available to the students, so they can make any look up in any time.

IRLUs' Examples

In this subsection we will consider some examples of the use of IRLUs in different learning contexts.

Physics:

Let's use an example from dynamics of the particle. The ultimate goal is to verify whether the student has got the skill to obtain the resultant of forces acting on a body. The wording of the problem would be:

Expresses the component of resultant force (in the positive direction of x- axis) acting on the mass on the plane inclined at angle b, in the figure, as a function on the mass ***m****, the gravity* ***g****, the angle* ***b*** *and the coefficient of friction* ***u****.*

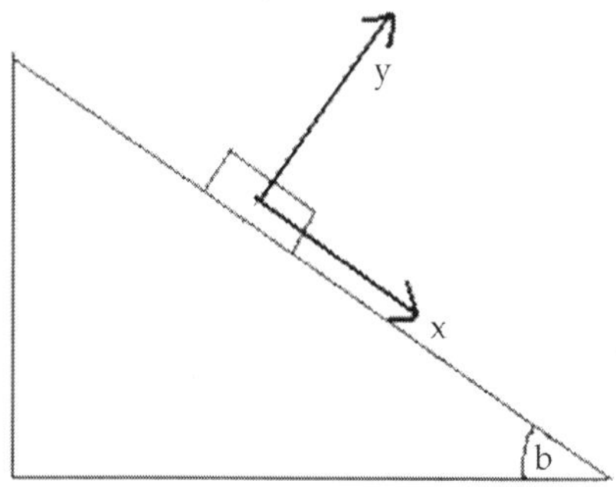

Figure 6: Mechanical example.

There are several different ways to write the correct answer. For example: m*g(Sin[b]-u*Cos[b]) or Cos[b](m*g)(Tan[b]-u), or g*m*Sin[b]*(1-u*Cot[b]), or..

Electrical network theory

Another example similar to that already indicated in this field, but where it is permissible to ask several things at once:

For the network in the figure, writes the right expressions (according to Ohm's Law) Vl=__and V2=____________Write the equation of loop (without using Ohm's law)__________=0

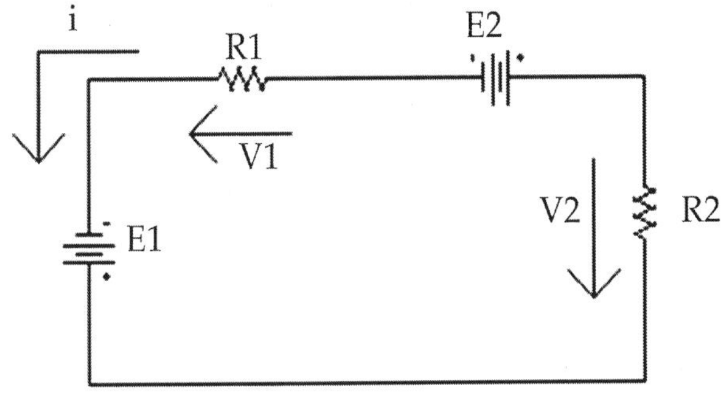

Figure 7: Electrical network example.

Control Systems

The most common technique used for model linear, time-invariant systems in control systems, is the block diagram, with the mathematical model represented as a transfer function. One of the first steps is to obtain the overall transfer function equivalent of the entire system. If the system is complex, the level of simplification is often different depending on the procedure (although the result should be unique). The way to verify the accuracy of the final result is to simplify previously compared with the correct solution. Two examples of use could be:

- In the following block diagram, you must to express the transfer function equivalent of the overall system, using algebraic techniques from block diagrams.

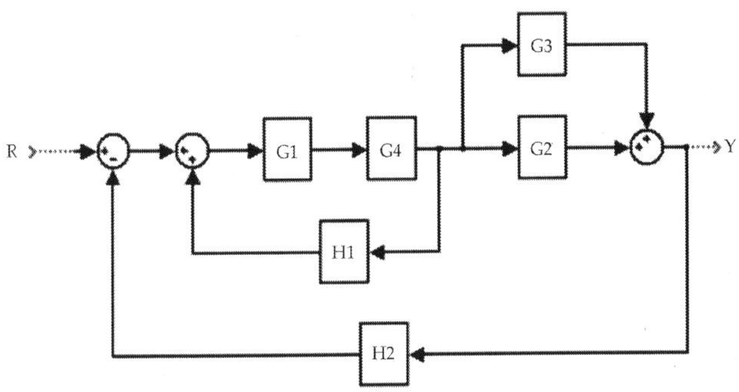

Figure 8: Control systems example 1.

- Write the final transfer function for the following system, using Mason's Rule.

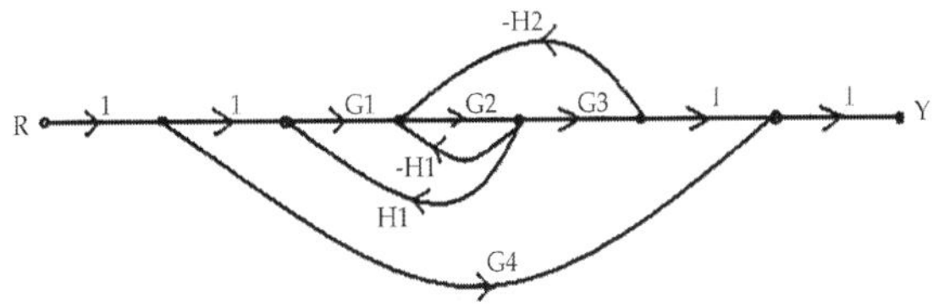

Figure 9: Control systems example 2.

As you can see, the scope is very extensive. In order to implement the computer program, the general procedure is to make a template-type, and change both the wording and the display-figure and of course, indicate the correct solution in each case. We can create different templates (e.g., based on the number of responses we want to ask) and then reused it in the subjects where we consider appropriate.

The examples discussed in this paragraph and others are available at the following URL: http:ffwww.poincare.pol-ab.uclm.esfwebMath

CONCLUSIONS

This chapter presents a model for the teaching assisted by network in the framework of connectivism as a learning theory in the digital age. Following are showed the main conclusions:

- The effort of building a computational kernel for a learning network based on symbolic calculus has been established inside of connectivism context.
- Connectivism has been considered as the best model to guide the process of creating a network learning.
- It has developed a computational core for a learning network through symbolic calculation and using webMathematica technology.
- The kernel of symbolic calculus in the learning network is considered as capable of fostering the interaction among teachers from different Scientific-technical subjects in the ECTS context.
- It has developed a set of interdisciplinary, interactive and reusable learning units, based on symbolic calculus and supported by the computational kernel of the learning network.

- A work in progress is to decide how to introduce the nodes: students, teachers and subjects, in the developed learning network, according to the connectivism paradigm.
- Another work in progress is to introduce the measures from complex networks topology, in order to evaluate the performance of network learning.

REFERENCES

1. Albert, R. & Barabasi, A.L. (2002). Statistical Mechanics of Complex Networks. *Rev. Mod. Phys.*, 74, 47-97
2. Babbage, C. (1836). *Scribbling books 2*. Science Museum Library, London (UK)
3. Menabrea, LF. (1961). *Sketch of the analytical engine invented by Charles Babbage. With notes upon the memoir by the translator, Ada Augusta, Countess of Lovelace, In: Charles Babbage and his calculating engines, part II*, Morrisson, P. & Morrison, E., (Eds.)., Cover Publications, New York
4. Barabasi, A.L. (2002). *Linked: The new science of networks*, Perseus Publishing, Cambridge (MA, USA)
5. Baumeister, P. (2005). Networked learning in the knowledge economy: A systemic challenge for universities. *European Journal of Open, Distance and E-Learning*. Retrieved on March 04, 2009, from http:ffwww.eurodl.orgfmaterialsf contribf2005fBaumeister.htm
6. Benkler, Y. (2006). *The wealth of networks: How social production transforms markets and freedom*, Yale University Press, New Haven (CT, USA)
7. Bernes Lee, T.; Cailliau, R.; Luotonen, A.; Frystyck Nielsen, H. & Secret, A. (1994). The world wide web. *Communications of the ACM*, 37, 76-82
8. Boccalettia, S.; Latorab, V.; Morenod, Y.; Chavezf, M. & Hwanga, D.U. (2006). Complex networks: Structure and dynamics. *Phys. Rep.*, 424, 175308 Bollobas, B. (1985). *Random Graphs*, Academic Press, London (UK)
9. Borgatti, S.P. & Foster, P. (2003). The network paradigm in organizational research: A review and typology. *Journal of Management*, 29, 991-1013
10. Bornholdt, S.; Schuster, G.H. (2003). *Handbook of Graphs and Networks: From the Genome to the Internet*, Wiley-VCH, Berlin (Germany)
11. Clergue, G. (1997). *L'apprentissage de la complexite*, Hermes, Paris (France)
12. Downes, S. (2004). Learning Networks, *International Journal of Instructional Technology and Distance Learning*, Perth (Australia),

October 2004. Retrieved on March 04, 2009, from http:ffwww.downes.cafcgi-binfpage.cgi?post=20

13. Downes, S. (2005). *An Introduction to Connective knowledge.* Retrieved on September.01, 2006, from. http:ffwww.downes.cafcgi-binfpage.cgi?post=33034
14. Downes, S. (2006). *Learning Networks and Connective Knowledge.* The University of Georgia. Retrieved on March. 01, 2009, from http:ffit.coe.uga.edufitforumfpaper92f paper92.html
15. Downes, S. (2007). An Introduction to Connective Knowledge, *Proceedings of the International Conference*. ISBN: 978-3-902571-67-0, June 25-26, 2007, Theo Hug (Hg.), 2008, Innsbruck University Press
16. Driscoll, M. (2000). *Psychology of Learning for Instruction*, Allyn & Bacon, Needham Heights (MA, USA)
17. Evans, W.R. (1950). Control system synthesis by root locus method. *Trans. AIEE*, 69, 66-69 Hill, W.F. (2002). *Learning: A survey of psychological interpretation*, Allyn & Bacon, Boston (MA, USA)
18. Larcombe, LP. (1999). *On Lovelace, Babbage and the origins of computer algebra, Computer algebra systems a practical guide*, Wister M, Chischester (UK)
19. Marin, I.; Miralles, J.J.; Artigao, M.M. & Martinez, J.A. (2008). Learning Network-Assisted by using symbolic computation: A useful technology in the learning of scientific-technological subjects, *Proceedings of the International Technology, Education and Development Conference 2008*, Valencia (Spain), March 2008
20. Mason, M. (2008). What is complexity theory and what are its implications for educational change. *Educational Philosophy and Theory*, 40, 35-49
21. Mercer, L.; Prusinkiewicz, P. & Hanan, J. (1990). *The Virtual Laboratory Environment*, 1990-91. Available: http:ffwww.cpsc.ucalgary.cafResearchfbmvfvlab.origfvlab.html
22. Mergel, B. (1998). *Instructional design and learning theories.* Retrieved January 15, 2009, from University of Saskatchewan, College of Education Web site: http:ffwww.usask.caf educationfcourseworkf802papersfmergelfbrenda.htm

23. Miralles, JJ.; Marin, I. & Fernandez, D. (2004). Redes de Aprendizaje con capacidades de calculo simb6lico (I): Proyecto Ada Augusta. *Mundo Linux*, 65, 22-28
24. Morin, E. (1999). *Le Moigne Jean-Louis, L'Intelligence de la Complexite*, L'Harmattan, Paris (France)
25. Morin, E. (2001). *La methode : L'identite humaine*, Editions du Seuil, Paris (France) Newman, M.E.J. (2003). The structure and function of networks. *SIAM Rev.*, 45, 167-256
26. Scharnhorst, A. (2003). Complex networks and the Web: Insights from nonlinear physics. *JCMC*, 8. Retrieved March 04, 2009, from http:ffjcmc.indiana.edufvol8f issue4fscharnhorst.html
27. Siemens, G. (2005). Connectivism: A Learning Theory for the Digital Age. *International Journal of Instructional Technology and Distance Leraning*, 2, (January 2005) 3-9, ISSN 1550-6908. Retrieved April 04, 2009 from http:ffwww.itdl.orgfJournalfJan_05f index.htm
28. Siemens, G. (2006). *Knowing Knowledge*, Lulu.com , ISBN-10: 1430302305.
29. Siemens, G. (2008). *Learning and knowing in networks: Changing roles for educators and designers*.
30. University of Georgia IT Forum. Retrieved January 22, 2009 from http:ff it.coe.uga.eduf itforumfPaper105fSiemens.pdf
31. Stephenson, K. (2002). *A quantum theory of trust. strategy+business*, 29. Retrieved March 04, 2006, from http:ffwww.strategy-business.comf39910
32. Strogatz, S.H. (2001). Complex Networks. *Nature*, 410, 268-276
33. Vaill, P.B. (1996). *Learning as a Way of Being*, Jossey-Blass Inc., San Francisco (CA, USA)
34. Wang, X.F. (2002). Complex Networks: Topology, Dynamics and Synchronization. *Int. J.Bifurcation and Chaos*, 12, 885-916
35. Watts, D.J.; Strogatz, SH. (1998). Collective dynamics of small-world networks. *Nature*, 393,440-442
36. Watts, D.J. (1999). Networks, dynamics, and the Small-World phenomenon. *American Journal of Sociology*, 13, 493-527

37. Watts, D.J. (2003). *Six degrees: The science of a connected age*, Norton, New York (USA) Wickham-Jones T. (2001). *webMathematica: A user guide (vl.0)*, Wolfram Research Inc. Wickham-Jones T. (2003). *webMathematica: A user guide (v2.0)*, Wolfram Research Inc. Wolfram, S. (1999). *The Mathematica Book*, Cambridge University Press, Cambridge (UK) Wright, A. (2007). *Glut: Mastering information through the ages*, Joseph Henry Press, Washington DC (USA)

Chapter

MULTIPLE FACTORIAL ANALYSIS OF SYMBOLIC DATA

7

Barnabé Tang Ahanda, Jean Gérard Aghoukeng Jiofack, Romain Germain Nzangué, Gilbert Hapi Mbiakop

Department de Mathematics, Faculty of Sciences, University of Yaounde 1, Yaounde, Cameroon

ABSTRACT

This document presents an extension of the multiple factorial analysis to symbolic data and especially to space data. The analysis makes use of the characteristic coding method to obtain active individuals and the reconstitutive coding method for additional individuals in order to conserve the variability of assertion objects. Traditional analysis methods of the main components are applied to coded objects. Certain interpretation aids are presented after the coding process. This method was applied to poverty data.

Citation: B. Ahanda, J. Jiofack, R. Nzangué and G. Mbiakop, "Multiple Factorial Analysis of Symbolic Data," Applied Mathematics, Vol. 3 No. 12A, 2012, pp. 2148-2154. doi: 10.4236/am.2012.312A295.

Keywords: Symbolic Object; Assertion Object; Multiple Factorial Analysis; Principle Components Analysis; Poverty

INTRODUCTION

Although the traditional analysis of data is based on a complete theory, it poses restrictive conditions for the resolution of problems: it requires that variables be single-valued (the value of a variable for an object is a unique and specific).

Recent evolutions in the information system have made it possible to find increasing and more complex data which need a more in-depth formalization than the one indicated in the usual rectangular table. The later does not take into account the possible variability of description as well as the uncertainty concerning certain variables or certain objects. A new formalism called symbolic objects has been developed to represent complex data like concepts, skills (see [1,2]). These objects concern complex data and provide new skills at the exit in the form of symbolic objects.

The techniques of factorial analysis (see [3,4]) which are data reduction and representation methods, are well known and well used. Certain stakeholders have extended these methods to symbolic objects by using the coding and decoding techniques (read [5,6]). The issue of multiple factorial analysis (MFA) is of particular interest to us (see [7,8]). A priori, this concerns the case where there is a group structure on variables. Its problem is enhanced by the characterization of groups and the search for a typology of groups of variables. To our knowledge, this method is not yet applied to symbolic objects.

In this article, we extend the MFA to a symbolic data base whose variables are structured into groups. It is a generalization of the MFA basic principle on the one hand, and the principle components analysis (PCA) techniques on the other hand (see [9]). We conclude this document with an application of the method presented to poverty data.

SYMBOLIC DATA BASE

General Context

A symbolic data base (SDB) is considered as represented in the Table 1 and made up of triplet $B=\{Y,C,A\}$ with $C=\{C_1,C_2,\cdots,C_I\}$ representing all the I objects, $Y=\{y_1,y_2,\cdots,y_J\}$ and $A=\{a^1,a^2,\cdots,a^I\}$ representing respectively the set of the J variables and the assertion objects used to describe I objects. We suppose that the SDB is associated with a measured space $(\Omega,\mathcal{F},\mu)$ an σ-algebra $\mathcal{C}$ on C, $O=\{O_1,O_2,\cdots,O_J\}$ where O_j is the space of observation of the variable $y_j, j=1,2,\cdots,J$. This space includes an tribe $\mathcal{O}_j$. Let $(X_{c_i,j})_{j=1,2,\cdots,J}$ be a set of measurable functions defined on Ω, deriving their value from the measurable spaces $(O_j,\mathcal{O}_j)$ and used to describe all the c_i objects. For I from 1 to I, $a^i=\bigwedge_{j=1}^{J}\left[y_j=X_{a^i,j}\right]$ a^i is a Boolean assertion so that from j from 1 to J, there is a measurable part $V_{a^i,j}\in\mathcal{O}_j$ so that we can have $X_{a^i,j}=X_{c_i,j}$ and $V_{a^i,j}=X_{a^i,j}(\Omega)$. In practice, it is written

$$a^i=\bigwedge_{j=1}^{J}\left[y_j=V_{a^i,j}\right]. \tag{1}$$

Table 1: Symbolic data base

	y_1		y_j		y_J
c_1	$[y_1=X_{a^1,1}]$	...	$[y_j=X_{a^1,j}]$	...	$[y_J=X_{a^1,J}]$
c_i	$[y_1=X_{a^i,1}]$	...	$[y_j=X_{a^i,j}]$	...	$[y_J=X_{a^i,J}]$
c_I	$[y_1=X_{a^I,1}]$	...	$[y_j=X_{a^I,j}]$	...	$[y_J=X_{a^I,J}]$

The observation space is therefore the product space $\Theta_j=\prod_{j=1}^{J}O_j$. This space includes an product tribe $\Xi_j=\prod_{j=1}^{J}\mathcal{O}_j$. To proceed, it is supposed that the descriptors are quantitative, i.e that O_j is a part of $\mathbb{R}$.

Thus, it is generalized that $X_{a^i,j}(\Omega)=V_{a^i,j}=\left[\underline{x_{ij}},\overline{x_{ij}}\right]$

where $\underline{x_{ij}},\overline{x_{ij}}$ represent respectively the minimum and maximum values per a^i object for the y_j variable. Consequently, we shall treat the Boolean assertions whose SDB is indicated in the following Table 2.

Group Structure

We shall assume that the J variables are divided into K groups with J_k as the number of k group variables; p_i the weight associated with the assertion object a^i and m_ℓ the weight of the y_ℓ variable. We shall designate I, K, J_k as the set and its cardinal. Let us consider $B_k = \{Y_k, C, A_k\}$ as the symbolic data base derived from B but reduced from the k group variables denoted Y_k. For i from 1 to I, the a^i object for the k group is represented by the assertion $a_k^i = \bigwedge_{j \subset J_k} \left[y_j = V_{a_k^i, j} \right]$. Formula (1) may be re-written as follows, while taking into consideration the existence of the following groups:

$$a^i = \bigwedge_{j \in J_1} \left[y_j = V_{a^i, j} \right] \wedge \cdots \bigwedge_{j \in J_k} \left[y_{j + \sum_{r<k} j_r} = V_{a_k^i, j + \sum_{r<k} j_r} \right]$$

$$\wedge \cdots \bigwedge_{j \in J_K} \left[y_{j + \sum_{r<K} j_r} = V_{a_K^i, j + \sum_{r<K} j_r} \right]$$

Objective

Like in the traditional case, the MFA based on symbolic data corresponds to the following three majors objectives:

1) Drawing up a typology of concepts, especially similar concepts from the point of view of all the variables after balancing the contribution of each group.
2) Comparing the typologies of individuals defined by groups of variables.
3) Exposing a typology of groups of variables and interpreting the proximities between them.

Table 2: Symbolic data base of boolean assertions

	y_1		y_j		y_J
c_1	$[\underline{x}_{11}, \overline{x}_{11}]$	...	$[\underline{x}_{1j}, \overline{x}_{1J}]$	...	$[\underline{x}_{1J}, \overline{x}_{1J}]$
c_i	$[\underline{x}_{i1}, \overline{x}_{i1}]$	...	$[\underline{x}_{ij}, \overline{x}_{iJ}]$	...	$[\underline{x}_{iJ}, \overline{x}_{iJ}]$
c_I	$[\underline{x}_{I1}, \overline{x}_{I1}]$	...	$[\underline{x}_{Ij}, \overline{x}_{IJ}]$	...	$[\underline{x}_{IJ}, \overline{x}_{IJ}]$

However, taking into consideration the variability of objects adds an interesting element to this problem because coherence warrants us to provide exit elements in keeping with the variability.

ANALYSIS AND INTERPRETATION

We shall present algorithm for a symbolic digital MFA. This approach essentially repeats the one used by the traditional MFA as well as that of the PCA on symbolic objects. The following stages shall be used:

1) Data coding;
2) PCA of groups;
3) Weighting of the data table;
4) PCA symbolic digital of all data;
5) Traditional MFA representation and interpretation;
6) Superimposed representation of symbolic objects and average cloud;
7) Representation of assertions of variables and symbolic factors of groups

Data Coding

Either Δ all the assertions in the SDB, the m coding order is any C application to Δ with values in $\left(\mathbb{R}^{j}\right)^{m}$.

In this work, we shall therefore use the centre and summit methods which is referred to as characteristic coding and reconstitutive coding respectively.

An assertion object is graphically represented by a hypercube. The characteristic coding represents each object by a point which is the centre of gravity of the assertion object. The reconstitutive coding represents an object by 2^{J} points which constitute the summit of the hypercube. Characteristic coding gives us the active elements of the analysis while reconstitutive coding gives the additional elements of the analysis that enable the reconstitution of the exit symbolic objects.

Weighting of Variables

Let X be the data table obtained after data coding. A J columns and $I\times\left[1+2^{J}\right]$ lines table is obtained where I individuals are active and the others in addition. For each group, a principle components analysis (PCA) on the

$X_k, k = 1 \cdots K$ tables where X_k is the X table reduced to the k group value. A series of $\lambda_i^k, i = 1, \cdots, k, k = 1, \cdots, K$. eigenvalues is obtained for each analysis. It is assumed that the real values in each group are classified in descending order. Each y_ℓ variable is weighted by $\frac{m_\ell}{\lambda_1^k}$ if the y_ℓ variable is derived from the k group. The weighted table is evaluated $X^{(p)}$.

Data Processing

Weighting conserves the internal structures of each group but changes the structure of the overall cloud (see [10]). In this case, it is referred to as the average cloud. As in the traditional case, we have represent equivalent objects by respective lines of the same K bases $\tilde{B}_k^{(p)}$ rank, with $B_k^{(p)}$ being the weighted B_k base.

$$\tilde{B}_k^{(p)} = \boxed{0} \boxed{B_k^{(p)}} \boxed{0} \Rightarrow \tilde{X}_k^{(p)} = \boxed{0} \boxed{X_k^{(p)}} \boxed{0}$$

The N_j^k K clouds may be simultaneously represented in the $\mathbb{R}^J$ space of variables by breaking down $\mathbb{R}^J$ into direct sum spaces isomorphic to $\mathbb{R}^{J_k}$ spaces.

The average cloud is obtained by a homothetic transformation of the $B^{(p)}$ $1/K$ relation weighted data base. It represents average individuals and average concepts in $\mathbb{R}^k$. The average cloud is obtained by a homothetic transformation of the $B^{(p)}$ $1/K$ relation weighted data base. It represents average individuals and average concepts in $\mathbb{R}^k$. Let us note B^* the SDB of the average cloud where B_k^* constitutes its restriction in the k group.

Proposition 1. The symbolic digital PCA of the B^* base leads to the same factors as the B base characteristic coding MFA.

Proof. It is immediate. The B^* base is constructed to this effect. □

Canonic Variable

The re-writing of table X^* in the principle components base $(F_1, \cdots, F_J)$ leads to a representation of objects by a new SDB $B = \{F, O, A_F\}$ whose assertion is written

$a_F^i = \bigwedge_{j=1}^{J}\left[F_j = W_{a_F^i,j}\right]$. Assertions of the variables associated with concepts are written $F^j = \bigwedge_{i=1}^{I}\left[F_j = W_{a_F^i,j}\right]$.

These factors are called canonic variables.

Superimposed Representation of Individuals and Concepts

Equivalent Assertions

A $B^{(p)}$ based a^i assertion is represented in each $B_k^{(p)}$ sub-base by a_k^i assertions which are called equivalent assertions associated with a^i. These are the different representatives in the $B^{(p)}$ sub-bases of the same assertion.

Superimposed Representation of Individuals

Here, individuals are described by the characteristic coding. An individual representing an object is represented by the expectation of unpredictable random. variable associated with the skill it represents. This representation no doubt has some shortcomings, but it gives us the first tendencies which are sometimes enough. In concrete terms, each group is projected with additional elements on the PCA of table X^*.

Proposition 2. Each individual (average individual) of table X^* is the centre of gravity of individuals observed from the point of view of all groups.

Proof. This is the property of traditional MFA. □

This proposition explains the notion of average cloud. The study of the attributes of variability of description indicated below will comfort us in this notion.

Superimposed Representation of Concepts

Symbolic digital PCA results will be used to obtain a superimposed representation of objects while taking into consideration variability. It based to the following proposition:

Proposition 3. Either $a^i = \bigwedge_{j=1}^{J}\left[y_j = V_{a^i,j}\right]$ an assertion of BDS B. We note $b^i = \bigwedge_{k=1}^{K}\left[u_k = V_{b^i,k}\right]$ the assertion representing the c_i objects in the space of the PCA principle components of this SDB. Let $G=\left(X_1^G,\cdots,X_J^G\right)$ the centre of gravity of the characteristic coding and u_{jk} the kth component of the jth factorial axis. Thus we have:

$$\min\left(W_{b^i,k}\right)=\sum_{j,u_{kj}<0}\left[\max\left(V_{a^i,j}\right)-X_j^G\right]u_{kj}$$

$$+\sum_{j,u_{kj}<0}\left[\min\left(V_{a^i,j}\right)-X_j^G\right]u_{kj}$$

$$\max\left(W_{b^i,k}\right)=\sum_{j,u_{kj}<0}\left[\min\left(V_{a^i,j}\right)-X_j^G\right]u_{kj}$$

$$+\sum_{j,u_{kj}<0}\left[\max\left(V_{a^i,j}\right)-X_j^G\right]u_{kj}$$

Proof. We note $\left(x_{ij}^G\right)_{j=1,\cdots,J}$ the coordinate of G^i centre of gravity of a^i assertion, and $\left(y_{kj}^G\right)_{j=1,\cdots,J}$ the coordinate of the image of G^i in the new basis. Let u_{jk} be the jth component of the kth factorial axis (denoted u_k), we note $x_r^i=\left(x_{ij}^r\right)_{j=1,\cdots,J}$ the coodinate of an elementary individual of a^i assertion for a point R^i, and $y_r^i=\left(y_{kj}^r\right)_{j=1,\cdots,J}$ his coordinate in the new basis.

We have:

$$y_{kj}^r=\overrightarrow{\boldsymbol{GR}}\cdot u_k=\sum_{j=1}^{J}\left(x_{ij}^r-x_{ij}^G\right)u_{jk}$$

then

$$\max\left(W_{b^i,k}\right)=\max\left\{y_{ki}^r/x_{ij}^r\in V_{a^i,j}\right\}$$

$$=\max\left\{\left[\sum_{j=1}^{J}\left(x_{ij}^r-x_{ij}^G\right)u_{jk}\right],x_{ij}^r\in V_{a^i,j}\right\}$$

as y_{kj}^r depend of x_{ij}^r, we have

$$\max\left(W_{b^i,k}\right)=\sum_{j,u_{kj}<0}\left[\left(\min\left\{x_{ij}^r,x_{ij}^r\in V_{a^i,j}\right\}\right)-X_j^G\right]u_{kj}$$
$$+\sum_{j,u_{kj}>0}\left[\left(\max\left\{x_{ij}^r,x_{ij}^r\in V_{a^i,j}\right\}\right)-X_j^G\right]u_{kj}$$

Therefore

$$\max\left(W_{b^i,k}\right)=\sum_{j,u_{kj}<0}\left[\min\left(V_{a^i,j}\right)-X_j^G\right]u_{kj}$$
$$+\sum_{j,u_{kj}>0}\left[\max\left(V_{a^i,j}\right)-X_j^G\right]u_{kj}.$$

With the similar methode, we have

$$\min\left(W_{b^i,k}\right)=\sum_{j,u_{kj}<0}\left[\max\left(V_{a^i,j}\right)-X_j^G\right]u_{kj}$$
$$+\sum_{j,u_{kj}>0}\left[\min\left(V_{a^i,j}\right)-X_j^G\right]u_{kj}.$$

This lead that the B_k representing group k will be projected successively with additional elements on the B^* base symbolic digital PCA. The superimposed representation on the same factorial plan enables us to generally observe the position of an object, and the point of view of each group by using variability. We then have the following result :

Proposition 4. Either a^i an assertion object represented in B^* by $\left(a^i\right)^*$. Let us note a_k^i the a^i assertion observed from the k group point of view. In the B^* base, the a^i assertion is written

$$a^i=\bigwedge_{j\in J_1}\left[y_j=\frac{1}{J\sqrt{\lambda_1}}V_{a_1^i,j}\right]$$
$$\wedge\cdots\bigwedge_{j\in J_k}\left[y_{j+\sum_{r<k}j_r}=\frac{1}{J\sqrt{\lambda_k}}V_{a_k^i,j+\sum_{r<k}j_r}\right]$$
$$\wedge\cdots\bigwedge_{j\in J_K}\left[y_{j+\sum_{r<K}j_r}=\frac{1}{J\sqrt{\lambda_K}}V_{a_K^i,j+\sum_{r<K}j_r}\right]$$

Either $a^i=\bigwedge_{j=1}^{J}\left[y_j=V_{a^i,j}^*\right]$ we note

$a_k^i = \bigwedge_{k-1}^{K}\left[F_k = W_{a_k^i,k}\right]$ the assertion representing the c_i object from the k group point of view in the space of the PCA principle components of this SDB. Either

$G = \left(X_1^G, \cdots, X_J^G\right)$ the centre of gravity of X^*, and F_{kj} the kth component of the jth factorial axis. Thus we have:

$$\min\left(W_{a_F^i,k}\right) = \sum_{j \in J_k, F_{kj} < 0}\left[\max\left(JV_{a^i,j}^*\right) - X_j^G\right]F_{kj} + \sum_{j \in J_k, F_{kj} > 0}\left[\min\left(JV_{a^i,j}^*\right) - X_j^G\right]F_{kj}$$

$$\max\left(W_{a_F^i,k}\right) = \sum_{j \in J_k, F_{kj} < 0}\left[\min\left(JV_{a^i,j}^*\right) - X_j^G\right]F_{kj} + \sum_{j \in J_k, F_{kj} > 0}\left[\max\left(JV_{a^i,j}^*\right) - X_j^G\right]F_{kj}$$

Proof. This is the consequence of the notations of the previous proposition. □

The representation of superimposed objects in the canonic base (that of factors) therefore respects the principle of centres of gravity. The hypercube representing an object is thus the centre of gravity of the hypercubes representing the same individual from the point of view of different groups. The MFA symbolic digital algorithm helps to preserve this essential property.

Representation of Variables and Assertions of Variables

The symbolic digital MFA enables us to define many representatives that contribute to the illustration of factors in the general analysis as well as in the different groups. The determination of the main factors of groups and the general analysis lead to many representations of variables and assertions of variables. The superimposed representation of these factors (as assertions of variables) on canonic variables then helps to find and interpret common factors and specific factors.

Representation of Variables and Assertions of Variables

The conduct of the MFA symbolic digital pre-supposes $K+1$ factorial analysis (the K groups and the average cloud). Two representations can be made for each of these analyses: a representation of variables (dual analysis of centres of gravity) and an analysis of assertions of variables taking into account variability. These representations clearly bring out the main variability factors of groups and the factors of general analysis.

Superimposed Representation of Factors (as Assertions of Variables) on Canonic Variables

This representation is the result of a MFA traditional representation which consists in highlighting the main variability factors of groups in the form of additional variables over canonic variables. But in symbolic data analysis, these variability factors are also expressed as assertions of variables and the representation of these assertions of variables over the canonic variables calls for an original observation. The position and collection of assertions of variables give us precious indications on their importance and the link between the different groups.

Link between Groups: Representation Quality of Groups

This is a traditional indicator in MFA. It is a representation of groups on canonic factors. The coordinates of an axis group is the combined inertia of the group's variables on the corresponding MFA axis of table X^*. For the two groups K_x. and K_y, we have

$\mathcal{L}(K_x, K_y) = \langle W_x^{(p)} D, W_y^{(p)} D \rangle$ where D is the diagonal

$I \times I$ matrix of the weight of individuals and $W_j^{(p)}$ is the matrix of scalar products between the individuals of group j for the X_j^p weighted matrix.

A symbolic representation may be proposed for it. Each assertion of variables can be coded into many variables whose projection over canonic variables results in a minimum value and a maximum value. The variability associated to a group will therefore be displayed by two values: the sum of the minimum inertia of the group's variables and the sum of its maximum inertia.

The representation of groups is done in the $\mathbb{R}^{I^2}$ space.

The Notion of Common Factor and Specific Factor

MFA helps to portray the characteristics of groups compared to the canonic variables (also referred to as factors). A common (or specific) factor to the MFA symbolic digital will be the common factor associated to the traditional MFA of centres of gravity.

POVERTY DATA

Presentation

The data is drawn from the first Cameroonian survey in households (ECAM I) conducted in 1996 by the National Institute of Statistics. The sample is made up of 1800 households with a total of 10230 individuals. The analysis is done on 13 continuous variables out of which 10 are active and 3 illustrative (see Table 3).

The average poverty line is used to establish the difference between poor and non poor individuals. We shall consider 10 classes of individuals with 5 poor classes organized in descending order from very poor to less poor individuals and 5 classes of non poor individuals organized in ascending order from less rich to very rich individuals. These classes constitute the concepts that we shall analyze. The groups of variables are regions. For presentation purposes, they shall be presented on a line and not in columns. An overview of the SDB for the Adamawa Region for three variables shows the following Table 4.

Analysis and Interpretation

After the algorithm application, the following results are obtained:

Inertia of the First Six Eigenvalues

Please see Table 5.

Relationship between General Variables and Groups

The F_1 variable is the variable which extracts the most important inertia from all groups. The other variables convey specific information to certain regions (see Table 6).

Table 3: Variables of ECAM1

N	Title	Description	Quality
1	DEPEAD	Expenditure per adult equivalent	Additional
2	DEPTET	Expenditure per head	Additional
3	DEPTOT	Total expenditure	Additional
4	DEPALI	Food expenditure	Active
5	DEPHAD	Clothing expenditure	Active
6	DEPMAI	Household expenditure	Active
7	DEPSAN	Health expenditure	Active
8	DEPTRA	Transportation expenditure	Active
9	DEPEDU	Education expenditure	Active
10	DEPSOP	Personal health care expenditure	Active
11	DEPLOI	Leisure expenditure	Active
12	DEPLOY	Rent expenditure	Active
13	DEPLOG	Lodging expenditure	Active

Table 4: An overview of SDB of ECAM1

		AD	
Concept	DEPALI	DEPHAB	DEPMAI
Poor of class 1	[96,464, 173,375]	[2000, 36,000]	[3950, 36,000]
Poor of class 2	[122,375, 573,832]	[15,000, 31,200]	[28,800, 34,800]
Poor of class 3	[109,500, 573,032]	[5700, 46,800]	[10,800, 28,800]
Poor of class 4	[157,210, 302,428]	[5700, 38,200]	[7200, 15,300]
Poor of class 5	[126,967, 358,482]	[0, 54,500]	[12,000, 53,700]
Non poor of class 1	[102,982, 536,028]	[0, 115,900]	[2450, 55,100]
Non poor of class 2	[220,303, 568,617]	[0, 82,000]	[11,950, 57,600]
Non poor of class 3	[95,421, 1,707,678]	[0, 297,300]	[23,000, 65,100]
Non poor of class 4	[146,782, 967,510]	[3500, 297,300]	[600, 71,300]
Non poor of class 5	[130,617, 813,428]	[8000, 485,200]	[22,800, 199,800]

Table 5: Histogram of eigenvalues

N	Eigenvalue	Inertia %	Combined %
1	6.80938	68.09	68.09
2	1.08859	10.89	78.98
3	0.60479	6.05	85.03
4	0.56497	5.65	90.68
5	0.31109	3.11	93.79
6	0.29054	2.91	96.69

Table 6: Relationship between general variables and groups

	F_1	F_2	F_3	F_4	F_5	F_6
AD	52.70	20.95	5.81	3.94	1.15	1.31
CE	77.57	9.29	0.83	0.93	0.35	0.26
ES	58.61	4.60	9.72	8.58	0.96	4.04
EN	62.35	10.50	3.13	5.98	2.81	1.40
LT	82.74	2.18	1.31	1.93	1.09	0.45
NO	51.03	6.12	7.38	8.56	10.40	4.10
NW	67.16	8.38	2.91	5.74	1.36	3.81
OU	46.08	10.14	6.78	9.28	6.78	5.22
SU	57.94	13.26	5.23	2.32	2.02	3.20
SW	57.84	12.42	11.17	3.22	1.13	2.62

Inter-Inertial/Total Inertia Relationship

Table 7 shows that on the first axis, there is a close proximity between equivalent classes. Broadly speaking, these equivalent classes have similar behaviours in their expenditure habits. The same phenomenon is observed in the descending values of r, on the 4th, 2nd, 6th and 3rd axes.

Degree of Resemblance between Groups and General Variables: Common and Specific Factors

The first MFA component is a factor common to all the Barnabé Tang Ahanda, Jean Gérard Aghoukeng Jiofack, Romain Germain Nzangué, Gilbert Hapi Mbiakop (see Table 8).

Table 7: Inter-inertial/total inertia relationship

Axis	F_1	F_2	F_3	F_4	F_5	F_6
Relationship	0.89	0.70	0.64	0.70	0.30	0.65

Table 8: Degree of resemblance between groups and general variables

	F_1	F_2	F_3	F_4	F_5	F_6
AD	0.87	0.58	0.65	0.13	0.05	0.09
CE	0.90	0.12	0.03	0.45	0.01	0.02
ES	0.90	0.15	0.31	0.66	0.15	0.80
EN	0.98	0.75	0.45	0.50	0.18	0.38
LT	0.98	0.28	0.11	0.33	0.02	0.08
NO	0.92	0.42	0.44	0.32	0.59	0.32
NW	0.97	0.57	0.12	0.78	0.28	0.35
OU	0.88	0.42	0.43	0.60	0.28	0.62
SU	0.93	0.47	0.60	0.55	0.28	0.70
SW	0.96	0.68	0.72	0.30	0.15	0.16

REFERENCES

1. E. Diday, "Introduction to the Symbolic Data Analysis," Cahiers du CEREMADE, No. 8823, 1988.
2. E. Diday, "From the Objects of the Data Analysis to Those of the Analysis of Knowledge, Symbolic and Numeric Induction from Data," Cépadues, 1991.
3. L. Lebart, A. Morineau and Mr. Piron, "Multidimensional Exploratory Statistics," 3rd Edition, DUNOD, Paris, 2000.
4. M. Volle, "Data Analysis," Economica, Paris, 1981.
5. J. F. Martin, "Fuzzy Coding and Its Applications in Statistics," Thesis, University of Pau, Pau, 1980.
6. F. J. Gallego, "Fuzzy Coding in Correspondence Analysis," Cahiers de l'Analyse des Données, Vol. 13, No. 2, 1980, pp. 413-430.
7. J. Pagès and B. Escofier, "Simple and Multiple Factorial Analysis," DUNOD, Paris, 1990.
8. B. Escofier and J. Pajes, "Multiple Factorial Analysis, Objectives, Methodology and Interpretation," 2nd Edition, Wiley, Paris, 2005.
9. P. Cazes, A. Chouakria, E. Diday and Y. Schektman, "Extension of Principal Component Analysis to Interval Data," Revue de Statistique Appliquée, Vol. 45, No. 3, 1997, pp. 5-24.
10. B. Ahanda Tang, "Extension of Factorial Analysis to Symbolic Data," Thesis, University of Paris, Paris, 1998.

Chapter

8

COMBINING SYMBOLIC TOOLS WITH INTERVAL ANALYSIS. AN APPLICATION TO SOLVE ROBUST CONTROL PROBLEMS

Inès Ferrer-Mallorquí, Josep Vehí

Institut d'Informàtica i Aplicacions, Universitat de Girona, Girona, Spain

ABSTRACT

Complex systems are often subjected to uncertainties that make its model difficult, if not impossible to obtain. A quantitative model may be inadequate to represent the behavior of systems which require an explicit representation of imprecision and uncertainty. Assuming that the uncertainties are structured, these models can be handled with interval models in which the values of the parameters are allowed to vary within numeric intervals. Robust control uses such mathematical models to explicitly have uncertainty into account. Solving robust control problems, like finding the robust stability or designing

Citation: Ferrer-Mallorquí, I. and Vehí, J. (2014), "Combining Symbolic tools with interval analysis. An application to solve robust control problems". American Journal of Computational Mathematics, 4, 183-196. doi: 10.4236/ajcm.2014.43015.

a robust controller, involves hard symbolic and numeric computation. When interval models are used, it also involves interval computation. The main advantage using interval analysis is that it provides guaranteed solutions, but as drawback its use requires the interaction with multiple kinds of data. We present a methodology and a framework that combines symbolic and numeric computation with interval analysis to solve robust control problems.

Keywords: Interval Analysis, Robust Control, Symbolic Computation

INTRODUCTION

The description of real systems using mathematical models has been usually concerned with the concept of uncertainty. The inaccuracies and uncertainties can be done by different causes: in the linearization of a system, for the fact that some parameters of the system can have different values for different operation conditions or because some parameters can vary with time. Moreover uncertainties are classified into two types: parametric or non-parametric. In case the origin and the structure of the uncertainty were known, they are called structured or parametric, otherwise, if they are unknown, the uncertainty is classified as unstructured or non-parametric. From a theoretical point of view, there are a lot of approaches related with parametric robust problems. Most of them use sparametric methodologies [1] -[3] . Some using QFT [4] , other Value Sets [5] .

Some of these methods have been integrated into packages as Matlab [6] or into frameworks as PARADISE [7] . All the cited approaches, manage these problems of robust control using parametric methods. Although these methods have the drawback that the solutions obtained are no guaranteed solutions.

Methods for assessing robust stability and robust design of parameter dependent linear systems, as Routh table or the small gain theorem,... give sufficient (but not always necessary) conditions to solve them. These theorems give a powerful methodology to transform robust control problems in simplified problems consisting of testing the positivity of a set of functions or conditions.

This paper presents a methodology that solves robust control problems combining symbolic computation with interval analysis [8] techniques. The main advantage of interval analysis is that it offers guaranteed solutions in case we want to solve robust control problems. Thus, its use means a really improvement in front parametric techniques.

Although the advantage of obtain guaranteed results is important, the rules and algorithms that must be generated to solve robust control problems using interval analysis, are difficult to be implemented for users. This difficulty discourage control engineers to use these techniques. This difficulty motivates us to offer the users an automatic application that uses interval techniques in a transparent way for the user. This application or framework makes a combination between symbolic tools and interval analysis methods. This paper presented his framework named IRCAD (Interval Robust Control framework for Analysis and Design) dedicated to the resolution of robust control problems applying techniques of interval analysis. This framework is composed by a set of tools of analysis and design that combine symbolic computation with the methodology of interval analysis. Thus, it is not necessary that the user have any special knowledge about interval analysis to use the functions of robust analysis and robust design that offer the framework.

The paper is organized as follows. Section 2 presents the formulation of the most typical robust parametric control problems. The use of interval analysis theory as a way to obtain guaranteed results on the solution of robust parametric control problems is the central point of Section 3. The problem of how to integrate all this diversity of data and all the involved methodologies of this kind of problems is discussed on Section 4. In Section 5 IRCAD is presented as a framework that allows the integration of the different data. Section 6 summarizes the framework and applies it to an illustrative example. Finally most emphasizing points and future intentions are put on the Section 7 of conclusions.

ROBUST PARAMETRIC CONTROL PROBLEM

To formulate the problem of parametric robust control we consider a system that is linear respect the variable s. This kind of systems can be expressed as a transfer function or in space state format, in continuous or in discrete domain.

In this paper we will consider a class of plants [9] with structured parametric uncertainties described by the following uncertain transfer function:

$$G(s,q)=\frac{\alpha_0(q)+\alpha_1(q)s+\cdots+\alpha_m(q)s^m}{\beta_0(q)+\beta_1(q)s+\cdots+\beta_n(q)s^n} \qquad (1)$$

depending on a structured perturbation characterized by the parameter vector

$$\boldsymbol{q}=\begin{bmatrix} q_1 & q_2 & \cdots & q_l \end{bmatrix}^{\mathrm{T}} \qquad (2)$$

where each parameter enters into the system description with polynomial dependency. We also consider a certain configuration of the feedback system with a fixed controller $C(s,\boldsymbol{k})$, where k is the design parameter vector. The motivation of this kind of system description is that the system parameters, q, can represent physical quantities that are known only to within a certain accuracy, or vary depending on operating conditions while the controller parameters, k, represent degrees of freedom available to the control system designer. Due to the physical interpretation of the uncertain parameters, each one can be considered independent from the other and their values lie between upper and lower bounds. Then, the uncertain domain can be defined as any hyperrectangle:

$$\boldsymbol{Q}=\left\{\boldsymbol{q}=\begin{bmatrix} q_1 & q_2 & \cdots & q_l \end{bmatrix}^{\mathrm{T}} \middle| q_i \in \left[q_i^-, q_i^+\right], i=1,\cdots,l\right\} \qquad (3)$$

With this kind of feedback system description, several robust control problems can be formulated. Problems concerning robust control analysis as testing the stability of the system or computing the stability margin. Also, problems related with control design as to find a set of controllers that fulfill one or more than one selected specifications.

Robust Control Problems: Analysis

In many control systems the plant parameters may vary over a wide range about a nominal value p^0. Robust parametric stability refers to the ability of a control system to maintain stability despite such large variations. Given an uncertain system and a controller, analysis of robust control problems consists in to check(calculate) some performance specifications:

Robust stability

The first problem formulated is the verification of the stability on uncertain systems. That is, how to find the necessary and sufficient conditions that allows to guarantee that a family of uncertain polynomial is stable or no.

In [10] , Ackermann presents different methods to check robust stability analysis.

-Boundary crossing theorem: It is based on Hurwitz determinants. Given a set of polynomials $P(s,Q)$, this set is robustly stable, if and only if:

1) There exists a stable polynomial $p(s,q)\in P(s,Q)$2) Any root cross imaginary edge: $jw\notin \text{Roots}\left[P(s,Q)\right]$for all $\omega\geq 0$

-Alternative boundary crossing theorem based on Hurwitz Determinants [11] : Giving a set of polynomials $P(s,Q)$ it is robustly stable, if and only if:

1) There exists a stable polynomial $p(s,q)\in P(s,Q)$
2) $H_n(q)\neq 0$for all $q\in Q$

This work uses the just presented boundary crossing theorem in a modified way. It gives necessary and sufficient conditions of absolute robust stability for a family of characteristic polynomials.

Stability margin

If the controller is given, the parametric stability margin is defined [12] as the maximal range of variation of the parameter p, measured in a suitable norm, for which closed-loop stability is preserved. It can be expressed as:

$$\rho_x := \sup\left\{\alpha : \delta(s,x,p)\ \text{stable}, \left\|p-p^0\right\|<\alpha\right\} \tag{4}$$

This margin is used as a quantitative measure of the robustness of the closed loop system with respect to parametric uncertainty evaluated at the nominal point p^0.

Parametric bode plots

Based on the experience with frequency domain analysis methods, one might conjecture that it is not necessary to check all ω values. Thus, it is common to construct Bode plots using only a grid of ω values. In the case of frequency domain methods for robustness analysis, the bottleneck is "singular frequencies".

In order to provide a systematic method of identifying singular frequencies, an algebraic definition is given on [10] . Before the definition some preliminary notation is introduced. When an uncertain polynomial $p(s,q)$ is evaluated at a frequency ω it equals a complex number

$p(j\omega,\boldsymbol{q}) = h(-\omega^2,\boldsymbol{q}) + j\omega g(\omega^2,q)$. Separating this complex number on real and imaginary parts:

$$h(-\omega^2,\boldsymbol{q}) = a_0(\boldsymbol{q}) - a_2(\boldsymbol{q})\omega^2 + a_4(\boldsymbol{q})\omega^4 - \cdots \tag{5}$$

$$\omega g(-\omega^2,q) = a_1(q)\omega - a_3(q)\omega^3 + a_5(q)\omega^5 - \cdots \tag{6}$$

Associated with these functions is a Jacobi an matrix.

$$J(\omega,\boldsymbol{q}) = \begin{pmatrix} \frac{\delta h(-\omega^2,\boldsymbol{q})}{\delta q_1} & \frac{\delta h(-\omega^2,\boldsymbol{q})}{\delta q_2} & \cdots & \frac{\delta h(-\omega^2,\boldsymbol{q})}{\delta q_l} \\ \frac{\delta \omega g(-\omega^2,\boldsymbol{q})}{\delta q_1} & \frac{\delta \omega g(-\omega^2,\boldsymbol{q})}{\delta q_2} & \cdots & \frac{\delta \omega g(-\omega^2,\boldsymbol{q})}{\delta q_l} \end{pmatrix}$$

In the Jacobian $J(\omega,\boldsymbol{q})$, the real part $h(-\omega^2,\boldsymbol{q})$, and the imaginary part $\omega g(\omega^2,q)$ are used to give algebraic definition of singular frequencies.

Definition. The nonnegative frequency ω_s is a singular frequency of the uncertain polynomial $p(s,\boldsymbol{q})$ if there exists a $\boldsymbol{q}^0 \in \mathbb{R}^l$ such that the three following conditions are simultaneously satisfied.

$$h(-\omega_s^2,\boldsymbol{q}^0) = 0 \tag{7}$$

$$\omega_s g(-\omega_s^2,\boldsymbol{q}^0) = 0 \tag{8}$$

$$rank\left[\boldsymbol{J}(\omega_s,\boldsymbol{q}^0)\right] < 2 \tag{9}$$

The last Equation (9) can be substituted by: $\det \boldsymbol{J}(\omega\boldsymbol{q}) = 0$

Frequency problems

-Value sets. Some gridding approaches have as a drawback the length of time it takes to compute the set of frequency plots. An alternative and faster technique, suggested in [5] , is to compute the frequency plot $p(j\omega,\boldsymbol{q}), \omega \geq 0$ for each q on a grid of Q. It is advisable to compute the value set for each ω on a grid of frequencies from 0 to +∞.

$$P(j\omega,Q) = \{p(j\omega,\boldsymbol{q}) \in \mathbb{C} | \boldsymbol{q} \in Q\} \tag{10}$$

In the case of a family of polynomials, the stability test is based on repeating the construction of the value set for a grid of frequencies to give the set of all possible frequency plots. The collection of value sets would then indicate stability or instability depending on how the zero-exclusion theorem is formulated.

Theorem 1. (zero-exclusion theorem). Given a polynomial family $P(s,Q)=\{p(s,q)|q\in Q\}$. This set is robustly stable if and only if

- A stable polynomial $p(s,q)\in P(s,Q)$ exists and
- $0\notin P(j\omega,Q)$ for all $\omega\geq 0$.

Discrete-time problems

Some problems concerned with discrete-time are:

- Model conversion. Conversion is needed in both senses, from a continuous time interval state-space model to a discrete-time interval model and also from a discrete-time uncertain system to a continuous-time uncertain model.
- Schur stability test. The problem of checking the stability of a discrete-time system is reduced to the determination of whether or not the roots of the characteristic polynomial of the system lie strictly within the unit disc, that is whether or not the characteristic polynomial is a Schur polynomial. If

$$P(z)=p_n z^n+p_{n-1}z^{n-1}+\cdots+p_1 z+p, \qquad (11)$$

where the z_i are the n roots of P(z) Then if P(z) is Schur, all these roots are located inside the unit circle $|z|<1$ so that when z varies along the unit circle $z=e^{j\theta}$, the argument of $P(e^{j\theta})$ increases monotonically. For a Schur polynomial of degree n, $P(e^{j\theta})$ has a net increase of argument of $2\pi n$, and thus the plot of $P(e^{j\theta})$ encircles the origin n times. This can be used as frequency domain test for Schur stability [13] [14] .

- Non-linear discrete-time control of uncertain systems. This problem can be formulated as:

Find one c

$$c\in S_c=\{c\in C|\forall p\in P, f(c,p)>0\} \qquad (12)$$

where f is a vector function that can be evaluated using algorithms based on interval analysis. This problem is both quite complex because it involves a quantifier and, at the same time, very simple because the only objective is to find one feasible vector. This makes considering a larger number of tuning parameters corresponding to the guaranteed-tuning problem [15] . This is evidence that it is possible to combine non-linearity, and structured uncertainty with guaranteed results.

Robust Control Problems Design

The requirement of robust design is that the design specifications must be satisfied over the entire parameter set. Given an uncertain system, the problem is to obtain a robust controller that fulfills some performance specifications.

As robust performance specifications we can enumerate stability, pole locations, steady-state, H_∞, etc. It is known that the solution of these problems can be reduced to the problem of checking the positivity of a set of rational functions over a given domain:

$$f_i(\alpha, \boldsymbol{q}, \boldsymbol{k}) > 0 \quad \forall \alpha \in A \quad \forall \boldsymbol{q} \in \boldsymbol{Q} \quad \forall \boldsymbol{k} \in \boldsymbol{K} \tag{13}$$

where α is the generalized frequency and $\boldsymbol{k} \in \boldsymbol{K}$ may be a single point (nominal controller k^0) or a certain domain in the parameter's space of the controller depending on the problem considered.

- Robustness analysis: to check if the controlled system achieves robust performances.

$$f_i(\alpha, \boldsymbol{q}) > 0 \quad \forall \alpha \in A \quad \forall \boldsymbol{q} \in \boldsymbol{Q} \tag{14}$$

- Robust control design: to find the robust K set such that the controlled system achieves robust performances.

$$f_i(\alpha, \boldsymbol{q}, \boldsymbol{k}) > 0 \quad \forall \boldsymbol{k} \in \boldsymbol{K} \quad \forall \boldsymbol{q} \in \boldsymbol{Q} \tag{15}$$

Most of the more used robust control design techniques are based on the worst possible scenario which may never occur in a particular control system. Some of these techniques are the following:

- H_∞control [16] . Robust performance specifications may be given as pole locations as well as H_∞ performances expressed in the frequency domain.
- l_1 control [17] . In the standard l_1 problem the design of an internally stabilizing controller such that the l_1 norm of the regulated output z due to the worst-case magnitude bounded disturbance ω is addressed. The Banach space of right-sided absolutely summable real sequences with the norm given by

$$\|x\|_1 := \sum_{k=0}^{\infty} |x(k)| \tag{16}$$

- μ synthesis. It minimizes the superior bound of the robust performance. The criterion is

$$J = \min_{t_p} \left\| \left[\min_{c(s)} \left\{ \sup \mu_{A(s)} \left[M(s) \right] \right\} \right] - 1 \right\| \tag{17}$$

where $\mu[0]$ is the structured singular value, A(S) is the disturbance matrix which indicates that $\mu[0]$ depends on the allowed structure for A(S), M(S) is a matrix obtained by rearranging the uncertain system into the M A-structure and C(S) indicates that the minimisation is function of the controller parameters [18] [19] .

- QFT (Quantitative feedback theory). While value sets are mainly concerned with analysis, when the aim is finding a compensator to satisfy design specifications, QFT is the most important approach. It can be considered as a natural extension of classical frequency-domain design approaches. One of the main objectives is to design a simple low-order controller where the bandwidth of the feedback controller being as small as possible. At a fixed frequency, the plant's frequency response set is called a template. In the bound generation step of QFT design procedure, the plant template is used to translate the given robustness specifications in domains in the Nichols chart where the controller gain-phase values are allowed to lie.

We are focused on the solution of robust control design problems. That is, we are testing if a controller K achieves the design specifications.

Some applications can be found in the design field. Parameter spaces of the controller are explored, finding the regions which fulfill a control specification in order to obtain a controller which satisfies it.

INTERVAL FORMULATION OF ROBUST PARAMETRIC CONTROL

The problem of robust parametric control, presented on the last section, is now formulated using interval analysis. Using the theory of interval analysis is possible to obtain guaranteed solutions to robust control problems. To formulate these problems on interval format, we formulate the problem of the parameter space approach considering the uncertain system of Figure 1, where k is the parameter vector of the controller,

$$\boldsymbol{k} = \begin{bmatrix} k_1 & k_2 & \cdots & k_l \end{bmatrix}^{\mathrm{T}} \tag{18}$$

q is the parameter vector of the process, and the uncertainty domain can be defined as a box:

$$K = \left\{ \boldsymbol{k} = \begin{bmatrix} k_1 & k_2 & \cdots & k_l \end{bmatrix}^{\mathrm{T}} \middle| k_i \in \left[\underline{k_i}, \overline{k_i} \right] \right\} \tag{19}$$

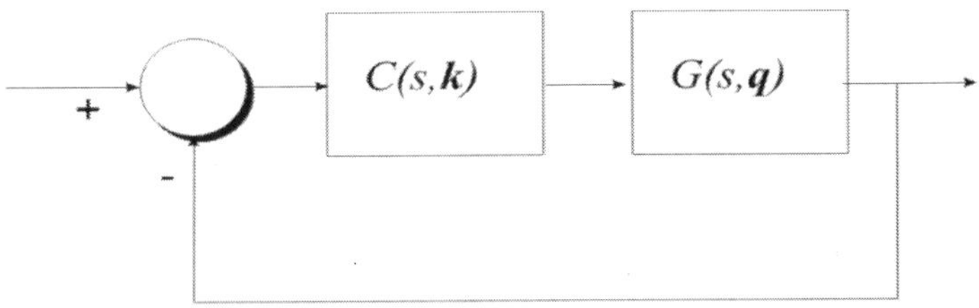

Figure 1: Uncertain system.

Design specifications are formulated in terms of closed-loop system stability and performances in the frequency domain. These specifications, such as bandwidth, resonance peak, control effort, etc., can be described as a set of N inequalities of the type:

$$f_i\left(\omega, \boldsymbol{q}, \boldsymbol{k}\right) > 0 \quad \omega \in \Omega \quad \boldsymbol{q} \in \boldsymbol{Q} \quad \boldsymbol{k} \in \boldsymbol{K} \quad i = 1, \cdots, N \tag{20}$$

where Ω is a subset of R^+ (usually an interval), Q an interval vector representing interval parameters and K; a nondegenerate set.

Then robust design can be formulated as follows: given a controller structure, C(s, K) the aim is to find the values of k which conform with the robust control specifications. Taking this formulation into account some control problems can be proposed [20] :

1) Performance checking. Given the uncertain system $G(s,Q)$ and the uncertain domain Q, the problem is to test if the designed controller $C\left(s, \boldsymbol{k}^0\right)$ achieve the robustness specifications obtained from Equation (20):

$$F_i\left(\omega, \boldsymbol{Q}, \boldsymbol{k}^0\right) > 0 \tag{21}$$

2) Performance margin computation. Taking set $\prod$ as a function of its radius ρ:

$$\prod(\rho) = \left\{ q : \left\| q - q^0 \right\|_\infty^\omega \leq \rho \right\} \tag{22}$$

find the maximal set $\prod(\rho^*)$ so that the designed controller $C(s, k^0)$ achieves robust performances for all ρ belonging to $\prod(\rho^*)$.

3) Robust controller design. Taking a particular structure for the controller and specifying the uncertain domain where the parameters of the system Q can vary, find a fixed structure controller $C(s, k^0)$ so the controlled closed loop system achieves the robust performance specifications:

$$F_i(\omega, Q, k^0) > 0 \tag{23}$$

4) Obtaining the set K for the robust controller problem. Given an uncertain plant $G(s, Q)$ the variation domain of the system parameter Q and a controller structure, find the robust set K which allows C(s, K) to achieve the robustness specifications:

$$F_i(\omega, Q, k) > 0 \tag{24}$$

5) Estimating the stability region problem, for a given k^0. This problem consists on building a set of all ρ 's achieves closed loop stability when given k^0.

In our approach two robust control problems, Design and Analysis, have been converted into a set of functions using symbolic computation. These functions have been intervalised resulting on a set of interval functions. At this point, to solve the problems it has only been necessary to test the positivity of these interval functions for all the variation of its parameters, Equation (14) and Equation (15).

INTEGRATION OF NUMERIC, INTERVAL AND SYMBOLIC COMPUTATION

As it has seen in Section 2, problems related with robust parametric control systems use symbolic data due these systems involves uncertain variables. Moreover in Section 3 is denoted that the interval formulation of the problems involves numeric and interval data. Thus, solving robust control problems using interval analysis tools, offer guaranteed results, but as a drawback, its use requires the interaction with multiple kind of data. These difficulties can discourage control engineers to use interval techniques although its benefits.

To make all these tools and techniques more available is interesting an environment that integrates the treatment of all these kind of data, allowing deal with robust control problems, solving them using a common methodology.

In this sense our work has found how to offer to the control engineer a set of friendly user interface tools that integrates the diversity of data that these kinds of problems involves and also allows to use interval tools in a transparent way for the user.

In Figure 2, a schema of the different tools used for these kinds of problems to manage numeric symbolic and interval computation is shown.

4As it can be seen in this schema, Matlab is the package used to centralize our environment. It is the selected package to develop the most routines of our environment due three main points:

- The high computational power with problems that involves matrix data.
- Its facility to deal with high level languages.
- Its easy connection with other commercial packages as Maple.

Our framework solves robust control problems, thus, problems that have implicitly high level degree of computation. This aspect would be a bottleneck if they would be solved writing scripts on a package like Matlab. Due this high cost of computation the routines on this framework have been implemented with the high level language C++. This language improves the building of procedures which require numeric-interval computation and allows to execute code in a faster way than using Matlab. This achievement was very decisive. In addition it allows to use tools from interval analysis libraries [21] , due its include instructions.

Once presented the two main packages used by the framework, it rests to say how Matlab procedures link with C++ critical routines. The deal of procedures and data among these two packages is done using code CMEX and DLL functions. The high level C++ functions are compiled using Borland, obtaining a set of dll function. There are executable functions that can be called from Matlab commands or from Matlab scripts.

Another aspect to consider is how to enter symbolic data, such as transfer functions or problem specifications. In our case it has been choose the Maple package. The extraction of these symbolic inputs from Matlab scripts and the pass of them to the Maple Package are implemented using the Symbolic Math Toolbox.

Therefore, Matlab is used to make the integration of numeric, interval and symbolic data, taking symbolic inputs, transforming them into a suitable format for the C++ routines, passing these transformed inputs to the C++ routines and finally collecting the data returned.

The last step, shown in the lower part of the diagram (user interface block) corresponds to the need to show results in a suitable way for control applications. The framework has implemented this part using the GUI-tools of Matlab. These tools allow to achieve a friendly interface that allows to present the results in an adequate format to be analyzed easily by the user.

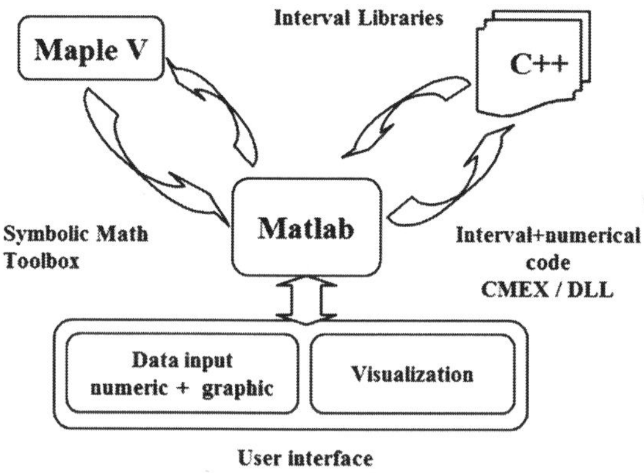

Figure 2: Cooperation numeric-interval-symbolic.

IRCAD: A FRAMEWORK FOR ROBUST CONTROL PROBLEMS

Once introduced the need to integrate the diversity of data and methods in section 4, now in this section we present the developed environment in order to achieve this integration. To obtain the kindly manipulation of numeric, interval and symbolic computation, our work has built a set of tools under the format of a framework called IRCAD (Interval Robust Control framework for Analysis and Design). To introduce the framework IRCAD (main screen is showed on Figure 3), we present the functions that it integrates. These functions or tools could be laid as three sets of problems.

The first set includes robust analysis problems like to test the stability of a system and to find the stability margin. The second set of problems are related to robust design problems and the main tool is an application that allows to select the design specifications that we want our control system achieve. The framework includes a third set corresponding to post-design tools, which allows that a selected controller can be applied to the system without leave the framework IRCAD.

The starting point to use the tools of the framework is to input the transfer function of the plant. Due the systems we consider are uncertain systems, the plant is defined as a family of transfer functions. To input them into the framework it is necessary to define a transfer function with undefined parameters, and also define their uncertain values range (interval values). So, as example, we can define a transfer function for a family of plants like $G(s,\boldsymbol{q})$ with the uncertain parameter $\boldsymbol{q} = [q_1, q_2]$.

Tools for Robust Analysis

The framework IRCAD has a set of tools addressed to solve robust analysis problems. Giving an uncertain system, where the transfer function is defined as a family of plants qi, from the menu Analysis, it is possible to make the following functions:

- Stability test. To calculate the absolute stability region, a branch-and-bound algorithm based on interval analysis is used. The result of this tool shows the regions stable and unstable. In case the depth of the algorithms was limited (in order to accelerate the end of the computation) can appear undefined regions. In the case there aren't problems of time computation, due the computer has high computation resources, it is possible to don't limit the depth of search and then the lonely regions that appear are estable and unestable regions.
- Stability margin computation. The parametric stability margin is defined as the length of the smallest perturbation Δq which destabilizes the closed loop. This margin is used as a quantitative measure of the robustness of the closed loop system with respect to parametric uncertainty evaluated at the nominal point q_0.
- Parametric Bode plots. It is common to construct Bode plots using only a grid of ω values. There are, however, cases when 'singular frequencies' occur. The framework considers these cases which require special attention in all frequency domain methods for robustness analysis.

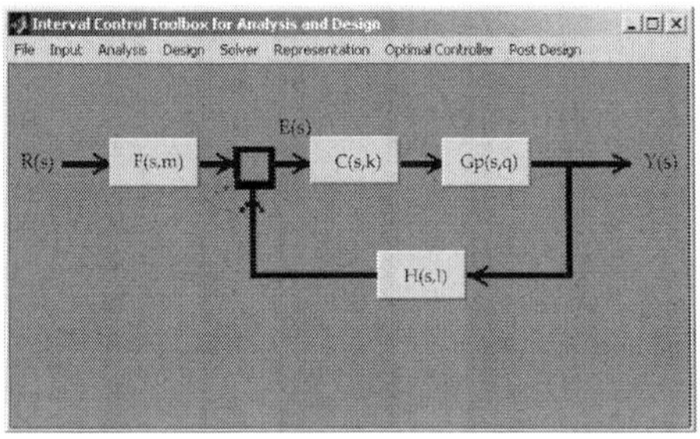

Figure 3: IRCAD framework.

Tools for Robust Design

The IRCAD framework has a powerful set of tools aimed at the design of controllers in the case of parametric control systems. Given an uncertain system, where the transfer function is defined as a family of plants q_i, from the menu design it is possible to select one or more than one specification. The specifications involve temporal and frequency domain and include: stability degree, absolute stability, resonance peak, velocity error, settling time, control effort and overshoot.

Once the specifications have been selected, each chosen specification is translated to an interval function, converting them into matrix structures and passing them to a solver (a C++ routine). The results of the solver routine, that is to say the results of the design problem, are returned on a matrix format as a set of controllers that fulfill the selected specification or the set of specifications. The resultant controllers are a set of interval values. Using the tools of Graphic User Interface from Matlab (GUI tools), the resultant matrix, which contains all the information about the valid controllers is passed to the user.

The framework allows to choose in which format we want to show the set of valid controllers. It is possible to visualize this set in a graphic way or in a numeric way. The decision often depends on the number of parameters selected for the controller. If the user wants to design a controller of two or less parameters, he can choose the graphical format, then the result will be a graph with the parameter space ranged by the parameters of the controller (k_1, k_2). In this graph the set of valid controllers are colored as a red region (Figure 5), the unfeasible controllers on a yellow region and the undefined regions on blue regions. If the user choose to obtain the set of valid controllers in a numeric format, then the framework shows on a table, the set of intervals for $k_1, k_2, \cdots, k_i$ that allows the system fulfill all the specifications.

Post-Design Tools

To help control engineer, IRCAD framework offers a very complete and power fulset of post-design functions. These functions allow the user, on the one hand to complete the design selecting one interval controller from the feasible controller set and in the other hand allow to verify that the selected controller achieve the specifications, without exit from application.

In order to offer a more reliable framework, the package gives the possibility to choose a criterion to select the optimal controller.

- Optimal Controller. The results of the robust control problem design are returned as a set of controllers that fulfill the specification or the set of specifications selected. To choose the optimal controller among this set is a task that can be not easy for the user, although the user was control engineer. To make the selection of the optimal controller from the given set of feasible controllers computed by the framework IRCAD, that offers the possibility to do it using three different criteria:
- Minimum norm: Given a graphical region of feasible controllers, in the parameter space delimited by the parameters of the controller K_1 and K_2, this criteria select the interval controller (K_1, K_2) that, fulfilling the specifications, has the minimum norm. This measure is calculated from the origin of the parameter space to the square defined by the interval(K_1, K_2).
- Maximum robustness: In this case the framework computes over each square (K_1, K_2) the value of the stability margin and selects as controller the square (K_1, K_2) that have the maximum value.
- Neighboring: Over each feasible square(K_1, K_2), the interval controller, the framework computes the number of neighbors. The square (K_1, K_2) with the maximum number of neighbors is selected.

When the user selects one of these criteria, then IRCAD computes one recommended controller following the method chosen. The framework shows the obtained controller by two ways at time: graphically and numerically.

Graphically, emphasizing this controller with a dark color on the graphic, that represents the set of feasible controllers (Figure 6), and Numerically, where is opened a window with the recommended values for the controller. The recommended value for the controller can be introduced on the controller box as an entry, or it can be introduced automatically as following is explained.

- Input optimal controller. The framework allows that the parameters of the commended controller obtained using one of the techniques exposed (minimum norm, maximum robustness or nearest neighbor) can directly be input. This automatic entry of the parameters of the controller allows to execute post-design tasks without exit from the framework.
- Post design tools. It is not necessary to go to any simulation environment to verify the selection, because the selected controller can be simulated using the same framework IRCAD.

All these features give to the user a friendly environment to work in the case of robust control problems, with the security that he has obtained guarantee solutions on the results.

The mainly tools of this block are concerning to the facilities to have tools to check that a designed controller fulfill a set of specifications.

ILLUSTRATIVE EXAMPLE

To illustrate how the solver created manages the data, an example suggested by [22] and later studied by [23] is used. The aim of the example is tuning a PI controller for an interval plant.

Example 1. Robust control design.

Given the transfer function of the plant:

$$G(s,\mathbf{q})=\frac{q_1}{1-\frac{s}{q_2}} \qquad (25)$$

whereqisthevectorofuncertainparameters, $\mathbf{q}=[q_1 \quad q_2];\ q_1=q_2=[0.8,1.25]$ and the PI controller is described by:

$$C(s,\mathbf{k})=\frac{k_1\left(1+\frac{s}{k_2}\right)}{s} \qquad (26)$$

where k is the vector of the design parameters, $\boldsymbol{k} = [k_1, k_2]$.

The design aim is to find the set K of controller parameters that fulfill the following performance specifications:

1. Absolute stability.
2. Velocity error lower than 2%.
3. Resonance peak lower than 3 dB.
4. Control effort lower than 20.

Given the Equation (26), the toolbox IRCAD manage each one of these specifications transforming them via the Extended Symbolic Toolbox into a set of polynomial interval functions [Equations (27)-(30)]. Matlab send these functions to the solver, transforming it previously on a matrix structure. This transformation is automatically done by IRCAD toolbox, thus it is transparent to the user. The role of the solver is to test the positivity of these functions over the parameter space, which is determined by the controller parameters. In this example the parameter space is determined by k_1 and k_2. IRCAD allows to present the results of the solver in two formats :

- -Graphically.
- -Set of intervals.

For this example the parameter space to be checked is limited giving the initial interval values to the parameters of the controller as $k_1 = [-200, 0]$ and $k_1 = [0, 10]$. In order to work with positive intervals the change $\overline{k_1} = -k_1$ has been made.

As is exposed at Section 2, in the subsection dedicated to robust control problems of design, to test if one specification or a set of specifications are fulfilled it is enough to test if a function or a set of functions are positive. In our example it is enough to test the positivity of the following functions:

1) For the specification of absolute stability the function obtained is:

$$f_1 = -k_2 + q_1 \overline{k_1} \tag{27}$$

2) For the specification of velocity error less than 2%:

$$f_2 = q_1 \overline{k_1} - 50 \tag{28}$$

3) For the specification of resonance peak lower than 3 dB:

$$f_3 = 2k_2^2\omega_1^4 - 4k_2^2\omega_1^2\overline{k_1}q_1q_2 + \overline{k_1}^2 q_1^2q_2^2k_2^2 + 2k_2^2\omega_1^2q_2^2 - 4k_2\omega_1^2q_2^2\overline{k_1}q_1 + \overline{k_1}^2 q_1^2q_2^2\omega_1^2 \tag{29}$$

4) And finally for control effort lower than 20:

$$f_4 = 400k_2^2\omega_1^4 - 800k_2^2\omega_1^2\overline{k_1}q_1q_2 + 400\overline{k_1}^2 q_1^2q_2^2k_2^2 + 400k_2^2\omega_1^2q_2^2 + 400\overline{k_1}^2 q_1^2q_2^2\omega_1^2 - \overline{k_1}^2 k_2^2q_2^2 - \overline{k_1}^2 \omega_1^4 - \overline{k_1}^2 k_2^2\omega_1^2 - \overline{k_1}^2 \omega_1^2q_2^2 - 800k_2\omega_1^2q_2^2\overline{k_1}q_1 \quad (30)$$

The resultant feasible regions for each specification are represented on Figure 4.

In this example, specification f_3 and specially specification f_4 have a hard computation cost due its high number of summand components and also for the high number of plant variables qi and controller variables k_i that they involve.

Although this, when IRCAD compute the four specifications at the same time the cost of computation is significantly reduced because some regions are discarded on the specification of lost cost f_1 and f_2.

Using IRCAD, the problem of design a controller that fulfills the four specifications gives the graphical classification of Figure 5.

The graphical result of Figure 5 corresponds to the following set of feasible intervals:

$k_1 = [-70, -65] \quad k_2 = [2.75, 3]$

$k_1 = [-75, -65] \quad k_2 = [3, 3.25]$

$k_1 = [-80, -65] \quad k_2 = [3.25, 3.5]$

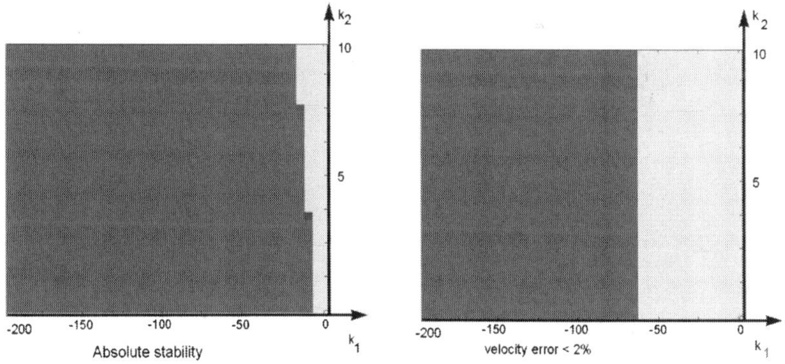

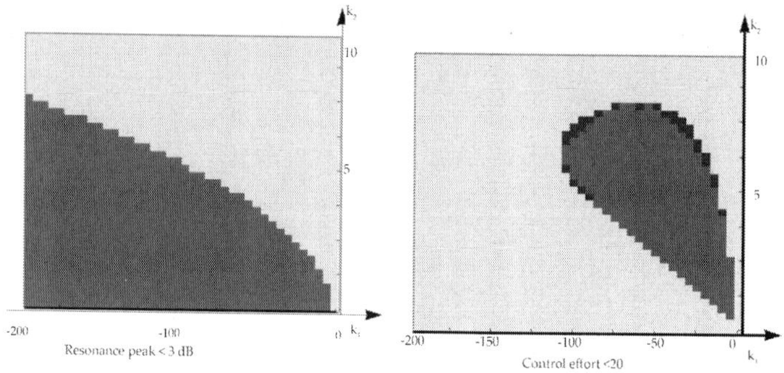

Figure 4: Feasible regions for each specification.

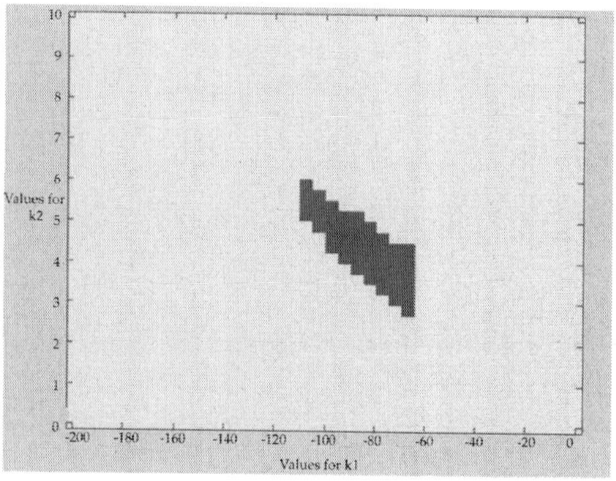

Figure 5: Feasible regions.

$k_1 = [-85, -65] \quad k_2 = [3.5, 3.75]$

$k_1 = [-90, -65] \quad k_2 = [3.75, 4]$

$k_1 = [-95, -65] \quad k_2 = [4, 4.25]$

$k_1 = [-100, -65] \quad k_2 = [4.25, 4.5]$

$k_1 = [-100, -75] \quad k_2 = [4.5, 4.75]$

$k_1 = [-105, -80] \quad k_2 = [4.75, 5]$

$k_1 = [-110, -85] \quad k_2 = [5, 5.25]$

$k_1 = [-110, -95] \quad k_2 = [5.25, 5.5]$

$k_1 = [-110, -100] \quad k_2 = [5.5, 5.75]$

$k_1 = [-110, -105] \quad k_2 = [5.75, 6]$

Finally, the solution of this robust control design example consists on the choice from this feasible region, following some criterion, of an interval pair k_1 and k_2 for the PI controller.

At this point IRCAD offers a set of post-design tools that allows to choose the optimal controller from the set of the proposed controllers on an automatic way. The framework chooses the controller following the criteria (see Section 5.3) that the user has selected.

In this example we suppose the user select the criterion of minimum norm. Then the controller suggested by the framework IRCAD is $[-67.5, 2.875]$ as is shown on Figure 6.

This action gives the optimal controller that fulfills the set of specifications. Using another post-design tool of the framework, the user can put this optimal controller on the closed-loop system. Finally, with this controller and this plant the framework offers some post design tools to simulate the features of the system.

CONCLUSIONS AND FUTURE WORK

An application to solve robust control problems was designed combining symbolic tools with interval analysis. The main advantage of using interval analysis is that guaranteed results are obtained. The drawback is the ne-

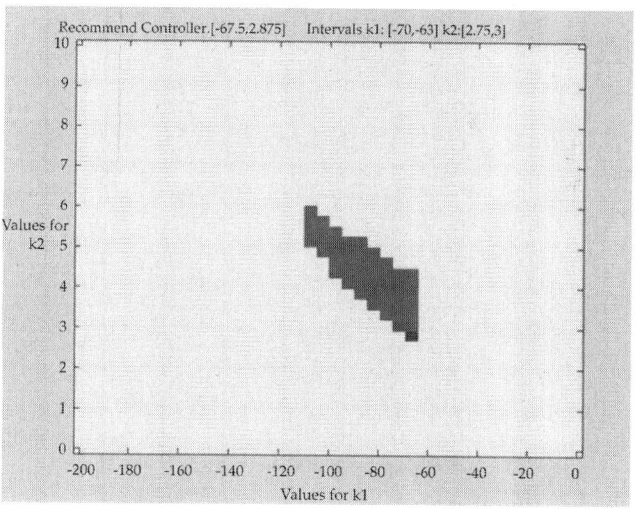

Figure 6: Minimum norm criteria.

Inès Ferrer-Mallorquí, Josep Vehí cessity to work with a variety of data, implicit on these kinds of problems. It is necessary to manage symbolic, interval and numeric data. The manipulation of these kinds of data needs to use the fitting software in each case. This diversity can discourage control engineers because most of them don't use these data in a mixed way. The framework created, allows users to manage on an easy way with problems that involves these kinds of data in a transparent way. The result is a most reliable framework for control engineers to solve problems that involves symbolic-interval-numeric data.

REFERENCES

1. The MathWorks Inc. (1999) The Polynomial Toolbox for Matlab, v. 2.0. Technical Report. www.polyx.com
2. Sakabe, K., Yanami, H., Anai, H. and Hara, S. (2004) A Matlab Toolbox for Robustcontrol Synthesis by Symbolic Computation. SICE Annual Conference in Sapporo, Hokkaido Institute of Tecnology, Japan, August 2004, 1968-1973.
3. Sienel, W., Bunte, T. and Ackermann, J. (1996) Paradise: Parametric Robust Analysisand Design Interactive Software Environment: A Matlab-Based Robust Control Toolbox. Proceedings of the 1996 IEEE International Symposium on Computer-Aided Control System Design, Dearborn, 15-18 September 1996, 380-385.
4. Nataraj, P.S.V. and Sardar, G. (2000) Computation of QFT Bounds for Robust Sensitivity and Gain-Phase Margins Specifications. Journal of Dynamic Systems Measurement and Control, 122, 528-834.
5. Ackermann, J. (1993) Robust Control: Systems with Uncertain Physical Parameters. Communications and Control Engineering Series. Springer-Verlag, Berlin.
6. Hyodo, N., Hong, M., Yanami, H. and Anai, H., et al. (2006) Development of a Matlab Toolbox for Parametric Robust Control: New Algorithms and Functions. SICE-ICASE International Joint Conference 2006, Busan, 18-21 October 2006, 2856-2861,
7. Ackermann, J. (2000) PARADISE—Parametric Robustness Analysis and Design Interactive Software Environment. Technical Report, Institute of Robotics and Mechatronics. www.op.dlr.de/FF-DRRR/paradise.
8. Moore, R.E. (1979) Methods and Applications of Interval Analysis. In: SIAM Studies in Applied Mathematics, SIAM, Philadelphia.
9. Vehí, J., Herrero, P., Sainz, M.A. and Jaulin, L. (2004) Quantified Set Inversion with Applications to Control. IEEE International Symposium on ComputerAided Control Systems Design, Taipei, 2-4 September 2004, 179-183.
10. Ackermann, J. (2002) Robust Control. The Parameter Space Approach. Springer, Berlin.
11. Frazer, R. and Duncan, W. (1929) On the Criteria for Stability of Small Motions. Proceedings of the Royal Society A, 124, 642-654.

12. Munro, N. (1999) Symbolic Methods in Control System Analysis and Design. IEEE Control Engineering Series, 56, 393 p.
13. Chapellat, H., Battacharyya, S.P and Keel, L.H. (1995) Robust Control: The Parametric Approach. In: Information and System Science Series, N.J. Prentice Hall, Englewood Cliffs.
14. Garloff, J. and Graft, B. (1999) Robust Schur Stability of Polynomials with Polynomial Parameter Dependency. Multidimensional Systems and Signal Processing, 10, 189-199. http://dx.doi.org/10.1023/A:1008402513296
15. Jaulin, L. and Walter, E. (1996) Guaranteed Tuning, with Application to Robust Control and Motion Planning. Automatica, 32, 1217-1221. http://dx.doi.org/10.1016/0005-1098(96)00050-7
16. Stoorvogel, A. (1992) The Control Problem. A State Space Approach. Prentice Hall International Series in Systems and Control Engineering, London.
17. Salapaka, M.V., Dahleh, M. and Voulgaris, P. (1997) Mixed Objective Control Synthesis: Optimal l1/h2 Control. SIAM: SIAM Journal on Control and Optimization (SICON), 35, 1672-1689.
18. Zhou, K., Doyle, J.C. and Glover, K. (1996) Robust and Optimal Control. Prentice Hall Inc., Upper Saddle River.
19. Gagnon, E., Pomerleau, A. and Desbiens, A. (1999) Mu-Synthesis of Robust Decentralized PI Controllers. IEEE Proceedings Control Theory Applications, 146, 289-294.
20. Vehí, J. (1998) Analysis and Design of Robust Controllers by Means of Modal Intervals. Ph.D. Thesis, Universitat de Girona, Spain. (in Catalan)
21. SIGLA/X. (1999) Modal Intervals. In: J. Vehí and M. Sainz, Eds., Application of Interval Analysis to Systems and Control, Girona, Spain, 157-221.
22. Fiorio, G., Malan, S., Milanese, M. and Taragna, M. (1993) Robust Performance Design of Fixed Structure Controllers for Systems with Uncertain Parameters. Proceedings of the 32nd Conference on Decision and Control, San Antonio, 15-17 December 1993, 3029-3031.
23. Malan, S., Milanese, M. and Taragna, M. (1997) Robust Analysis and Design of Control Systems Using Interval Arithmetic. Automatica, 33, 1363-1372. http://dx.doi.org/10.1016/S0005-1098(97)00028-9

Chapter 9

SYMBOLIC AND GRAPHICAL COMPUTATIONS OF A CLASS OF SLIGHTLY PERTURBED EQUATIONS

Mohammed A. Sharaf[1], Abdel-naby S. Saad[2], Zainab A. Mominkhan[3]

[1]Department of Astronomy, Faculty of Science, King Abdulaziz University, Jeddah, Saudi Arabia

[2]Department of Astronomy, National Research Institute of Astronomy and Geophysics, Cairo, Egypt

[3]Department of Mathematics, College of Science for Girls, King Abdulaziz University, Jeddah, Saudi Arabia

ABSTRACT

In this paper, a class of slightly perturbed equations of the form $F(x)=\xi-x+\alpha\Phi(x)$ will be treated graphically and symbolically, where $\Phi(x)$ is an analytic function of x. For graphical developments, we set up a simple graphical method for the real roots of the equation $F(x)=0$ illustrated by four

Citation: M. Sharaf, A. Saad and Z. Mominkhan, "Symbolic and Graphical Computations of a Class of Slightly Perturbed Equations," Applied Mathematics, Vol. 4 No. 5, 2013, pp. 817-824. doi: 10.4236/am.2013.45112.

transcendental equations. In fact, the graphical solution usually provides excellent initial conditions for the iterative solution of the equation. A property avoiding the critical situations between divergent to very slow convergent solutions may exist in the iterative methods in which no good initial condition close to the root is available. For the analytical developments, literal analytical solutions are obtained for the most celebrated slightly perturbed equation which is Kepler's equation of elliptic orbit. Moreover, the effect of the orbital eccentricity on the rate of convergence of the series is illustrated graphically.

Keywords: Symbolic Computation; Small Perturbations; Graphical Representations; Transcendental Equations

INTRODUCTION

Symbolic Computation is a modern area of research of interdisciplinary character that is placed in the common area of action of mathematics and the sciences of computation. Symbolic computing is concerned with the representation and manipulation of information in a symbolic form. It is based on defining objects not as numerical quantities but as entities that have certain mathematical properties. The representation of mathematical objects in a symbolic rather than numeric computational form has existed since the early days of computer science [1]. The 1970s and 1980s have seen the development, however, of environments that place a greater emphasis on computation with mathematical objects in an implicit or symbolic form. A brief history of symbolic mathematical computation is given in [2] and references therein.

Symbolic and graphical computations utilized with nowadays existing symbols used for manipulating digital computer programs, they are definitely invaluable for obtaining solutions with any desired accuracy [3,4].

In the present paper, a class of slightly perturbed equations of the form

$$F(x)=\xi-x+\alpha\Phi(x), \qquad (1)$$

will be treated graphically and symbolically where $\Phi(x)$ is an analytic function of x.

Advances in graphical technology have now made it possible to interact with information in innovative ways, most notably by exploring multimedia environments and by manipulating three-dimensional virtual worlds. In many situations, the graph offers much more insight into the problem than does the algebra. Moreover it makes easy to understand and interpret data at a glance and helps to make comparisons among many things. For the iterative solution of equations, graphical solution usually provides excellent initial conditions.

The above mentioned advantages motivated us introducing graphical and symbolic developments to a class of slightly perturbed equations that would have some applications in the future.

GRAPHICAL COMPUTATIONS

In this section, simple graphical method for the real roots of the equation

$$F(x)=\xi-x+\alpha\Phi(x)=0,\ |\alpha|<1, \tag{2}$$

will be developed, where ξ and α are given and x is required.

Take a rectangular system of axes and construct the curve and the straight line whose equations are

$$\left.\begin{aligned} y&=\Phi(x), \\ y&=\frac{1}{\alpha}(x-\xi). \end{aligned}\right\} \tag{3}$$

The abscissa of their point of intersection is the value of x satisfying the equation; for, eliminating y, Equation results (1). Instead of drawing the straight line a straightedge can be drawn between the two points $(\xi,0)$ and $(\xi+\beta\alpha,\beta)$, with $\beta=100$ (say).

The accuracy of the computed root (or roots) x_0 (say) may checked by the condition that $F(x_0)=\varepsilon$, where ε is of order 10^{-3} to 10^{-1} at most.

The following are two sets of examples, for the first set we considered four equations listed together with the values of α and ξ in Table 1. While the second set of examples is four cases of the most celebrated slightly perturbed equation which is Kepler's equation of elliptic orbit given as

$$M=E-e\sin E \tag{4}$$

where

$$M = \sqrt{\frac{\mu}{a^3}}(t-\tau) = n(t-\tau) \quad (5)$$

The angle M, called the mean anomaly by Kepler, E the eccentric anomaly, e the eccentricity of the orbit $(e<1)$, μ the gravitational parameter, n the mean motion, a the semi-major axis of the orbit, t the time, and τ is the time of passage through pericenter (the closet point to the focus of the orbit). The solution of Kepler's equation is to find E when M and e are given.

The four cases of Kepler's equation are listed in Table 2.

Figures 1 and 2 give the solutions of the two sets of equations mentioned in Tables 1 and 2 respectively.

Table 1: Equation (2) for some Φ, α and ξ

No	$\Phi(x)$	α	ξ
1	$0.8\sin x + 0.5\cos x$	0.73	1.3439
2	Laguerre polynomial: $L_3(x) = \frac{1}{6}\left[-x^3 + 9x^2 - 18x + 6\right]$	0.06	5.04
3	$x^2 e^{-x}$	0.8	0.9
4	$\cos^4 2x$	0.5	1.5708

Table 2: Kepler's equation for some e and M

No	e	M^{o}
1	0.34	20
2	0.56	80
3	0.2	46
4	0.65	87

SYMBOLIC COMPUTATIONS

Lagrange's approach to solving Kepler's equation in 1770s, led to useful expansion of the Equation (1) as

$$y = x + \alpha\Phi(y),\ |\alpha| < 1.$$

where α is to be considered a small parameter-originally identified with a planetary eccentricity. Then y could be expanded in terms of x and α [5] as

$$y = x + \sum_{n=1}^{\infty} \frac{\alpha^n}{n!} \frac{d^{n-1}}{dx^{n-1}} \left[\phi(x)\right]^n \quad . \quad (6)$$

Lagrange's series is, of course, the Taylor series representation of the root of the functional equation $y - x - \alpha\phi(x) = 0$. Sufficient conditions for a unique root are obtained by a direct application of Rouche's theorem for analytical function of a complex variable [6].

In what follows we shall illustrate the symbolic computations by considering the applications of Equation (6) on the two sets of equations mentioned in Section 2.

Symbolic Computations of the First Set of Table 1

Because of space limitations, only three terms for the symbolic expressions of the functions of Table 1 are given without specification of the values of α and ξ.

We can write each of these expansions as a third degree polynomial in α as

$$b_0 + b_1\alpha + b_2\alpha^2 + b_3\alpha^3$$

The b's coefficients are listed in for each function. In what followsFor the function: $0.8\sin x + 0.5\cos x$

$$b_0 = x$$

$$b_1 = 0.5\cos[x] + 0.8\sin[x]$$

$$b_2 = 0.4\cos[2x] + 0.39\cos[x]\sin[x]$$

$$b_3 = -0.055625\cos[x] + 0.313125\cos[3x] \\ -0.089\sin[x] - 0.033\sin[3x].$$

For the Laguerre polynomial $L_3(x)$

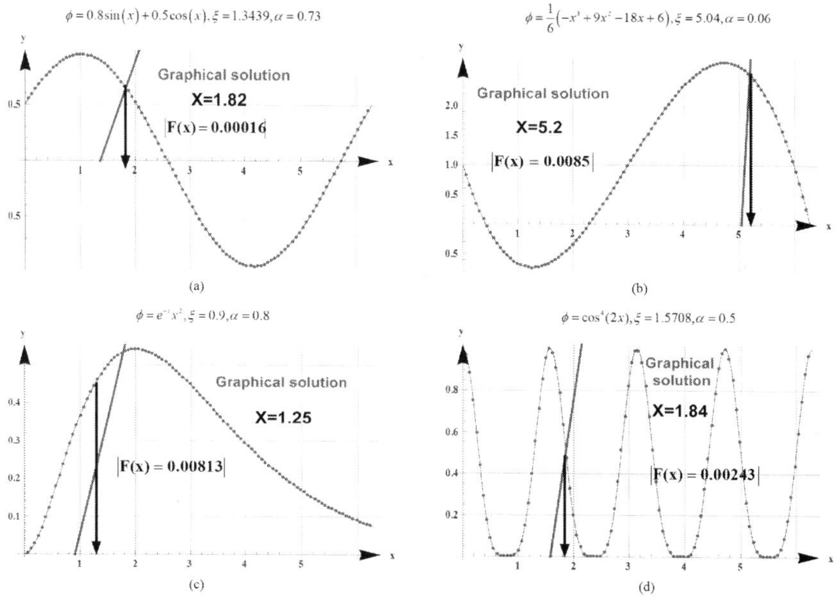

Figure 1: Graphical solutions of the functions of Table 1.

$$b_0 = x$$

$$b_1 = 1 - \frac{1}{6}(-6+x)(-3+x)x$$

$$b_2 = \frac{1}{12}(6+(-6+x)x)(-6+(-6+x)(-3+x)x)$$

$$b_3 = -\frac{1}{12}(6+(-6+x)(-3+x)x)$$
$$\times(126+x(-276+x(189+4(-12+x)x))).$$

For the function x^2e^{-x}

$$b_0 = x$$

$$b_1 = e^{-x}x^2$$

$$b_2 = -e^{-2x}(-2+x)x^3$$

$$b_3 = -\frac{1}{2}e^{-3x}x^4(10+3(-4+x)x).$$

For the function $\cos^4 2x$

$$b_0 = x$$

$$b_1 = \cos^4[2x]$$

$$b_2 = -8\cos^7[2x] + \sin[2x]$$

$$b_3 = 40\cos^{10}[2x] - 48\cos^{10}[2x]\cos[4x].$$

Symbolic Computations for Kepler's Equation of Table 2

Comparing Equation (1) with Equation (4) we get:

$$y \equiv E;\ x \equiv M;\ \alpha \equiv e < 1;\ \phi(y) \equiv \sin y,$$

then applying Lagrange's expansion theorem as given by Equation (5) we obtain:

$$u = E - M = \sum_{n=1}^{\infty} \frac{e^n}{n!} \frac{\mathrm{d}^{n-1}}{\mathrm{d}M^{n-1}} (\sin M)^n , \tag{6}$$

it could be shown that [6],

$$\frac{\mathrm{d}^{n-1}}{\mathrm{d}M^{n-1}} \sin^n M$$
$$= \sum_{k=0}^{\left[\frac{n-1}{2}\right]} (-1)^k \left(\frac{n-2k}{2}\right)^{n-1} \binom{n}{k} \sin(n-2k)M, \tag{7}$$

Example 2: $e = 0.56 : M = 80^0 \Longrightarrow$ **Graphical solution=** $E = 1.92^r$.

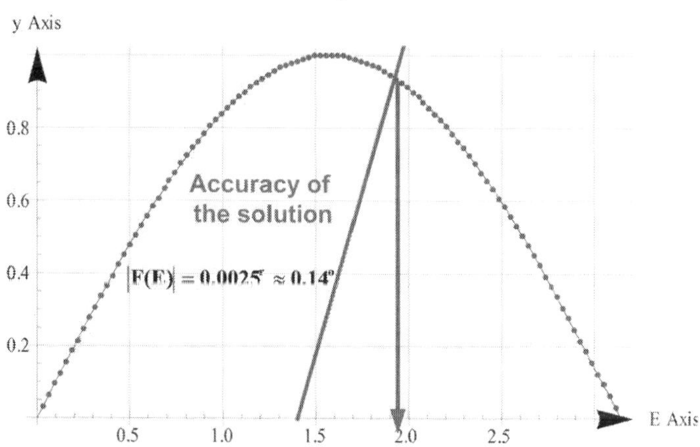

(a)

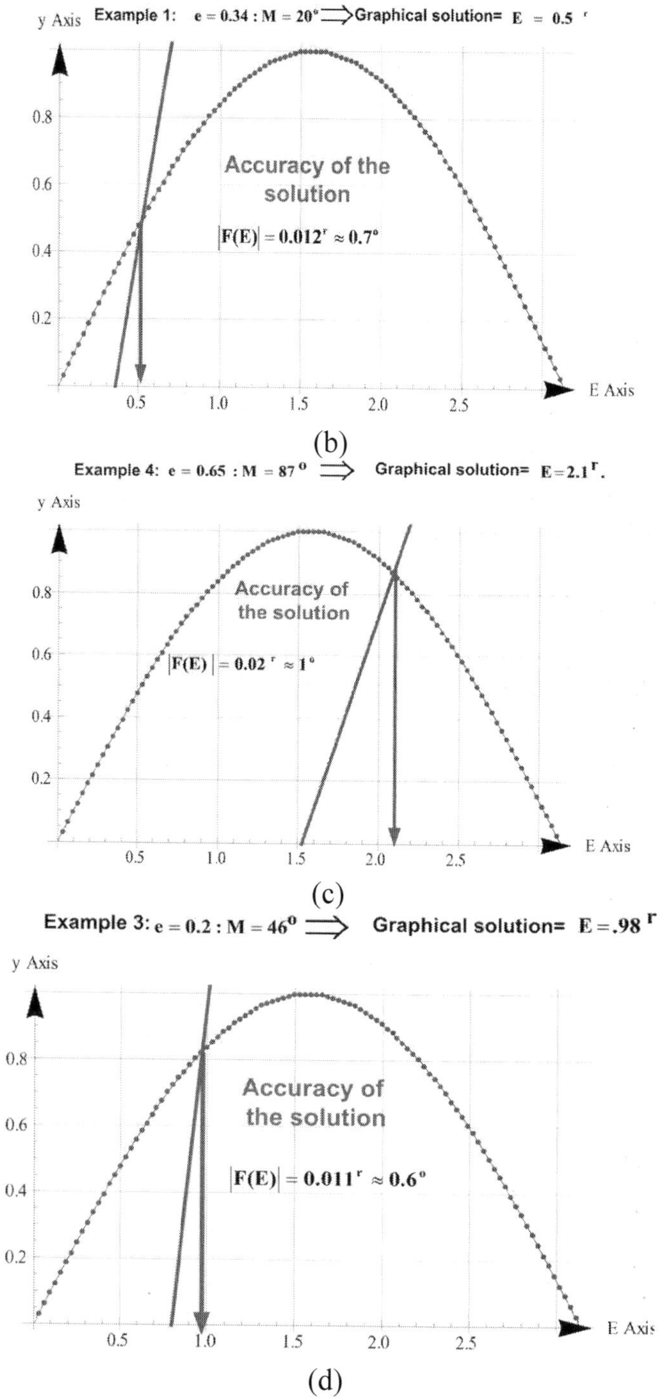

Figure 2: Graphical solutions of Kepler's equations of Table 2.

where [z] is the integer part of z. Using Equation (7) into Equation (6) we get:

$$u = \sum_{n=1}^{\infty} e^n \sum_{k=0}^{\left[\frac{n-1}{2}\right]} A_{n,k} \sin(n-2k)M, \tag{8}$$

where

$$A_{n,k} = \frac{(-1)^k \left(\frac{1}{2}n-k\right)^{n-1}}{k!(n-k)!} \tag{9}$$

Equation (8) gives the solution of Kepler's equation as power series in the eccentricity e with the coefficient which turned out to be linear combinations of sine functions of integer multiples of the mean anomaly M. It was shown [5] that $e < 0.66274341\cdots$ is the requirement for the series (8) to represent the unique root of Kepler's equation for all values of M. Laplace is the first to show that if e exceeds this critical value, then the series will diverge for some values of M.

Lagrange rearrange the terms in the series (8) to obtain a form which we would now call a Fourier sine series. The Fourier coefficients were infinite power series in e which converge for all elliptic orbits. This series could be written as:

$$u = \sum_{m=1}^{\infty} H_m \sin mM, \tag{10}$$

where

$$H_m = \frac{2}{m}\sum_{k=0}^{\infty} \frac{(-1)^k \left(\frac{1}{2}me\right)^{m+2k}}{k!(m+k)!} = \frac{2}{m} J_m(me), \tag{11}$$

$J_n(z)$ the Bessel function of the first kind of order n.

In 1824, Bessel, attempted the direct solution of Kepler's equation as a Fourier series and obtained the coefficients in an integral forms. The power series expansions of these integrals produced, of course, the same collection of series that Lagrange had obtained [Equation (10)] almost fifty years earlier-but, because Bessel made such an extensive study of these functions for many years, they bear his name not of Lagrange.

From the above discussions we can write u in the compact form

$$u = \sum_{j=1}^{\infty} C_j \begin{Bmatrix} e^j \\ \sin jM \end{Bmatrix} . \qquad (12)$$

Of course, the analytical expression of any of the C's coefficients is computed from the truncated series of Equation (12).

In what follows, we shall find the analytical expression for the coefficient C_s o $\sin sM$ or of e^s for the series solution of Kepler's equation for elliptic orbits represented by m terms. To do so, let m (integer) the maximum number of terms for the series solution of Kepler's equation and s (integer) $\le m$ is number of the required term of the series.

Series Solution of the Kepler's Equation in the Form: $\boldsymbol{E = M + \sum_{i=1} C_i \sin iM}$

$$s = 2, m = 5 \Rightarrow C_2 = \frac{e^2}{2} - \frac{e^4}{6}$$

$$s = 2, m = 10 \Rightarrow C_2 = \frac{e^2}{2} - \frac{e^4}{6} + \frac{e^6}{48} - \frac{e^8}{720} + \frac{e^{10}}{17280}$$

$$s = 8, m = 10 \Rightarrow C_8 = \frac{128e^8}{315} - \frac{2048e^{10}}{2835}$$

$s = 1, m = 15$

$$\Rightarrow C_2 = e - \frac{e^3}{8} + \frac{e^5}{192} - \frac{e^7}{9216} + \frac{e^9}{737280} - \frac{e^{11}}{88473600} + \frac{e^{13}}{14863564800} - \frac{e^{15}}{3329438515200}$$

$$s = 7, m = 7 \Rightarrow C_7 = \frac{16807e^7}{46080}$$

$$s = 31, m = 30 \Rightarrow C_{31} = 0$$

Series Solution of the Kepler's Equation in the Form: $E = M + \sum_{i=1} C_i e^i$

$s = 2, m = 5 \Rightarrow C_2 = \frac{1}{2}\sin[2M]$

$s = 8, m = 15$

$$\Rightarrow C_8 = -\frac{1}{720}\sin[2M] + \frac{4}{45}\sin[4M] - \frac{243}{560}\sin[6M] + \frac{128}{315}\sin[8M]$$

$s = 1, m = 15 \Rightarrow C_1 = \sin[M]$

$s = 7, m = 7$

$$\Rightarrow C_7 = \frac{\sin[M]}{9216} + \frac{243\sin[3M]}{5120} - \frac{3125\sin[5M]}{9216} + \frac{16807\sin[7M]}{46080}$$

$s = 10, m = 30$

$$\Rightarrow C_{10} = \frac{\sin[2M]}{17280} - \frac{16}{945}\sin[4M] + \frac{2187\sin[6M]}{8960} - \frac{2048\sin[8M]}{2835} + \frac{78125\sin[10M]}{145152}$$

$s = 31, m = 30 \Rightarrow C_{31} = 0$

$s = 11, m = 1 \Rightarrow C_1 = \sin[M]$

Graphical Representations of the C's' Coefficients

First: If C_j are the coefficients of e_j $\forall j$.

In appendix some graphics showing the effect of the eccentricity e on the coefficients, that is, the effect of e on the convergence rate of the series.

Second: If C_j are the coefficients of $\sin jM$ $\forall j$.

Graphical representations of the C's coefficients $\sin jM$ $\forall j$ are also given in appendix.

CONCLUSION

Mohammed A. Sharaf, Abdel-naby S. Saad, Zainab A. Mominkhan (a)-(d) show the graphical representations of the transcendental equations given in Table 1. The literal analytical solution of Kepler's equation of elliptic orbit is described by setting up simple graphical representations. The four cases of slightly perturbed equation (Kepler's equation) listed in Table 2 are shown by Figures 2(a)-(d). While the graphical representations of the C›s coefficients e_j and $\sin jM$, $\forall j$ are given by Figures 3 and 4 in the Appendix. The effect of the orbital eccentricity on the rate of convergence of the series is illustrated graphically. One merit for applying graphical solution is providing excellent initial conditions for iterative solutions of equations.

Appendix

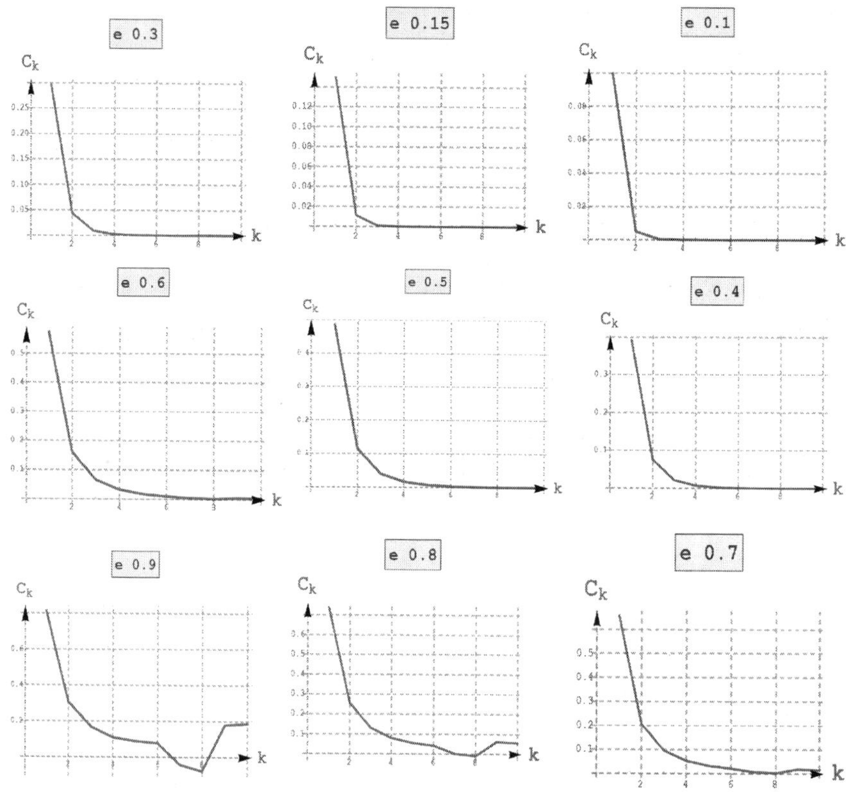

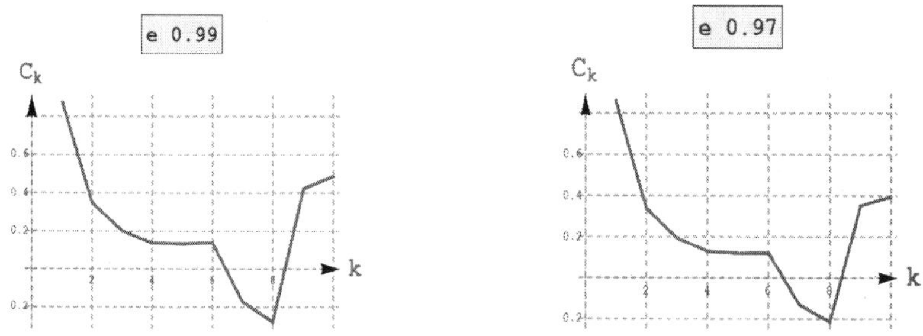

Figure 3: Graphical representations of the C's coefficients $^{e_j \forall j}$. Note that, the series diverge for e ≥ 0.7 as proved by Laplace.

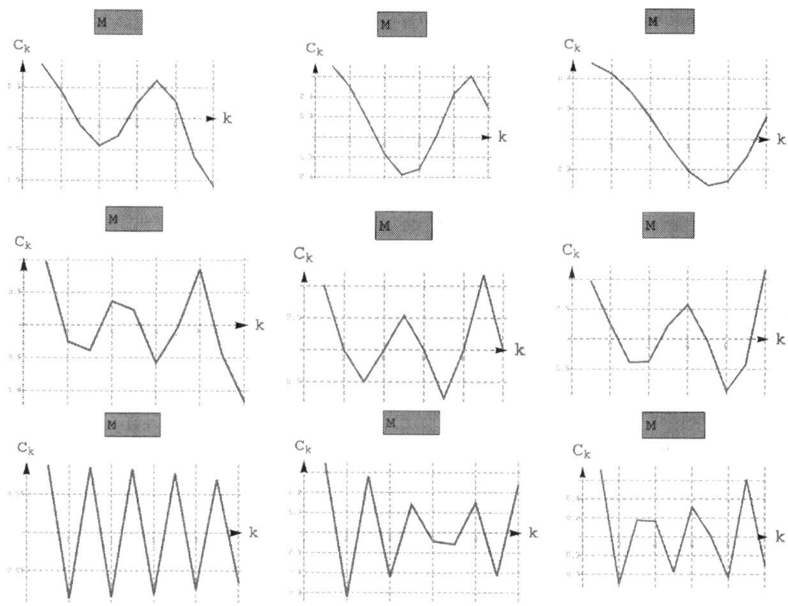

REFERENCES

1. E. Fiume, "An Introduction to Scientific, Symbolic, and Graphical Computation," A K Peters/CRC Press, London, 1995.
2. E. Kaltofen, "Challenges of Symbolic Computation," Journal of Symbolic Computation, Vol. 29, 2000, pp. 891-919. doi:10.1006/jsco.2000.0370
3. V. A. Brumberg, "Analytical Techniques of Celestial Mechanics," Springer-Verlag, Berlin, Heidelberg, 1995. doi:10.1007/978-3-642-79454-4
4. M. A. Sharaf, A.-N. S. Saad and A. A. Sharaf, "Unified Symbolic Algorithm of Gauss Method for Near-Parabolic Orbits," Celestial Mechanics and Dynamical Astronomy, Vol. 70, No. 3, 1998, pp. 201-214. doi:10.1023/A:1008339612666
5. M. P. Fitzpatrick, "Principles of Celestial Mechanics," Academic Press, New York and London, 1970.
6. R. H. Battin, "An Introduction to the Mathematics and Methods of Astrodynamics," AIAA, Education Series, Reston, 1999. doi:10.2514/4.861543

SECTION 3: NUMERICAL COMPUTING APPLICATIONS

Chapter 10

NON-NEGATIVITY PRESERVING NUMERICAL ALGORITHMS FOR PROBLEMS IN MATHEMATICAL FINANCE

Yuan Yuan

Hydroforce (Beijing) Technology Co., Ltd., Beijing, China

ABSTRACT

We give a study result to analyze a rather different, semi-analytical numerical algorithms based on splitting-step methods with their applications to mathematical finance. As certain subsistent numerical schemes may fail due to producing negative values for financial variables which require non-negativity preserving. These algorithms which we are analyzing preserve not only the non-negativity, but also the character of boundaries (natural, reflecting, absorbing, etc.). The derivatives of the CIR process and the Heston model are being extensively studied. Beyond plain vanilla European

Citation: Yuan, Y. (2018), "Non-Negativity Preserving Numerical Algorithms for Problems in Mathematical Finance". Applied Mathematics, 9, 313-335. doi: 10.4236/am.2018.93024.

options, we creatively apply our splitting-step methods to a path-dependent option valuation. We compare our algorithms to a class of numerical schemes based on Euler discretization which are prevalent currently. The comparisons are given with respect to both accuracy and computational time for the European call option under the CIR model whereas with respect to convergence rate for the path-dependent option under the CIR model and the European call option under the Heston model.

Keywords: Stochastic Differential Equations, Numerical Methods, Splitting-Step Algorithm, Non-Negativity, Convergence, Simulation

INTRODUCTION

Stochastic differential equations (SDEs) are fundamental in mathematical finance. Particularly, they serve as models for describing the evolution of certain financial variables, such as the stock price, interest rates or volatility of an asset. Although there are extensive studies on SDEs, explicit solutions of SDEs are rarely known or very difficult to obtain, so numerical approximations are relied on. Actually a wealth of numeric methods have been proposed and tested (see for example [1] [2] [3] [4]). In our work, we mainly dedicate to applying rather different numerical algorithms to evaluate financial derivatives with comparison to methods based on Euler discretization. We concentrate on the Cox-Ingersoll-Ross (CIR) model [5] and the Heston model [6] as they are two of the fundamental models in mathematical finance.

Problem Statement and Motivation

In some cases, the Itô-type SDEs of the form

$$\mathrm{d}X(t)=f\left(X(t)\right)\mathrm{d}t=\sigma\left(X(t)\right)\mathrm{d}W(t),\ X(0)=X_0\in D \tag{1.1}$$

are well-defined only with certain boundary conditions. For example, the Cox-Ingersoll-Ross (CIR) [5] model and the Heston model [6] which were proposed to describe the short rate of interest and stock price dynamics respectively are well-defined in the domain $D=[0,\infty)$, where $X(t)=0$ implies an absorbing or reflecting state. Precisely, the CIR process can be expressed in the form

$$\mathrm{d}V(t)=\kappa\left(\theta-V(t)\right)\mathrm{d}t-\sigma\sqrt{V(t)}\mathrm{d}W^{V}(t) \tag{1.2}$$

where $W^V(t)$ is a Wiener process and κ,θ,σ are positive constants. The Heston model is a two-factor model of the form:

$$\mathrm{d}S(t)=\mu S(t)\mathrm{d}t+\sqrt{V(t)}S(t)\mathrm{d}W^{S}(t) \tag{1.3}$$

$$\mathrm{d}V(t)=\kappa\left(\theta-V(t)\right)\mathrm{d}t+\sigma\sqrt{V(t)}\mathrm{d}W^{V}(t) \tag{1.4}$$

where $W^S(t),W^V(t)$ are two correlated Wiener processes with $dW^S(t)$ $dW^V(t)=\rho dt,\rho\in(-1,1)$, and μ,κ,θ,σ are positive parameters. The component S describes the evolution of a financial variable such as stock index or exchange rate, and V describes the stochastic variance of its returns. One can notice that the component V in the Heston model evolves according to the CIR process (1.2).

First let's mention that the SDE (1.2) for CIR process is not explicitly solvable but its transition probability is known; it can be represented by a non-central chi-square density. Depending on the number of degrees of freedom $d:=4\kappa\theta/\sigma^2$, there are significant differences in the boundary behavior of CIR process. According to Feller's classification [7], if $d\geq 2$, the boundary $V(t)=0$ is unattainable and the strong solution is strictly positive; if $d<2$, the boundary is attracting and attainable but strongly reflecting, i.e., each sample path will reflect instantaneously into the positive domain once it reaches 0 (see e.g. Karlin & Taylor [8]). The reflecting behavior in the latter case is particularly difficult to capture numerically. During our work, we only focus on the case that the boundary is attainable, i.e. $d<2$.

Second, as the square-root term in CIR process avoids the possibility of negative values, Euler-Maruyama [1] and higher order Taylor-type methods [1] [2] cannot be applied in this case as they lead to negative values of X(t) in practice. The numerical methods which preserve non-negativity are preferred. Actually various numerical schemes have been devised to meet this requirement, among which the schemes based on Euler discretization normally gives high efficiency, but their strong convergence rate and discretization errors are difficult to establish. In contrast to the Euler-type methods giving strong convergence of order $\frac{1}{2}-\varepsilon$, Moro and Schurz [9] provide the splitting-step methods of strong convergence at least of order 1. Although they are more costly with respect to computational time, they give higher accuracy and convergence rate. With this motivation, we would like to evaluate some important financial instruments using splitting-step methods with comparison to Euler-type numerical schemes. We apply them to option pricing including a path-dependent option valuation comparing both accuracy and computational cost.

Third, exact simulation methods exist for both the CIR process (see Glasserman [10]) and the Heston model (see Broadie and Kaya [11] [12]). The drawback of these exact simulation methods is that they are very time-consuming. They are competitive when one just simulates the process at one time (or few times), for example to price a European option with a Monte-Carlo algorithm. On the contrary, they are too slow when one has to simulate the process along a time-grid, for example to compute a path-dependent European option. Thus, the splitting-step methods may have advantage on pricing path-dependent options as they are more efficient.

Review of Research Status

For the CIR process, a number of numerical schemes based on implicit time-stepping integrators have been devised for the case of unattainable boundary condition, see for example Alfonsi [13], Kahl and Schurz [14] and Dereich et al., 2012 [15] . There also are some splitting methods that serve for both boundary conditions without any restriction of parameters, such as Alfonsi [16], Moro and Schurz [9] . The latter paper gives a splitting-step method which highly relied on the knowledge of explicit solution of SDEs. This method is what we mainly use throughout our project.

Other direct approaches that can be applied to both attainable and unattainable zero boundary cases are based on modification of Euler-Maruyama approximations. See for example Deelstra and Delbaen [17], Bossy and Diop [18], Berkaoui, Bossy and Diop [19], Higham and Mao [20], Lord et al., 2008 [21] . These methods all are proved to converge to the exact solution as the time step tends to 0, but the strong convergence rate and discretization errors are difficult to establish. The full truncation method in Lord et al., 2008 [21] has been shown in practice to be the leading method in this class.

The exact simulation method for CIR model exists, see Glasserman [10], but it requires more computational time than discretization schemes (up to a factor 10). This was analyzed in Alfonsi [13] . For the Heston model, the exact simulation scheme (see Broadie and Kaya [11] [12]) has the same drawback of highly time-consuming and were analyzed in Broadie and Kaya [11] [12] and Lord et al., 2008 [21] . After comparing to various numerical schemes, Kahl and Jäckel [22] conclude that combining the balanced Milstein method (BMM) for the variance process with their bespoke IJK method for the logarithm of the stock price gives the best results with respect to strong convergence measure. The BMM method actually preserves positivity for the variance process if $4\kappa\theta > \sigma^2$: But this restriction is rarely satisfied in practice, and one typically finds that the sampling scheme for V produces negative values with substantial probability.

Anderson [23] considered two new discretization schemes: the truncated Gaussian scheme (TG) and the quadratic-exponential scheme (QE) with additional martingale-corrected versions labeled by "TG-M" and "QE-M" respectively. Both TG and QE schemes were attested to outperform the full truncation method proposed by Lord et al., 2008 [21] with respect to biases and only cost marginally more computational time than the Euler scheme. The TG scheme is more natural to use than the QE scheme in multi-asset applications that involve several correlated variance processes, but generally performs somewhat worse than QE at practically relevant time steps. Thus they concluded that the QE scheme should be the default choice of the schemes considered due to its simplicity and strong performance; martingale correction (the QE-M scheme) is optional.

Further, for pricing financial derivatives under Heston model, closed form semi-analytical formulae for plain vanilla option prices have been derived in [6] . However, these formulae require the evaluation of logarithms with complex arguments during the involved inverse Fourier integration step. Carr & Madan [24] pioneered the use of the fast Fourier transform (FFT) technique by mapping the Fourier transform directly to option prices via the characteristic function. Lewis [25] considered the same type of approach, except that there the transform was taken with respect to the log-spot price. An important step in Lewis [25] was made by considering the resulting integral as a contour integral in the complex plane. Lee [26] generalizes Carr & Madan [24] and unifies it with extensions of some relevant elements and proved an error analysis for these FFT methods. Kahl & Jächel [27] propose a new approach which enables the use of Heston's analytics for practically all levels of parameters and even maturities of many decades since most implementations of Heston's formulae are not robust for moderate to long dated maturities or strong mean reversion. Lord & Kahl [28] present the optimal contour of the Fourier integral, taking into account numerical issues such as cancellation and explosion of the characteristic function. Fang &

Oosterlee [29] proposed the COS method which is based on the Fourier and Fourier-cosine expansion.

Summary of the Paper

In the remaining paper, we introduce the general structure and properties of splitting-step algorithm proposed by Moro and Shurtz in [9] in Section 2, including an efficient sample manner for generating a non-central chi-square distribution random number.

In Section 3, we present numerical results of applications to CIR model, evaluating a European plain vanilla call option and a path-dependent option with comparisons to the methods based on Euler discretization. The comparisons are with respect to both accuracy and computational time. All of our numerical results are generated on a MacBook Pro with an Intel Core i7 2.3 GHz processor, 16 GB 1600 MHz DDR3 memory, using Xcode 6.4 and R 3.1.0 in a Mac OS X Yosemite environment.

In Section 4 we study the option valuation under the Heston model. The results show that the splitting-step method gives the best convergence rate for pricing a European plain vanilla call option among the five algorithms utilized in our project.

Finally we conclude and point out the future work in Section 5.

GENERAL STRUCTURE OF SPLITTING-STEP ALGORITHMS

Construction of the Algorithms

The splitting-step algorithm [9] has general structure as followed. Suppose that the more general equation to be integrated is of the form

$$\mathrm{d}X(t)=\left[\alpha(X(t),t)+\beta(X(t),t)\right]\mathrm{d}t+\sigma(X(t),t)\mathrm{d}W(t), \tag{2.1}$$

where W(t) is a standard Wiener process? Then decompose the above equation into two subsystems

$$\mathrm{d}X_1(t)=\beta(X_1(t),t)\mathrm{d}t+\sigma(X_1(t),t)\mathrm{d}W(t), \tag{2.2}$$

$$\mathrm{d}X_2(t)=\alpha(X_2(t),t)\mathrm{d}t \tag{2.3}$$

where it's required that we know the exact strong solution for X1(t) or the conditional probability $P[X_1(t)|X_1(0)]$. Afterwards, approximate the solution of (2.1) along the time interval [t,Δt] using the following two-step algorithm for each Δt.

- Step 1: Knowing the value X_t, taking it as an initial data of (2.2), i.e. $X_t = X_1(t)$, we obtain an intermediate value $\tilde{X}_t = X_1(t+\Delta t)$ through the exact integration of (2.2), or alternatively, through the conditional transition probability $P[X_1(t+\Delta t)|X_1(t)]$.
- Step 2: Using X_t obtained in step 1 as the initial condition for (2.3), i.e. $\tilde{X}_t = X_2(t)$, integrate (2.3) by any converging deterministic numerical algorithm to get X,(t+Δt) (at least of deterministic order 1). Then $X_{t+\Delta t} = X_2(t+\Delta t)$.

Easily speaking, the procedure is as follow:

1), Solve Equation (2.2) exactly for $X_1(t+\Delta t) = \Phi(t, X(t))$;
2), Approximate numerically the equation (2.3)
For $X_2(t+\Delta t)$ with initial condition
$X_2 = X_1(t+\Delta t)$, then
$X_2(t+\Delta t) = X_1(t+\Delta t) + \alpha(X_1(t+\Delta t), t)\Delta t + \cdots$
3) Take $X(t+\Delta t) = X_2(t+\Delta t)$. Thus
$X(t+\Delta t) = \Phi(t+\Delta t, X(t)) + \alpha(\Phi(t+\Delta t, X(t)), t)\Delta t + \cdots$

We adapt the example in [9] to see how to obtain $X_1(t)$ through the conditional transition probability $P[X_1(t) | X_1(0)]$. Suppose that $\beta(X_1, t) = 0$ $\sigma(X_1, t) = \sqrt{X_1}$, then the conditional probability distribution is given by

$$P[X_1(t) | X_1(0)] = \frac{2}{t}\left(\frac{X_1(0)}{X_1(t)}\right)^{\frac{1}{2}} I_1\left(\frac{4}{t}\sqrt{X_1(t)X_1(0)}\right) e^{-\frac{2}{t}[X_1(t)+X_1(0)]} + e^{-\frac{4}{t}X_1(0)}\delta(X_1(t))$$

where δ(x) is the Dirac delta function and I is the modified Bessel function of the first kind with index n which is given by

$$I_v(x) = \sum_{n=0}^{\infty} \frac{1}{n!(v+n+1)}\left(\frac{x}{2}\right)^{v+2n}, x > 0$$

with gamma function $\Gamma(t) := \int_0^{\infty} x^{t-1} e^{-x} dx, t > 0$ specially $\Gamma(n) = (n-1)!$ for $n = 0, 1, 2, \cdots$. Then we can sample the conditional probability distribution function using the rejection or inverse methods but it's computational expensive. There is a more simple way to obtain $X_1(t)$. Noticing that the variable $Z(t) = \frac{4}{t} X_1(t)$ follows a non-central χ^2-distribution, that is

$$P[Z \mid Z_0] = \sum_{j=1}^{\infty} \frac{(\lambda/2)^j e^{-\lambda/2}}{j!} P_{\chi_{2j}^2}(Z) + e^{-\frac{t}{4}\lambda} \delta(Z)$$

where $\lambda = \frac{4}{t} Z_0$ and $P_{\chi_{2j}^2}(x)$ is χ^2-pdf with 2j degrees of freedom. Then

$$X_1(t) = \frac{1}{2k} \begin{cases} 0 & \text{if } K = 0, \\ \sum_{i=1}^{2K} z_i^2 & \text{if } K \neq 0, \end{cases}$$

where $k = \frac{2}{t}$ and K is chosen from a Poisson distribution with mean $\lambda/2$ and z_i are independent Gaussian random numbers with zero mean and unit variance.

Generally, a non-central chi-square distribution random variable $\chi_d'^2(\lambda)$ with d degree of freedom and non-centrality parameter λ whose probability density function is given by

$$P[\chi_d'^2(\lambda) = x] = p(x; d, \lambda) = \frac{e^{-(\lambda+x)/2}}{2} \left(\frac{x}{\lambda}\right)^{(d-2)/4} I_{(d-2)/2}\left(\sqrt{\lambda x}\right) \tag{2.4}$$

see Johnson et al., 1994 [30] . This distribution is properly defined for d positive, and was extended to the case d=0 by Siegel [31], afterwards was extended to the case for d=−2,−4,··· by Moro and Schurz [9] . They used the fact that the distribution of (0.1) can be expressed as a mixture of central χ^2 variables with Poisson weights, then it can be sample as follow: Choose K from a Poisson distribution with mean = 2, then

$$\chi_d'^2(\lambda) = \chi_{d+2K}^2$$

And

$$\chi_d'^2(\lambda) = \begin{cases} 0 & \text{if } d+2K \le 0 \\ \chi_{d+2K}'^2 & \text{if } d+2K > 0 \end{cases} \qquad d = 0, -2, -4, \cdots$$

where χ_{d+2K}^2 can be sampled using any standard random number generator of the χ_d^2 distribution. This sampling is used specially when λ is small. For the case of large λ, other approximations (see e.g. Johnson et al., 1994 [30]) might be taken into account for improving the speed of the splitting-step algorithms.

Strong and Weak Convergence of Splitting-Step Methods

Let $C^{i,j}(\mathbb{R}^d \times [0,T])$ denote the vector space of continuous functions f=f(x,t) which are i times continuously differentiable with respect to the space

Coordinate $x_k \in \mathbb{R}$, $(k = 1,2,\cdots,d)$ and j times continuously differentiable with respect to the time coordinate t∈[0,T] . Let

$0 < t_0 < t_1 < \cdots < t_n < t_{n+1} < \cdots < t_N = T$

be any random partition of the given time interval [0,T] with sufficiently small maximum step size $\Delta = \max_{i=1,2,\cdots,N} |t_i - t_{i-1}| \le 1$. Then the time discretized approximation $\tilde{X}_\Delta$ of a continuous-time process X, is said to be of general strong order of convergence γ to X at time T if there exists a positive constant C, which does not depend on Δ, and a $\delta_0>0$ such that the following strong error ε(Δ) satisfies

$\varepsilon(\Delta) = \mathrm{E}\left(\left|X(T) - \tilde{X}_\Delta(T)\right|\right) \le C\Delta^\gamma$

for each $\Delta \in (0,\delta_0)$.

Along with the strong convergence, the weak convergence can be defined. A discrete-time approximation Y_Δ is said to converge with weak order β>0 to X at time T as Δ→0 if for each smooth function g of polynomial growth there exists a constant C_g, which does not depend on D and $\Delta_0 \in [0,1]$ such that the following weak error ϵ(Δ) satisfies the estimate

$\epsilon(\Delta) = \left|E_g(X(T)) - E_g(Y_\Delta(T))\right| \le C_g \Delta^\beta$

for each Δ∈(0,Δ0).

The splitting-step algorithm has been shown that has strong and weak order 1.0 of convergence under certain assumption of the coefficient functions in [9]. We quote the convergence theorem here:

Recall that the original SDE is

$$\mathrm{d}X(t)=\left[\alpha(X(t),t)+\beta(X(t),t)\right]\mathrm{d}t+\sigma(X(t),t)\mathrm{d}W(t). \tag{2.5}$$

We refer to the splitting

$$\mathrm{d}X_1(t)=\beta(X(t),t)\mathrm{d}t+\sigma(X(t),t)\mathrm{d}W(t) \tag{2.6}$$

$$\mathrm{d}X_2(t)=\alpha(X(t),t)\mathrm{d}t \tag{2.7}$$

Theorem Assume that the coefficient functions $\alpha,\beta\in C^{2,1}\left(\mathbb{R}^d\times[0,T]\right)$ and $\sigma\in C^{3,2}\left(\mathbb{R}^d\times[0,T]\right)$ with exclusively uniformly bounded derivatives are such that

$$\sup_{0\le t\le T} E\left[\left|X(t)^2\right|+\left|\alpha(X(t),t)\right|^2+\left|\beta(X(t),t)\right|^2+\left|\sigma(X(t),t)\right|^2\right]<+\infty$$

for a fixed finite, nonrandom terminal time T>0T>0 . Then the splitting-step algorithm with step 1 and 2 (see section §2.1) has (global) strong and weak order 1.0 of convergence on the interval [0,T] (in the worst case).

The proof and a general theorem on L^2-convergence based on variation-of-constants formula (VOP) see Moro and Schurz [9] . We check the weak convergence with numerical experiment by taking g the identity function in section 3, where we present numerical studies of various strategies for integrating SDEs.

NUMERICAL STUDIES WITH THE CIR MODEL

CIR Model and Its Properties

The famous Cox-Ingersoll-Ross (CIR) model was introduced in 1985 by John C. Cox, Jonathan E. Ingersoll and Stephen A. Ross in [5] as an extension of the Vasicek model. It serves to describe the evolution of interest rates. This model specifies that the interest rate dynamics follows the stochastic differential equation, which is also called CIR Process:

$$\mathrm{d}V(t)=-\kappa(V(t)-\theta)\mathrm{d}t+\sigma\sqrt{V(t)}\mathrm{d}W(t),\quad \forall t\in\mathbb{R}^+ \tag{3.1}$$

with k, θ and σ strictly positive parameters and W(t) a Wiener process. The parameter k determines the speed of adjustment, θ is the long-run mean and σ is the so-called volatility to volatility. The drift factor, κ(θ−V(t)) is exactly the same as in the Vasicek model. It ensures mean reversion of the interest rate towards the long-term value θ, with speed of adjustment governed by the strictly positive parameter k. According to Feller's boundary criteria, see Feller [32], different k, θ and σ values may produce distinct behavior at the boundary V(t)=0

- If $\kappa\theta \geq \sigma^2/2$, the upward drift is sufficiently large to make the boundary unattainable, i.e., the solution is always positive V(t)>0V(t)>0 if V0>0.V0>0.
- If $\kappa\theta < \sigma^2/2$, there are infinite many values of t>0 for which V(t)=0 . The boundary becomes attainable, but it is strongly reflecting. That is, when a sample path reaches 0, then it returns immediately to the positive domain in a reflecting manner.

Since for all positive values of k and θ, the standard deviation factor $\sigma\sqrt{V(t)}$ rejects the possibility of negative interest rates, integration strategies must preserve non-negativity. Otherwise, it not only lacks any possible interpretation in the context of finance but also could induce severe errors in option valuation.

Simulation Schemes for the CIR model

We now turn to the simulation of CIR model (12). The exact simulation method for CIR model exists, see Glasserman [10], but it's very time-consuming. A simple Euler discretization of the CIR process may produce negative values of V(t) . Practitioners often choose either a absorption or reflection fix whenever the process attains a negative value. That is to say, they deal with it by either setting V(t)=0 or setting V(t)=−V(t) whenever V(t)<0 (see e.g. [33]). Other direct approaches base on forced Euler-Maruyama approximations by slight modification of the model are also prevalent, see for example Higham and Mao [20], Deelstra and Delbaen [17], Lord et al., 2008 [21], which the full truncation method in Lord et al., 2008 [21] in practice has been proved to be the leading method of this class. Apart from Euler-type schemes, the splitting-step methods we introduced in the foregoing section can be applied to CIR model without any modification of the SDE. Moreover, the non-negativity is preserved during all the time interval.

Numerical Schemes Based on Euler Discretization

Here we outline the numerical schemes based on Euler discretization which we mentioned above. Let

$$0 < t_0 < t_1 < \cdots < t_n < t_{n+1} < \cdots < t_N = T$$

be any random partition of the given time interval [0,T].

- Higham and Mao [20] considered an Euler discretization of the CIR model with a novel fix, for which they prove strong convergence. Their scheme deals with the squart-root term in (12) by taking absolute value of the inside component, i.e.

$$V(t_{n+1}) = V(t_n) - \kappa\Delta t\left(V(t_n) - \theta\right) + \sigma\sqrt{|V(t_n)|}\Delta W_n, \tag{3.2}$$

$$V(t_0) = V(0), \tag{3.3}$$

where $\Delta W_n = W(t_{n+1}) - W(t_n)$.

Although it can be applied to the simulation, it leads to negative values of V(t) in practice as we can see from Figure 1

- In Lord et al., 2008 [21], when they referred the method Higham and Mao, they had one more step which was taking the absolute value of the right side of (3.2), we call this method Higham and Mao complemented, which is as follow

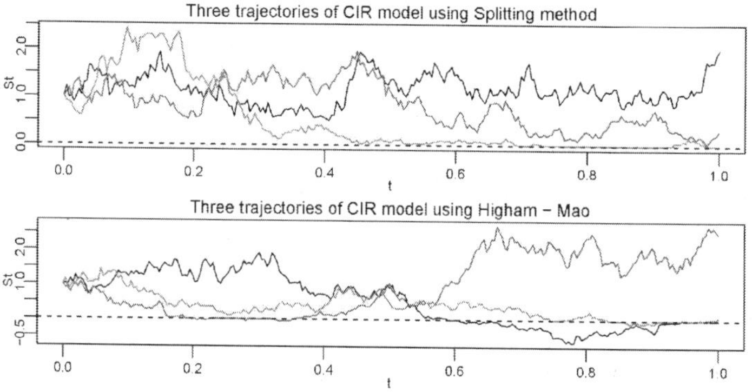

Figure 1: Trajectories of CIR model with V_0=1,T=1,κ=1,θ=1,σ=2. The splitting-step algorithm preserves non-negativity for positive initial data while Higham-Mao produces negative values occasionally.

$$\tilde{V}(t_{n+1}) = \tilde{V}(t_n) - \kappa\Delta t\left(\tilde{V}(t_n) - \theta\right) + \sigma\sqrt{|\tilde{V}(t_n)|}\Delta W_n, \tag{3.4}$$

$$V(t_{n+1}) = \left|\tilde{V}(t_{n+1})\right|, \tag{3.5}$$

$$V(t_0) = V(0). \tag{3.6}$$

- Deelstra and Delbaen [17] proposed an Euler-type scheme by taking positive part of the component inside the squared-root, this scheme is called partial truncation in Lord et al., 2008 [21] and has the form as follow

$$\tilde{V}(t_{n+1}) = \tilde{V}(t_n) - \kappa\Delta t\left(\tilde{V}(t_n) - \theta\right) + \sigma\sqrt{\left(\tilde{V}(t_n)\right)^+}\Delta W_n \tag{3.7}$$

$$\tilde{V}(t_{n+1}) = \left(\tilde{V}(t_{n+1})\right)^+ \tag{3.8}$$

$$V(t_0) = V(0) \tag{3.9}$$

- Full truncation was devised by Lord, Koekkoek & Van Dijk in [21] which is the leading method of these classes. The difference between full truncation and partial truncation is the treatment of the drift term, where the full truncation method has one more x and can be expressed as follow

$$\tilde{V}(t_{n+1}) = \tilde{V}(t_n) - \kappa\Delta t\left(\tilde{V}(t_n)^+ - \theta\right) + \sigma\sqrt{\left(\tilde{V}(t_n)\right)^+}\Delta W_n \tag{3.10}$$

$$\tilde{V}(t_{n+1}) = \left(\tilde{V}(t_{n+1})\right)^+ \tag{3.11}$$

$$V(t_0) = V(0) \tag{3.12}$$

We refer the above four numerical schemes as Higham and Mao, Higham and Mao complemented, partial truncation and full truncation respectively.

Splitting-Step Algorithms Applied to CIR Process

According to the general structure of splitting-step algorithm in Section 2, we split the SDE (3.1) into two equations as follows.

$$\mathrm{d}V_1(t) = \kappa\theta\mathrm{d}t + \sigma\sqrt{V_1(t)}\mathrm{d}W(t) \tag{3.13}$$

$$\mathrm{d}V_2(t) = -\kappa V_2(t)\mathrm{d}t \tag{3.14}$$

We note that the process defined by (3.13) is a kθ-dimensional squared Bessel process (BESQ) with index of the process = $\nu = \dfrac{2\kappa\theta}{\sigma^2} - 1$ (see Makarov and Glew [34]). The boundary $V_1(t)=0$ is entrance if $\nu \geq 0$, regular if $-1<\nu<0$, or exit if $\nu \leq -1$. For the regular diffusion on $(0,\infty)$ the transition probability density function (PDF) is given by

$$P\left(V_1(t+\Delta t)\,|\,V_1(t)\right) = \left(\frac{V_1(t+\Delta t)}{V_1(t)}\right)^{\frac{\nu}{2}} \frac{e^{-2(V_1(t+\Delta t)+V_1(t))/\sigma^2 \Delta t}}{\sigma^2 \Delta t/2} I_\nu\left(\frac{4\sqrt{V_1(t+\Delta t)V_1(t)}}{\sigma^2 t}\right) \tag{3.15}$$

in the case of a regular boundary, where I_v is the modified Bessel function of the first kind with index n. Then we notice that the transition density $P[V_1(t+\Delta t)|V_1(t)]$ can be written in terms of a non-central χ^2 distribution. Recalling the probability density function (PDF) of a non-central chi-square distribution random variable with d degree of freedom and non-centrality parameter λ:

$$P\left[\chi_d'^2(\lambda) = x\right] = p(x;d,\lambda) = \frac{e^{-(\lambda+x)/2}}{2}\left(\frac{x}{\lambda}\right)^{(d-2)/4} I_{(d-2)/2}\left(\sqrt{\lambda x}\right),\ x>0$$

Comparing the equation above and (3.15), we have

$$P\left(V_1(t+\Delta t)\,|\,V_1(t)\right) = \frac{4}{\sigma^2 \Delta t} p\left(\frac{4V_1(t+\Delta t)}{\sigma^2 \Delta t}; \frac{4\kappa\theta}{\sigma^2}, \frac{4V_1(t)}{\sigma^2 \Delta t}\right), \tag{3.16}$$

as $\nu = \dfrac{d}{2} - 1$. The Equation (3.16) shows that the random process $\dfrac{4V_1(t+\Delta t)}{\sigma^2 \Delta t}$ can be represented by a non-central chi-square distribution with $d = \dfrac{4\kappa\theta}{\sigma^2}$ degree of freedom and non-centrality parameter $\lambda = \dfrac{4V_1(t)}{\sigma^2 \Delta t}$. Hence we have

$$V_1(t+\Delta t) = \frac{\sigma^2 \Delta t}{4} \chi_d'^2(\lambda) \tag{3.17}$$

where

$$\lambda = \frac{4V_1(t)}{\sigma^2 \Delta t}, \quad d = \frac{4\kappa\theta}{\sigma^2} \tag{3.18}$$

Then along the time interval $[t,t+\Delta t]$, given $V(t)$, first we take it as an initial data of (3.13) and integrate the SDE though the transition conditional probability (3.15) to obtain $V_1(t+\Delta t)$ which can be done by sampling a $\chi_d'^2(\lambda)$ random number in (3.17). Second we regard $V_1(t+\Delta t)$ has the initial data of

(3.14) and integrate it by any deterministic numerical method of at least order 1. In our project, we practically sample the $\chi_d'^2(\lambda)$ random number by the aid of the function rnchisq() inside the Rmath library from R and integrate (3.14) with deterministic Euler method.

In Figure 1, three trajectories for the CIR process are shown with a comparison to Higham-Mao which is simply taking absolute value of V(t) in the squart-root term. Here we choose $\kappa=1, \theta=1, \sigma=2$, so that the boundary is attainable.

The figure depicts that the splitting-step algorithm preserves non-negativity for positive initial data while Higham-Mao produces negative values occasionally. Thus splitting method is preferable as the CIR process is not defined for negative values.

Weak Convergence

In this section we verify that the splitting-step algorithm has weak convergence of order 1.0 on the interval [0,T] for a fixed finite, nonrandom terminal time T>0.

We take g the identity function and use the model

$$\mathrm{d}V(t) = \left(a + bV(t)\right)\mathrm{d}t + \sigma\sqrt{V(t)}\,\mathrm{d}W(t) \tag{3.19}$$

Then split it in this way:

$$\mathrm{d}V_1(t) = a\mathrm{d}t + \sigma\sqrt{V_1(t)}\,\mathrm{d}W(t) \tag{3.20}$$

$$\mathrm{d}V_2(t) = bV_2(t)\,\mathrm{d}t \tag{3.21}$$

Similarly to (3.17) (3.18), we have

$$V_1(t+\Delta t) = \frac{\sigma^2 \Delta t}{4}\chi_d'^2(\lambda) \tag{3.22}$$

where

$$\lambda = \frac{4V_1(t)}{\sigma^2 \Delta t}, \quad d = \frac{4a}{\sigma^2} \tag{3.23}$$

We use (3.22) (3.23) to obtain $V_1(t+\Delta t)$ which is corresponding to (3.20) and deterministic Euler method to integrate (3.21).

Now, we calculate the expected value of exact solution of (3.19) with the initial value $V(0)=V_0 \geq 0$.

Taking the expected value of both sides of (3.19) reads:

$$\mathrm{d}\mathrm{E}\left[V(t)\mid V(0)=V_0\right] = \left(a+b\mathrm{E}\left[V(t)\mid V(0)=V_0\right]\right)\mathrm{d}t+\mathrm{E}\left[\sigma\sqrt{V(t)}\mathrm{d}W(t)\mid V(0)=V_0\right] \tag{3.24}$$

Noticethatthesecondtermontherightside $\mathrm{E}\left[\sigma\sqrt{V(t)}\mathrm{d}W(t)\mid V(0)=V_0\right]=0$ as W(t) is a standard Wiener process. We define

$$y(t)=\mathrm{E}\left[V(t)\mid V(0)=V_0\right]$$

Then y(t) is a deterministic function of t. Equation (3.24) becomes

$$\mathrm{d}y(t)=\left(a+by(t)\right)\mathrm{d}t$$

with $y(0)=V_0$. Solving the above equation we have

$$y(t)=y(0)\mathrm{e}^{bt}+\frac{a}{b}\left(\mathrm{e}^{bt}-1\right)$$

Therefore,

$$\mathrm{E}\left[V(t)\mid V(0)=V_0\right]=y(t)=V_0\mathrm{e}^{bt}+\frac{a}{b}\left(\mathrm{e}^{bt}-1\right) \tag{3.25}$$

For a=1,b=1 , we have

$$\mathrm{E}\left[V(t)\mid V(0)=V_0\right]=\mathrm{e}^{t}\left(1+V_0\right)-1 \tag{3.26}$$

Then we apply the splitting-step method to obtain the numerical approximation $\tilde{V}(t)$. Figure 2 depicts the error

$$\epsilon_{\Delta t}=\left|\mathrm{E}\left[V(t)\right]-\mathrm{E}\left[\tilde{V}(t)\right]\right| \tag{3.27}$$

Versus decreasing uniform step size Dt, which shows that the method has weak order 1.0 while using constant step sizes Dt. Moreover, we compare to Higham-Mao complemented, partial truncation and full truncation for the same equation. Figure 2 shows that splitting methods has the highest precision among these four algorithms.

Option Valuations

Among the variety of financial derivatives, the option is one of the most important financial instruments. In current financial markets, there are mainly four kinds of options: American option, European option, Asian option, and Barrier option. In our work, we only focus on pricing European options, which can only be exercised on the maturity date whereas an American option can be exercised at any time before expiration.

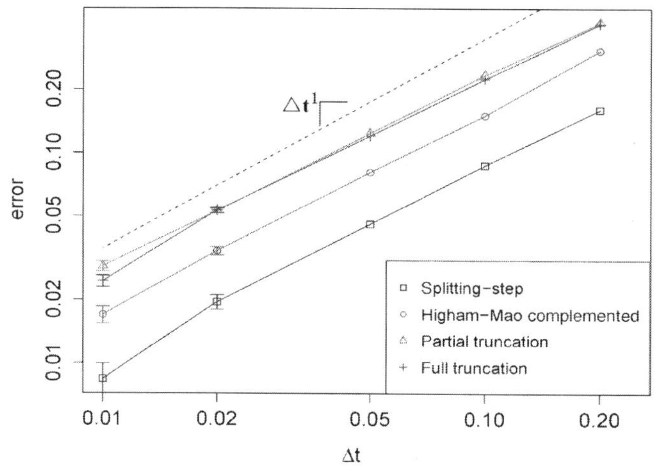

Figure 2: Value of weak convergence error (3.27) as a function of Dt for T=1.0 and S(0)=1.0, a+b=1 , σ=2 using the splitting-step, Higham-Mao complemented, partial truncation and full truncation algorithms in log-log scale for 10^7 realizations. The dashed line is proportional to Dt.

A European Call Option

A European call option on an asset is a contract that allows the buyer to buy (but not obligate) this asset at the price K at time T. The number K>0 is called the exercise (or strike) price and T>0 the maturity (or expiration) time. We pay K to buy the asset at time T if V(T)≥K and sell it immediately at V(T) making a profit of V(T)−K. The option is worthless if V(T)<K as we can buy it cheaper than K.

Thus at time T, the call option has the value

$$C_T = \mathrm{E}\left[\left(V(T)-K\right)^+\right] = \mathrm{E}\left[\max\left(V(T)-K,0\right)\right] \tag{3.28}$$

Suppose that C_T is the numerical approximation of C_T, then the error

$$\epsilon_{call} = \left|\tilde{C}_T - C_T\right| \tag{3.29}$$

reveals the precision of an algorithm. Since the exact solution C_T no explicitly known, we compute a numerical reference solution with a very small t and regard it as C_T. Here we use the numerical approximation of C_T with $t=10^4$ computed by splitting method as reference. Figure 3(a) depicts the comparison of the four algorithms with respect to the precision of calculating call option for each step size t and Figure 3(b) shows the computational cost versus precision among them. We observe that Higham-

Mao complemented, partial truncation and full truncation work better on this call option pricing as they have more precision and higher speed.

A Path-Dependent Option

Apart from the typical European call option in the previous section, more exotic options exist in the market (see, for example [35] [36]). Here we consider one of path-dependent options: up-and-out call option (hereafter referred as U&O option). An up-and-out call option pays off the usual max(V(T)−K,0) at expiry unless at any time during the life of the option the underlying asset reaches certain level (from below, obviously) then it is said to knock out and become worthless. The Figure 4 explains the payoff of each underlying asset. We assume that a underlying will be knocked-out once it reaches L from below at any time prior to expiry time T. Then the underlying asset corresponds to the blue one has payoff 0 as it's knocked-out.

As we say that an U&O option expires worthless if the asset price touches some barrier L from below, say, at any time prior to expiry, where L is larger than the present asset value V_0, the price of U&O call option at expiry time T is given by

$$C^{U\&O}(T) = \mathrm{E}\left[\left(V(T)-K\right)^{+} 1_{\{0<V(t)<L,0\leq t\leq T\}}\right] \tag{3.30}$$

where $1_{\{x\}}$ is one if x is true and zero otherwise.

The exact option price $C^{U\&O}(T)$ is not available since the exact solution of U&O call option under CIR model is not explicitly known. Then instead of evaluating biases we work on the convergence properties of the algorithms.

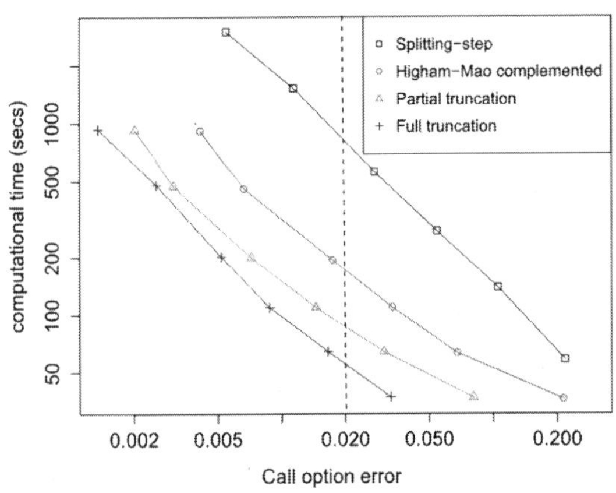

(a)

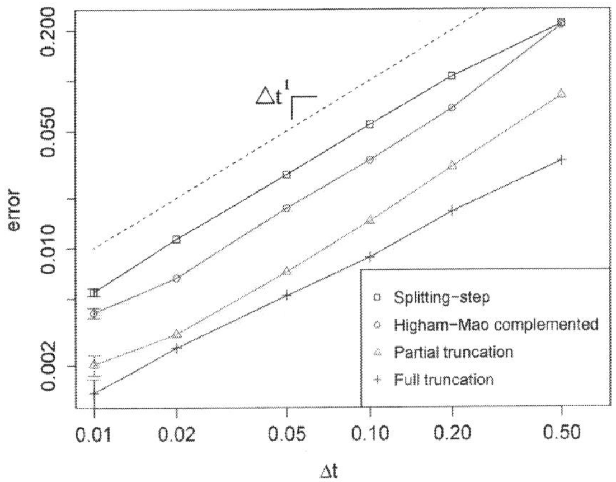

(b)

Figure 3: (a) The call option valuation error (3.29) as a function of Dt for V(0)=1.0 , κ=θ=1 , σ=2 , K=2 , T=10 using the splitting-step, Higham-Mao complemented, partial truncation and full truncation algorithms in log-log scale for 10^7 realizations. The dashed line is proportional to Dt. (b) Computational time in seconds as a function of the call option valuation error for the same parameter setting.

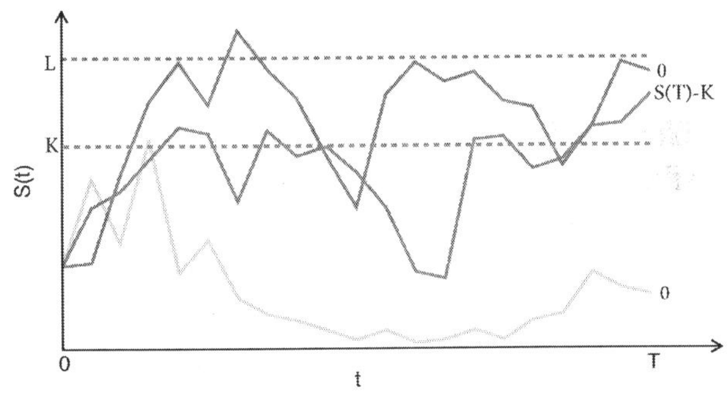

Figure 4: Payoff of each sample path. The underlying asset corresponds to the blue one has payoff 0 as it's knocked-out.

Let $\tilde{u}_{\Delta t}$ which depends on the time step Dt be a numerical approximation of an exact value u. The numerical method said to be of order p means that there exists a number C independent of Dt such that

$$\left|\tilde{u}_{\Delta t} - u\right| \leq C\Delta t^{p}$$

At least for sufficiently small Dt. It's also said that the convergence rate of the method is Δt^p. Normally the error $\tilde{u}_{\Delta t} - u$ depends smoothly on Dt. Then

$$\tilde{u}_{\Delta t} - u = C\Delta t^p + O\left(\Delta t^{p+1}\right)$$

i.e.,

$$\tilde{u}_{\Delta t} = u + C\Delta t^p + O\left(\Delta t^{p+1}\right)$$

An approach of p is to check the relative differences between $\tilde{u}_{\Delta t}$ computed for different Dt. In most cases we compare solutions where Dt is halved successively. Then we get

$$\begin{aligned}\tilde{u}_{\Delta t} - \tilde{u}_{\Delta t/2} &= u + C\Delta t^p - u - C\left(\Delta t/2\right)^p + O\left(\Delta t^{p+1}\right)\\ &= C\Delta t^p\left(1 - \frac{1}{2^p}\right) + O\left(\Delta t^{p+1}\right)\\ &= C'\Delta t^p + O\left(\Delta t^{p+1}\right)\end{aligned}$$

with $C' = C\left(1 - \frac{1}{2^p}\right)$ Hence, we can get an estimate of the order of accuracy p after computing $\tilde{u}_{\Delta t}$ and $\tilde{u}_{\Delta t/2}$ for different Dt.

The relative differences $\tilde{u}_{\Delta t} - \tilde{u}_{\Delta t/2}$ indicate the convergence rate of an algorithm. The faster converges a scheme (has higher p value), the faster relative differences reduce to 0 as Dt tends to 0. Now we define the relative difference of U&O call option value as:

$$\epsilon_{\Delta t}^{U\&O} = \left|\tilde{C}_{\Delta t}^{U\&O}(T) - \tilde{C}_{\Delta t/2}^{U\&O}(T)\right| \tag{3.31}$$

where $\tilde{C}_{\Delta t}^{U\&O}(T)$ is the numerical approach of $C_{\Delta t}^{U\&O}(T)$ using time step Dt. Thus $\epsilon_{\Delta t}^{U\&O} = C\Delta t^p + O\left(\Delta t^{p+1}\right)$, with C a constant independent of Dt. We show the relation between $\epsilon_{\Delta t}^{U\&O}$ and the time step Dt to get an estimate of p for the four algorithms. We also analyze the computational cost with respect to the relative difference $\epsilon_{\Delta t}^{U\&O}$. For sufficient small Dt, the relative differences $\epsilon_{\Delta t}^{U\&O}$ computed by splitting method are much less than standard error of the mean (SEM) which suggests to increase substantially the number of sample paths for the sake of validity of these $\epsilon_{\Delta t}^{U\&O}$. But roughly $\text{SEM} \sim \frac{1}{\sqrt{N}}$ where N is size of the sample, which means that the SEM will reduce to 10% if we increase N to 100N. As we have already used $N=10^7$ sample paths, it's impossible in practice to increase it to at least $N=10^{11}$ to make $\epsilon_{\Delta t}^{U\&O}$ larger than SEM. Then for sufficient small Dt, we estimate $\epsilon_{\Delta t}^{U\&O}$ computed by splitting by the fit of existing data. This has implied that splitting convergent the fastest as it's the first one to reach the SEM level. Figure 5(a) depicts that the splitting-step method has the highest order of accuracy p since p corresponds to the slope of these lines. The parameter what we used are $V(0)=1.0$, $\kappa=\theta=1$, $\sigma=2$, $T=10$, $K=2$, $L=10$ and 10^7 sample paths are generated. Figure 5(b) portrays the computational time in seconds with respect to the relative difference $\epsilon_{\Delta t}^{U\&O}$. At the beginning, splitting performs the worst as it's the most time-consuming one, see for example when $\epsilon_{\Delta t}^{U\&O}=0.02$. But along the trend of smaller relative differences, splitting performs better and better and costs the least since certain value of relative difference. For example, with the same parameter setting, we can deduce that splitting will cost the least for the cases of $\epsilon_{\Delta t}^{U\&O}<0.001$ from the tendency of these four algorithms.

OPTION PRICING WITH HESTON'S STOCHASTIC VOLATILITY MODEL

Heston Model Basics

The Heston model [6] is defined by the coupled two-dimensional stochastic differential equation (SDE):

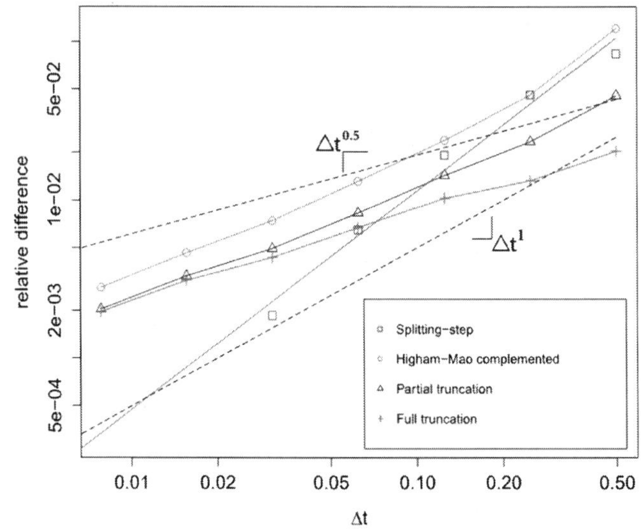

(a)

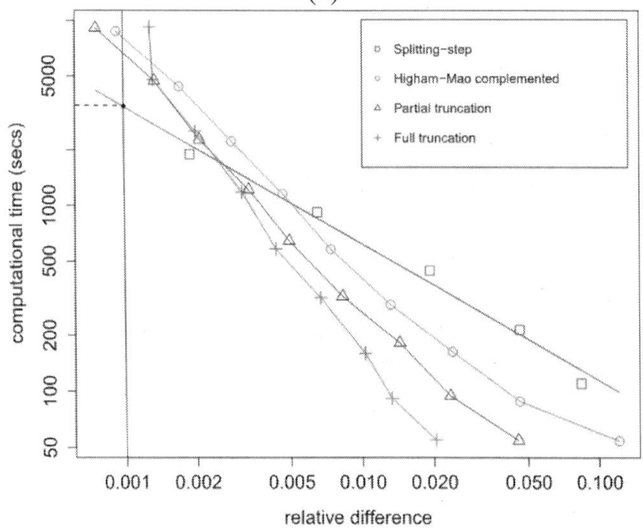

(b)

Figure 5: (a) The relative difference (3.31) of values of up-and-out option at time T as a function of Dt for T=10 , T=10 , T=10 , σ=2 , K=2 , L=10 , using the splitting-step, Higham-Mao complemented, partial truncation and full truncation algorithms in log-log scale for 10^7 realizations. Dash lines are p power laws of Dt and $\Delta t - \sqrt{\Delta t}$. (b) Computational time in seconds as a function of the relative difference (3.31) for the same parameter setting.

$$dS(t) = \mu S(t)dt + \sqrt{V(t)}S(t)dW^S(t) \tag{4.1}$$

$$dV(t) = \kappa(\theta - V(t))dt + \sigma\sqrt{V(t)}dW^V(t) \tag{4.2}$$

where κ,θ,σ are positive constants, $W^S(t)$ and $W^V(t)$ are Wiener processes with correlation, i.e., $dW^S(t)dW^V(t)=dt,\rho\in(-1,1)$. The parameters μ is the rate of return of the asset. θ is the long variance, which means as t tends to infinity, the expected value of V(t) tends to θ. k is the rate at which V(t) reverts to θ σ is the so-called vol of vol, which determines the variance of V(t) . We note that the variance (4.2) follows a CIR process.

For V(t) it has the same boundary behavior as we mentioned in §3.1.

1) 0 is an attainable boundary when $\sigma^2>2\kappa\theta$. The boundary is strongly reflecting;
2) ∞ is an unattainable boundary.

There is an additional condition for S(t) which is iii) S(t) has an absorbing barrier at 0.

Path Simulation

There are plenty of methods can be used to simulate Heston model. Broadie and Kaya [11] [12] derived an exact simulation method without bias, but highly time-consuming. This is analyzed in Broadie and Kaya [21], and Lord et al., 2008 [21] . The exact scheme is competitive when one simulates the process just at one time (or few times), for example to price European options with a Monte-Carlo algorithm. On the contrary, they are drastically too slow for simulating the process along a time-grid, which occurs when computing path-dependent options prices. Kahl and Jäckel [22] state that combining the balanced Milstein method (BMM) for the variance process with their bespoke IJK method for the logarithm of the stock price gives the best results with respect to strong convergence measure. The BMM method actually preserves positivity for the variance process if $4\kappa\theta>\sigma^2$: But this restriction is rarely satisfied in practice, and one typically finds that the sampling scheme for V produces negative values with substantial probability. The TG and QE schemes proposed by Anderson [23] were attested to outperform the full truncation method proposed by Lord et al., 2008 [21] with respect to biases and only cost marginally more computational time than the Euler scheme.

Comparing to the methods above, Euler discretization and splitting-step are very simple to implement. For simulating the variance process (4.2) we use Splitting-step, Higham-Mao, partial truncation and full truncation methods introduced in §3. Then we switch to logarithms for the asset price S(t) , as in Lord et al., 2008 [21], i.e.

$$\ln S(t+\Delta t) = \ln S(t) + \left(\mu - \frac{1}{2}V(t)\right)\Delta t + \sqrt{V(t)}\left(\rho \Delta W_V(t) + \sqrt{1-\rho^2}\Delta W_S(t)\right) \tag{4.3}$$

with $W^S(t), W^V(t)$ two independent Wiener processes. For Higham-Mao, as it may produce negative values of V(t) with substantial probability which we have seen in §3.2.1, we follow their spirit to take the absolute value of V(t) in (4.3), i.e.,

$$\ln S(t+\Delta t) = \ln S(t) + \left(\mu - \frac{1}{2}|V(t)|\right)\Delta t + \sqrt{|V(t)|}\left(\rho \Delta W_{V(t)} + \sqrt{1-\rho^2}\Delta W_S(t)\right) \tag{4.4}$$

A European Call Option with Heston Model

Closed form semi-analytical formulae for plain vanilla option prices have been derived in [6] . However, these formulae require the evaluation of logarithms with complex arguments during the involved inverse Fourier integration step. Carr & Madan [24], Lewis [25] considered the use of the fast Fourier transform (FFT) technique by mapping the Fourier transform directly to option prices via the characteristic function. Lee [26] generalizes Carr & Madan [24] and unifies it with extensions of some relevant elements and proved an error analysis for these FFT methods. Kahl & Jächel [27] propose a new approach which enables the use of Hestons analytics for practically all levels of parameters and even maturities of many decades since most implementations of Hestons formulae are not robust for moderate to long dated maturities or strong mean reversion. Lord & Kahl [28] present the optimal contour of the Fourier integral, Fang & Oosterlee [29] proposed the COS method which is based on the Fourier and Fourier-cosine expansion.

For simplicity and efficiency, in this section we mainly work on the convergence performance of the algorithms we mentioned in §4.2 and demonstrate that besides the well performance in convergence rates, splitting has superiority in computational efficiency.

We define $C^{Hes}(T)$ to be the exact European call option price at maturity time T based on Heston model, i.e.

$$C^{Hes}(T) = \mathrm{E}\left[\left(S(T)-K\right)^+\right] = \mathrm{E}\left[\max\left(S(T)-K\right),0\right]$$

and $\tilde{C}_{\Delta t}^{Hes}$ is an numerical approximation of $C^{Hes}(T)$ with time step Dt.

Then the relative difference is defined as

$$\epsilon_{\Delta t}^{Hes} = \left| \tilde{C}_{\Delta t}^{Hes} - \tilde{C}_{\Delta t/2}^{Hes} \right| \qquad (4.5)$$

The same technique with §3.3.2 gives that $\epsilon_{\Delta t}^{Hes} = C\Delta t^p + O\left(\Delta t^{p+1}\right)$, with C a constant independent of Dt, where p is the order of accuracy defined in §3.3.2. We estimate p by computing $\epsilon_{\Delta t}^{Hes}$ for different Dt. The slope of the plots of $\epsilon_{\Delta t}^{Hes}$ with respect to Dt gives the approximation of p.

Figure 6 are drawn with parameters S(0)=100 , V(0)=0.04 , μ=0.02 , κ=1.5, θ=0.06 , σ=0.7 , ρ=0 , K=120 , T=10 and 10^7 sample paths. Obviously the splitting-step method has the highest order of accuracy p. Although the splitting method is more time-consuming than the others for each Dt, and with respect to relative differences, it costs the most at the beginning as for the case $\epsilon_{\Delta t}^{Hes} > 2$ illustrated in Figure 6(b). But it costs the least computational time if one restricts that the relative differences should be less than a certain value. For example, if one requires that the relative difference should be less than 0.10 with the same parameter setting, splitting practically works much faster than the others. The Figure 6(b) gives the evidence.

More precisely, we analyze the computational cost by taking the case $\epsilon_{\Delta t}^{Hes}$ =0.065 as an example. As partial truncation and full truncation give fluctuant points (see Figure 6(b)), we choose Higham-Mao to compare with. The fitted line of Higham-Mao is extended to intersect with the vertical line $\epsilon_{\Delta t}^{Hes}$ =0.065 . From Figure 7, one can find that for reducing $\epsilon_{\Delta t}^{Hes}$ to 0.065, splitting needs around 10^4 seconds » 2.8 hours while Higham-Mao needs much more than 10^6 seconds » 277.8 hours which is more than 100 times than splitting.

CONCLUSIONS

A number of existing numerical algorithms for integrating SDEs cannot be used to simulate certain financial models which require non-negativity, such as the CIR model, the Heston model that we have seen. A slight modification of Euler-Maruyama [1] method conducts to various Euler-type numerical schemes. They are simple to implement and efficient enough, but have lower accuracy compared to exact simulation schemes. Sometimes they also give

(4.5). The vertical dash line corresponds to $\epsilon_{\Delta t}^{Hes}$ =0.065 .

Less precision is compared to semi-analytical numerical schemes such as the splitting-step methods which we are analyzing in our project. Since the exact simulation schemes are thoroughly too slow for simulating a process along a time-grid, splitting-step methods preponderate them with respect to efficiency.

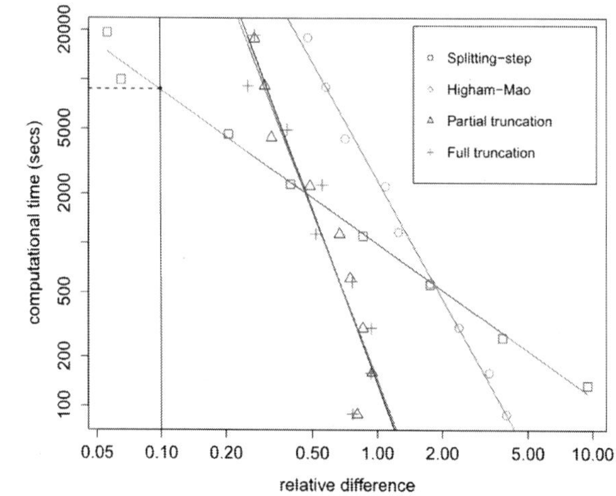

(a)

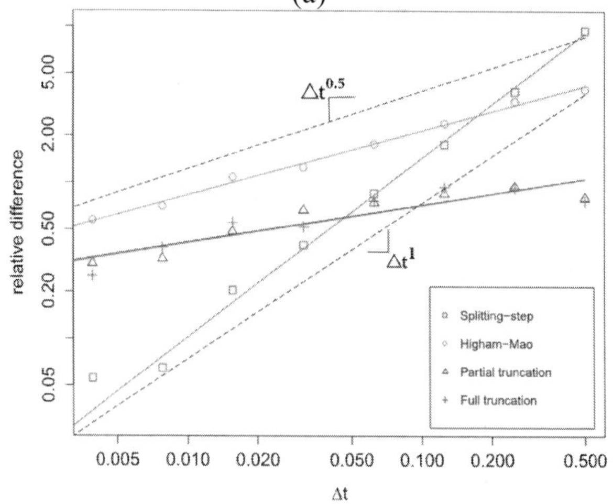

(b)

Figure 6: (a) Relative differences of call option (4.5) as a function of Dt in log-log scale with S(0)=100 , V(0)=0.04 , μ=0.02 , κ=1.5 , θ=0.06 , σ=0.7 , ρ=0 , K=120 , T=10 , and 10^7 sample paths. Dash lines are power laws of Dt and $\Delta t - \sqrt{\Delta t}$. (b) Computational time in seconds as a function of the call option valuation relative differences for the same parameter setting.

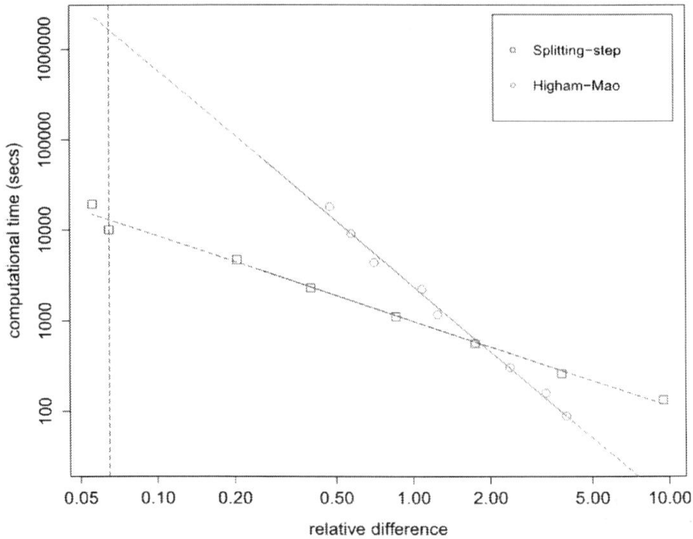

Figure 7: Computaional cost in seconds as a function of relative difference $\epsilon_{\Delta t}^{Hes}$.

The splitting algorithms heavily rely on the exploitation of the specific structure of original system (2.1). One can decompose the original system into two subsystems appropriately for which either one knows the explicit solution or the conditional transition probability of one of the subsystems and integrate the remaining one by deterministic numerical methods of at least order 1. In this way, it preserves the non-negativity and a maximum of convergence order 1.0 both in strong and weak sense. For the analytical solution of SDEs, an extensive list of known one can be found in textbooks such as Kloeden and Platen [1] .

Among the numerical schemes in §3, splitting-step method gives the best convergence rate and normally highest accuracy. We observe that it doesn't perform better than other four schemes on call option valuation with CIR model. Despite it costs more computational time to generate sample paths, it may cost even less for certain × errors. For example, if one requires that the relative difference of A European call option under the Heston model should be less than 0.10 with the parameter setting in ×4.3, the splitting method costs the least computational time than others.

FUTURE WORK

Since derivatives are one of the three main categories of financial instruments where the other two being stocks (i.e., equities or shares) and debt (i.e., bonds and mortgages), and as we said that the splitting methods are much more efficient but has less accuracy than exact simulation methods which are not suitable for pathwise simulations, then one direction of our interest is applying splitting-step method to other path-dependent options, probably with comparison to other efficient algorithms like what we have done in §3.3.2.

Another direction is the application of splitting-step models to portfolio problems which are aiming to find the optimal investment strategy of an investor. Following the optimal portfolio strategy leads to the maximum expected utility of the terminal wealth. For example the classical Merton's problem [37] and mixed stock & bond portfolio problem like in [38] are under consideration.

As Moro and Schurz [9] have mentioned that the splitting-step scheme is applicable to multi-dimensional problems, we intend to encounter some suitable splitting manners to deal with them which in particular can be SDEs with linear, decoupled, commutative or diagonal noise terms. Furthermore, non-linear partial differential equations, for instance, Super-Brownian motion interest us to verify its properties by implement splitting-step methods.

ACKNOWLEDGEMENTS

I would take the opportunity to thank my tutor Esteban Moro Egido who gives the support during our project.

REFERENCES

1. Kloeden, P.E. and Platen, E. (1998) Numerical Solution of Stochastic Differential Equations. Springer, New York.
2. Kloeden, P.E., Platen, E. and Schurz, H. (1994) Numerical Solution of SDEs through Computer Experiments. Springer, New York. https://doi.org/10.1007/978-3-642-57913-4
3. Schurz, H. (1997) Stability, Stationarity, and Boundedness of Some Implicit Numerical Methods for Stochastic Differential Equations and Applications. Logos-Verlag, Berlin.
4. Milstein, G.N. (1995) Numerical Integration of Stochastic Differential Equations. Kluwer, Dordrecht. https://doi.org/10.1007/978-94-015-8455-5
5. Cox, J.C., Ingersoll, J.E. and Ross, S.A. (1985) A Theory of Term Structure of Interest Rates. Econometrica, 53, 385-407.
6. Heston, S. (1993) A Closed-Form Solution for Options with Stochastic Volatility with Applications to Bond and Currency Options. Reviews of Financial Studies, 6, 327-343. https://doi.org/10.1093/rfs/6.2.327
7. Feller, W. (1951) Two Singular Diffusion Problems. Annals of Mathematics, 54, 173-182. https://doi.org/10.2307/1969318
8. Karlin, S. and Taylor, H.M. (1981) A Second Course in Stochastic Processes. Academic Press, New York.
9. Moro, E. and Schurz, H. (2007) Boundary Preserving Semi-Analytic Numerical Algorithms for Stochastic Differential Equations. SIAM Journal on Scientific Computing, 29, 1525-1549. https://doi.org/10.1137/05063725X
10. Glasserman, P. (2003) Monte Carlo Methods in Financial Engineering. Springer Verlag, New York. https://doi.org/10.1007/978-0-387-21617-1
11. Broadie, M. and Kaya, O. (2004) Exact Simulation of Option Greeks under Stochastic Volatility and Jump Diffusion Models. In: Ingalls, R.G., Rossetti, M.D., Smith, J.S. and Peters, B.A., Eds., Proceedings of the 2004 Winter Simulation Conference, INFORMS, Washington, 1607-1615. https://doi.org/10.1109/WSC.2004.1371506
12. Broadie, M. and Kaya, O. (2006) Exact Simulation of Stochastic Volatility and Other Affine Jump Diffusion Processes. Operations Research, 54, 217-231. https://doi.org/10.1287/opre.1050.0247

13. Alfonsi, A. (2005) On the Discretization Schemes for the CIR (and Bessel Squared) Processes. Monte Carlo Methods and Applications, 11, 355-384. https://doi.org/10.1515/156939605777438569
14. Kahl, C. and Schurz, H. (2006) Balanced Milstein Methods for Ordinary SDEs. Monte Carlo Methods and Applications, 12, 143-170. https://doi.org/10.1515/156939606777488842
15. Dereich, S., Neuenkirch, A. and Szpruch, L. (2012) An Euler-Type Method for the Strong Approximation of the Cox-Ingersoll-Ross Process. Proceedings of the Royal Society of London A, 468, 1105-1115. https://doi.org/10.1098/rspa.2011.0505
16. Alfonsi, A. (2010) High Order Discretization Schemes for the CIR Process: Application to Affine Term Structure and Heston Models. Mathematics of Computation, American Mathematical Society, 79, 209-237. https://doi.org/10.1090/S0025-5718-09-02252-2
17. Deelstra, G. and Delbaen, F. (1998) Convergence of Discretized Stochastic (Interest Rate) Processes with Stochastic Drift Term. Applied Stochastic Models and Data Analysis, 14, 77-84. https://doi.org/10.1002/(SICI)1099-0747(199803)14:1<77::AID-ASM338>3.0.CO;2-2
18. Bossy, M. and Diop, A. (2004) An Efficient Discretization Scheme for One Dimensional SDEs with a Diffusion Coefficient Function of the Form. INRIA Working Paper No. 5396.
19. Berkaoui, A., Bossy, M. and Diop, A. (2008) Euler Scheme for SDEs with Non-Lipchitz Diffusion Coefficient: Strong Convergence. ESAIM Probability and Statistics, 12, 1-11.
20. Higham, D.J. and Mao, X. (2005) Convergence of Monte Carlo Simulations Involving the Mean-Reverting Square Root Process. Journal of Computational Finance, 8, 35-62.
21. Lord, R., Koekkoek, R. and van Dijk, D. (2008) A Comparison of Biased Simulation Schemes for Stochastic Volatility Models. Journal of Quantitative Finance, 10, 177-194.
22. Kahl, C. and Jackel, P. (2006) Fast Strong Approximation Monte-Carlo Schemes for Stochastic Volatility Models. Quantitative Finance, 6, 513-536. https://doi.org/10.1080/14697680600841108
23. Andersen, L. (2008) Simple and Efficient Simulation of the Heston Stochastic Volatility Model. Journal of Computational Finance, 11, 1-42. https://doi.org/10.21314/JCF.2008.189

24. Carr, P. and Madan, D. (1999) Option Valuation using the Fast Fourier Transform. Journal of Computational Finance, 2, 61-73. https://doi.org/10.21314/JCF.1999.043
25. Lewis, A.L. (2001) A Simple Option Formula for General Jump-Diffusion and Other Exponential Levy Processes. http://www.optioncity.net/pubs/ExpLevy.pdf
26. Lee, R.W. (2004) Option Pricing by Transform Methods: Extensions, Unification and Error Control. Journal of Computational Finance, 7, 51-86. https://doi.org/10.21314/JCF.2004.121
27. Kahl, C. and Jackel, P. (2005) Not-So-Complex Logarithms in the Heston Model. Wilmott Magazine, 19, 94-103.
28. Lord, R. and Kahl, C. (2007) Optimal Fourier Inversion in Semi-Analytical Option Pricing. http://papers.tinbergen.nl/06066.pdf
29. Fang, F. and Oosterlee, C.W. (2008) A Novel Pricing Method for European Options Based on Fourier-Cosine Series Expansions. SIAM Journal on Scientific Computing, 31, 826-848. https://doi.org/10.1137/080718061
30. Johnson, N.L., Kotz, S. and Balakrishnan, N. (1994) Continuous Univariate Distributions. Wiley, New York, 2.
31. Siegel, A.F. (1979) The Noncentral Chi-Squared Distribution with Zero Degrees of Freedom and Testing for Uniformity. Biometrika, 66, 381-386. https://doi.org/10.1093/biomet/66.2.381
32. Feller, W. (1971) Probability Theory and Its Applications. 2nd Edition, Wiley, New York.
33. Gatheral, J. (2006) The Volatility Surface: A Practitioners Guide. John Wiley and Sons, New York.
34. Makarov, R.N. and Glew, D. (2010) Exact Simulation of Bessel Discussions. Monte Carlo Methods and Applications, 16, 283-306. https://doi.org/10.1515/mcma.2010.010
35. Shreve, S.E. (1996) Stochastic Calculus and Finance. http://citeseerx.ist.psu.edu/viewdoc/download?doi=10.1.1.137.6951&rep=rep1&type=pdf
36. Wilmott, P., Howison, S. and Dewynne, J. (1995) The Mathematics of Financial Derivatives. Cambridge University Press, New York. https://doi.org/10.1017/CBO9780511812545

37. Merton, R.C. (1969) Lifetime Portfolio Selection under Uncertainty: The Continuous-Time Case. Review of Economics and Statistics, 51, 247-257. https://doi.org/10.2307/1926560

38. Korn, R. and Kraft, H. (2002) A Stochastic Control Approach to Portfolio Problems with Stochastic Interest Rates. SIAM Journal on Control and Optimization, 40, 1250-1269. https://doi.org/10.1137/S0363012900377791

Chapter

NUMERICAL SIMULATIONS OF MICROSCALE GAS FLOWS: CONTINUUM APPROACH

11

A. Chaudhuri[1], A. Hadjadj[1], C. Guha[2] and T. K. Dutta[2]

[1]CORIA-UMR CNRS 6614, Avenue de l'Universite, 76801 St Etienne du Rouvray France

[2]Chemical Engineering Department, Jadavpur University, Kolkata-700032 India

INTRODUCTION

Enormous technical applications of micro and nanosciences have evolved as one of the most active interdisciplinary field of research and development during last two decades. The journey of scientific innovations in micro and nanoscale has just begun and yet plenty of rooms at the bottom remain to be explored. It is the imagination and vision of Nobel Laureate, Professor Richard Feynman that have been implemented in the field of

Citation: A. Chaudhuri, A. Hadjadj, C. Guha and T. K. Dutta, "Numerical Simulations of Microscale Gas Flows: Continuum Approach", IntechOpen Recent Advances in Technology, 2009, DOI: 10.5772/7422.

micro and nanodevices. Micro and nanodevices can be widely used in aerospace, automotive, biotechnology, medicine, instrumentation, robotics manufacturing and many other applications. In coming decades, Micro-Electro-Mechanical-Systems (MEMS)/Nano-Electro Mechanical-Systems (NEMS) technology are going to become part of our daily life. MEMS refer to devices that have typical characteristic lengths lying between 1 μm and 1 mm that combine electrical and mechanical components. In this chapter, we will restrict our discussion to only microscale gas flows in continuum regime together with the application of wall-slip models and numerical methods that are commonly used in this field. The dynamics of fluids and their interaction with surfaces in microsystems are very much different from those in large-scale systems. In microsystems, the flow becomes granular for liquids and rarefied for gases, and the walls tend to move. Phenomena such as viscous heating, thermal creep, electro-kinetics, anomalous diffusion, and even quantum and chemical effects appear important. Design and optimization of effective micro and nanodevices require a deeper physical understanding of hydrodynamic and kinetic phenomena in single and multiphase flows in small dimensions. The material of the wall, the quality of the surface as well as the energy exchange in small dimensions play the crucial roles for microscale flow characteristics. Since surface to volume ratio of microstructures is very high surface effects dominate over volumetric effects. The local Knudsen number is a measure of the degree of rarefaction of gases encountered in MEMS devices. The popular methods for analyzing the fluid-flow and heat transfer characteristics of gases in microscale include the particulate method of Boltzmann equations, Direct Solution Monte Carlo (DSMC) and Burnett equation models. Several investigators have used the DSMC approach where the Knudsen number is very high involving high computation cost and time requirement. The standard continuum approach of considering the bulk of the fluid with no-slip wall conditions can no longer predict the momentum and heat transfer accurately in microdevices. Navier-Stokes simulations of gas flow in microdevices with wall-slip models have evolved as an alternative solution technique to treat this miniaturization effect in the slip flow regime.

In section 2, we will briefly highlight the governing equations of fluid-flow with slip models and the applicability of the continuum approach. The following sections will provide the discussion on simulations of non-reactive and reactive compressible flows in microscale. In section 3, we will present the behavior of non-reactive fluid-flow and heat transfer in microchannels followed by the comparison of fluid-flow behavior in micronozzles and the conventional full-scale nozzles. The effect of miniaturization on flow separation pattern will also be addressed in this section. In section 4, we will illustrate few reactive flow studies in microchannels. Keeping in view the useful application of shock waves at microscale we will demonstrate the studies of shock-wave attenuation in microchannels in section 5. Finally, the results will be summarized in section 6 and concluding remarks will be provided as the basis of future work in this area.

FLOW REGIMES AND MODELS

Knudsen number is defined as the ratio of microscopic length scales (mean free path) and characteristic length of the physical domain. This is the key factor determining the gas flow regimes. The effect of miniaturization in microscale gas flows can be attributed to rarefaction, giving rise to strong analogy between low-pressure flows and microscale flows. Based on the ratio of Mach number and Reynolds number, the Knudsen number can also be interpreted as

$$Kn = \sqrt{\gamma\pi/2}\,\frac{M}{\mathrm{Re}} \qquad (1)$$

where γ is the ratio of specific heats. Equation (1) relates rarefaction effects to compressibility and viscous effects through Mach and Reynolds numbers respectively. As Knudsen number increases, the rarefaction effects become more and more pronounced and eventually the continuum assumption breaks down. Traditionally, gas flows are divided into four regimes as i) $Kn \leq 10^{-3}$: continuum regime, ii) $10^{-3} < Kn \leq 10^{-1}$: slip regime, iii) $101 < Kn \leq 10$: transitional and iv) $Kn > 10$: free-molecular regimes.

The effect of wall in microdimension is of utmost importance which plays crucial role in microscale gas flows affecting the momentum and energy transfer. Classical continuum regime is essentially governed by compressible Navier-Stokes equations with conventional boundary conditions. On the other hand, in slip flow regime, Knudsen layer (zone of non- equilibrium near the wall) must be taken into account with modified velocity slip and temperature jump at the wall. The continuum approach is thus applicable taking sufficient precautions using appropriate slip models. Most of the Microsystems using gases as working fluid fall in slip flow regime or early transition regime (M. Gad-el-Hak, 1999; Karniadakis & Beskok, 2002). Solution of Navier-Stokes equations with wall-slip models thus evolved as relatively cheaper choice compared to DSMC and Burnett equation models in terms of computational costs and complexity.

Compressible Navier-Stokes system of equations with Newtonian stress tensor, Fourier law of thermal diffusion, Fick's law of mass diffusion and calorically perfect/thermally perfect gas/ gas-mixture behavior are usually assumed as governing equations for defining compressible fluid flow. The validity of these assumptions is widely accepted and has been addressed by many authors in standard books and literature. Specifically, in the following section, we wish to emphasize on the modified slip boundary conditions applicable in accordance with the continuum approach.

Wall-Slip Models

In microscale gas-flows, the interaction of the fluid molecules with the surface material (chemical affinity) plays an important role. Momentum and energy transfer can thus be correlated to the interaction of impinging gas molecules to the surface of the wall material. On the basis of kinetic theory, Maxwell postulated and proved the existence of wall-slip. In order to formulate the velocity slip, Maxwell considered details of momentum transfer mechanism of the molecules and the wall, introducing TMAC (Tangential Momentum Accommodation Coefficient, σ_v). This lies between zero (no momentum transfer to the wall) and unity (gas molecules transmit entire tangential momentum to the wall) representing specular and diffusive reflection. Later Smoluchowski in 1898 experimentally confirmed the hypothesis of Poisson to define the temperature jump corresponding to velocity slip at the wall containing thermal accommodation coefficient σ_T . Detail discussions about accommodation coefficients have been reported in Gad-el-Hak, (1999); Karniadakis, (2000) and Arkilic et al., (2001). In most of the numerical simulations, it is usual practice to set accommodation coefficients to unity. It is really a tremendous challenge to get the desired accuracy to measure the precise mass flux experimentally with imposed pressure gradients. Arkilic et al., (2001) experimentally determined sub-unity TMACs for nitrogen, argon and carbon dioxide in contact with single-crystal silicon. Ewart et al., (2007) reported their recent experimental determination of TMAC based on mass flow measurements. Based on findings of Porodonov et al., (1974); Arkilic et al., (1997b); Maurer et al., (2003); Colin et al., (2004) and Ewart et al., (2007), it can be summarized that TMAC lies between 0.81-0.981 for nitrogen, 0.7-0.942 for argon and 0.895-0.986 for helium. Suitable sub-unity values of the TMAC thus can be utilized in numerical simulations. Maxwell's first order velocity slip with thermal creep and corresponding temperature jump relation can be written as:

$$u_{slip} - u_{wall} = \frac{2-\sigma_v}{\sigma_v}\lambda\left.\frac{\partial u}{\partial n}\right|_w + \frac{3}{4}\frac{\mu}{\rho T}\left.\frac{\partial T}{\partial s}\right|_w \tag{2}$$

$$T_{slip} - T_{wall} = \frac{2-\sigma_T}{\sigma_T}\frac{2\gamma}{(\gamma+1)}\frac{k}{\mu C_p}\lambda\left.\frac{\partial T}{\partial n}\right|_w \tag{3}$$

where n and s denote the normal and tangential directions of the derivative terms.

λ, u, μ, ρ, T, k and C_p are mean free-path, velocity, viscosity, density, temperature, thermal conductivity and specific heat at constant pressure, respectively.

Significant discrepancies have been observed between experimental data and DSMC calculations based on first order slip models, for Knudsen number greater than 0.1. Several authors have proposed higher-order boundary conditions to extend the validity of the slip- flow regime. In non-dimensional form, the generalized second-order slip velocity relation (Sreekanth, 1969) can be expressed as:

$$\hat{u}_{slip} - \hat{u}_{wall} = \xi_1 Kn\left.\frac{\partial \hat{u}}{\partial \hat{n}}\right|_w + \xi_2 Kn^2\left.\frac{\partial^2 \hat{u}}{\partial \hat{n}^2}\right|_w \tag{4}$$

The tabulated coefficients ξ_i , used by various researchers for the above general equation, are reported in Colin, (2005). Nevertheless, the implementation of higher-order boundary conditions of this form is particularly difficult. The recent work of Dongari et al., (2007) has dealt with analytical solution of Navier-Stokes equations using second-order slip model for isothermal gas flow in long microchannel. The change in curvature of pressure versus the streamwise direction has been reported and the presence of Knudsen's minima has also been correctly predicted by their developed theory.

In recent studies, implementation of the first-order slip boundary conditions on various reacting and non-reacting test-cases has been carried out after the successful validation of the developed explicit Navier-

Stokes solver. We have utilized robust shock capturing schemes including AUSM (Advection Upwind Splitting Methods, Liou & Steffen, 1993) and subsequent modified version of these schemes as well as high-order WENO (Weighted- Essentially Non-Oscillatory) schemes (Jiang & Shu, 1996) for convective flux calculations and central differencing discretization of diffusive fluxes. Usually, choice of shock capturing schemes is essential to deal with flow discontinuities over a wide range of Mach number. Detail of these schemes is well known to computational community and widely available in literature. In this chapter, we will mainly focus on the physics of the microscale thermal- fluid behavior setting aside the details of the numerical tools.

FLUID FLOW AND HEAT TRANSFER IN MICROCHANNELS AND MICRONOZZLES

Microchannels

Microchannels are currently being used in many areas such as those of medicine, biotechnology, avionics, telecommunications, metrology, computer technology, office equipment, home appliances, environment protection etc. Experimental investigations of flow and heat transfer through microchannels of various cross sectional geometries have been carried out by many researchers, (Wu & Little, 1984; Pafler et al., 1990; Choi et al., 1991; Pong et al., 1994; Liu et al., 1995; Harley et al., 1995; Yu et al., 1995; Shih et al., 1996; Bayt & Breuer, 1998), mainly by focusing on the measurements of pressure drop and bulk fluid temperatures at the channel ends. Numerous discrepancies have been encountered compared to the macroscale behavior. In the review of Papautsky et al., (2001), it is reported that the data of the slip-flow gas indicate an approximate 60% reduction in friction factor compared to macroscale theory for a single-phase laminar flow. Rostami et al., (2002) made a detail review of the aforementioned experimental findings along with the numerical findings of Beskok et al., (1996); Arkilic et al., (1997) and several other contributors. For microchannels and microtubes, Reynolds analogy and Colburn analogy were found invalid (Wu & Little, 1984; Choi et al., 1991). Relatively recent experimental work of Celata et al., (2007) reported the determination of local friction factor (tube diameters ranging from 30 - 254 μm) for helium taking compressibility effects into account. Their findings are in close agreement with incompressible and quasi-incompressible correlations. The existence of nonlinear pressure drop across the channel for higher-Knudsen number flows and dependency of product of friction factor and Reynolds number for fully developed laminar flow on Reynolds number have been reported in various experimental studies. Application of first-order slip models to account for the non-continuum effects evolved as favorable choice by the numerical findings of the literature. Chen, (2004) utilized 3D numerical simulation to predict the steady compressible laminar flow in microchannel and reported that the slip effect reduces the wall friction and curvature of the pressure distribution. Later, Jain & Lin, (2006) also emphasized about the need of accounting 3D effects for the prediction of friction coefficients in microchannels.

Arkilic et al., (1997a) reported a theoretical model for slip flow in long isothermal channels. The corresponding nonlinear pressure distribution (normalized by outlet pressure) along streamwise direction (normalized by the length of the channel) is given by

$$\hat{P}(\eta) = -\chi + \sqrt{\chi^2 + (1+2\chi)\eta + (\Pi^2 + 2\chi\Pi)(1-\eta)} \tag{5}$$

where

$$\chi = 6\frac{2-\sigma_v}{\sigma_v}Kn_o \tag{6}$$

In the above equation, is the ratio of inlet and outlet pressure and Kn_0 is the outlet Knudsen number.

We carried out a standard benchmark test of isothermal slip flow through microchannel for validation of first-order slip model using finite volume method (FVM). The dimension and other flow-field parameters are shown in Figure 3.1.1a, where nitrogen gas is being considered as the working fluid. It can be seen from Figure 3.1.1b that the agreement between the FVM and analytical solution (Arkilic et al., 1994) is excellent. Computational results are also in good agreement with experimental data of Pong et al., (1994).

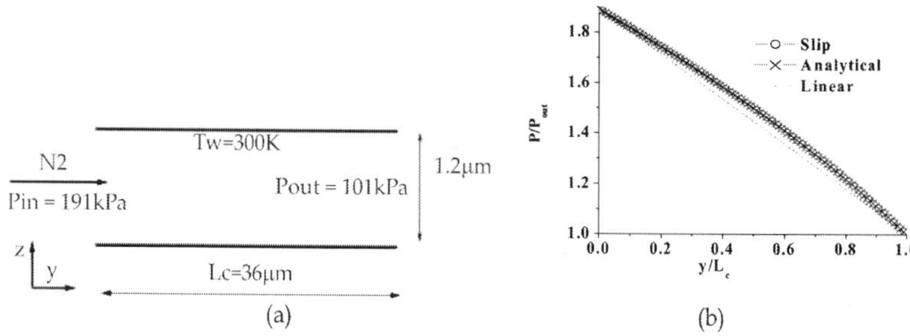

Fig. 3.1.1. Validation of first-order slip model: (a) problem description, (b) pressure profile.

The study of microchannel flow is partly in response to the need of thermal control in the operation of MEMS. Yan & Farouk, (2002) investigated high Knudsen number low-speed flow through partially heated parallel plates using DSMC. Chaudhuri et al., (2007b) studied the effects of Knudsen number and gas species at atmospheric pressure levels using similar configuration for partially heated microchannel (resembling classical Graetz problem). The dominant molecular diffusion and nonlinear pressure distribution for higher Knudsen number are found to be similar for both rarefied (Yan and Farouk, 2002) and microscale behavior.

In comparison to experimental and numerical studies of subsonic flow inside microchannels, studies on flow-thermal characteristics in high-speed flows are not abundantly found in literature. Liou & Fang, (2001) reported that the magnitude of the temperature jump at the wall increases with increasing Knudsen number. The heat transfer at the isothermal wall is found to increase significantly with Knudsen number. Raju & Roy, (2005) investigated the heat transfer characteristics of high-speed compressible flows and compared DSMC results of Liou & Fang, (2001) with their finite element simulations. Chaudhuri et al., (2007a) compared finite volume simulation with results of the DSMC (Liou & Fang, 2001; Le & Hassan, 2006) and FEM results (Raju & Roy, 2005) for inlet Mach number 4.15. Acceptable agreement has been achieved in terms of detached bow shock patterns and flow field inside the microchannel (Figure 3.1.2). Chaudhuri et al. further extended hypothetical test-cases with higher inlet Mach number (hypersonic). The jump in near wall temperature appeared higher for higher inlet Mach numbers (Figure 3.1.2). Position of fully developed thermal boundary-layer shifted further downstream with increasing inlet Mach number.

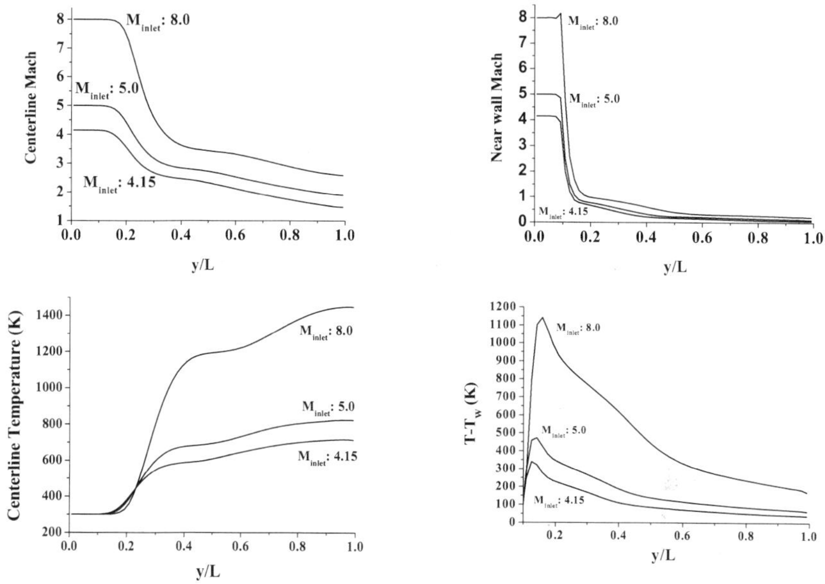

Figure 3.1.2: Comparison of Mach number and temperature distribution (streamwise) for high- speed gas flow through microchannels with inlet Knudsen number 0.062.

Micronozzles

Variety of aerospace missions and application systems have been benefited from micro- engineering technology due to reduced size, mass and power requirements (Janson et al., 1999). Among them, micropropulsion devices provide an active field of applied research. Electrical and chemical microthrusters are developing in recent years in order to fulfil the required design and fabrication of microspacecraft with high accuracy. The main application of the microthruster is the micropropulsion system for microsatellites weighing typically 20 to 100 kg or nanosatellites typically less than 20 kg. Among the propulsion systems which use MEMS technology or hybrid technology for the miniaturization are the cold gas thrusters, the subliming solid thrusters, the field emission electric propulsion (FEEP), or the hall-effect thrusters. Unique thrust requirements have initiated the need for optimal design and fabrication concept for a MEMS-based microthruster. Due to the inherent complexity of the fluid dynamics in the micrometer scale, experimental and numerical studies were proposed for better understanding of the flow properties and optimal design (Kujawa et al., 2003). The performance of a micropropulsion system is primarily determined by the performance of micronozzles. The nozzle converts the stored energy in a pressurized gas into kinetic energy through an expansion. The nozzle efficiency is characterized by the amount of kinetic energy leaving the nozzle, and is governed by the exit Mach number. Many experimental and numerical studies were carried out on microthruster/ micronozzle behavior for varying geometries and propellants. Bayt et al., (1997) experimentally and numerically studied a micron scale Laval nozzle. For the numerical study, they used a 2-D finite volume NS solver. Bayt, (1999) fabricated an extruded 2-D converging-diverging nozzle with minimum throat widths averaging 19 μm and 35 μm etched by deep reactive ion-etching. Their experiment reveals that supersonic flow can be achieved in contoured micronozzles (outlet Mach number 3.8 for expansion nozzle ratio of 17) for micropropulsion systems. Reed, (2003) carried out experiments on planar micronozzles for nitrogen and helium flows under ambient temperature to study the nozzle performance with different exit to throat aspect ratios. Lewis, (2000) dealt with design and fabrication of initial prototypes of microthruster chips. From relatively more recent work of Chen et al., (2005), it can be realized that thrust is not only a function of pressure difference but also depends on the flow history for miniature nozzles. In spite of existence of discrepancies on thrust and mass flow measurements among the theoretical and experimental findings the following points can be summarized for mini/microscale nozzle behavior (Rossi et al., 2000; Rossi et al., 2001; Hammel, 2002; Alexeenko et al., 2002; Hao et al., 2005; Liu et al., 2006; Morffiigo et al., 2007; Xu & Zhao, 2007; Lin & Gadepalli, 2008). (1) Applicability of slip models in DSMC/NS solutions is valid in slip flow regimes for successful prediction of micronozzle flow behavior. (2) Precise low thrust in the order of mN to J.N can be produced for micropropulsion systems. (3) There exists shock free supersonic flow in the diverging section compared to conventional nozzle under similar constraints. In addition, the sonic point usually appears further downstream of the throat location. (4) Flow-separation with high expansion ratio decreases with decreasing the nozzle size.

In our previous work (Chaudhuri et al., 2006), numerical results for flow through microthruster are compared with the 1-D numerical study by Rossi et al. (2000), and 2D FEM results by Raju & Roy, (2002) as a no-slip benchmark problem predicting essential 2-D features of the fluid-flow inside the micronozzle. In this section, we will present three test- cases with different outlet pressure (65 kPa, 55, kPa and 10 kPa) to investigate the internal micronozzle flow having the same inlet pressure of 100 kPa. Figure 3.2.1 shows the schematic of the nozzle configuration. The effect of the nozzle size with different scales is also studied for outlet pressure of 10 kPa. It can be seen from Figure 3.2.2 that the flow inside the micronozzle remains entirely subsonic for outlet pressure 65 kPa. When the outlet pressure is kept 55 kPa, (Fig 3.2.2) the first sonic point appears further downstream of the throat. The existence of the shock free supersonic zone with gradual dissipation to subsonic flow is also predicted by the simulation and is similar to the findings of Hao et al., (2005). On the other hand, the fully supersonic flow is observed downstream of the throat (Figure 3.2.3) for outlet pressure of 10 kPa.

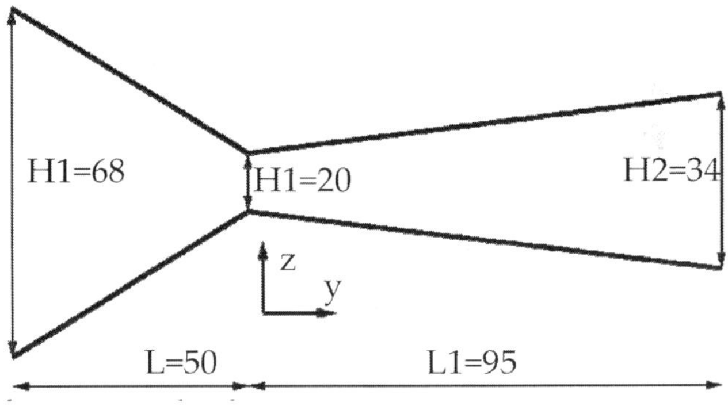

Figure 3.2.1: Schematic of micronozzle, dimensions are in μm.

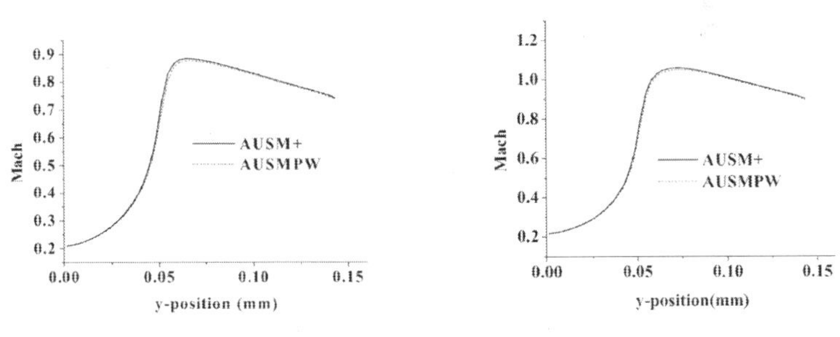

Figure 3.2.2: Centreline Mach distribution, case: 65kPa (left), case: 55kPa (right).

The dominant surface-to-volume ratio effect drastically changes the flow field in microscale compared to the normal-scale nozzles. Hao et al., (2005) reported the presence of shock wave in a larger nozzle (1000 times).

The higher temperature at the throat is correlated to the lower value of the Mach number at the micronozzle throat (Hao et al., 2005). It is evident from Figure 3.2.2 that both AUSM+ and AUSMPW predict the essential flow characteristics with similar level of accuracy. Figure 3.2.3 illustrates the effect of size on the micronozzle flow. Five different micronozzles with throat dimension 4 μm, 8 μm, 20 μm, 100 μm and 200 μm are further studied in this work (varying scales with geometrical similarity). The outlet Mach number increases with the increase of nozzle scale and the position of sonic point moves away from the throat towards the outlet of the nozzle with the decrease in size of the micronozzle. The predicted exit Mach numbers with no-slip condition are 2.02, 1.91, 1.57, 1.2 and 1.001 for nozzle throat sizes of 200 μm, 100 μm, 20 μm, 8 μm and 4 μm respectively. Hao et al., (2005) measured pressures at two different positions near the throat of the micronozzle having 20 μm throat dimension. The Knudsen number for 65 kPa, 55 kPa and 10 kPa in this dimension marginally falls in the slip flow regime. The results are in good agreement with experimental data of Hao et al., (2005). From Figure 3.2.4, it can be seen that while decreasing the dimension of nozzle throat, the deviation between the two solutions (with and without slip boundary conditions) becomes more and more prominent. This is in accordance with the increase in Knudsen number varying from 0.002 to 0.034 in smaller dimensions. Recent numerical investigations of San et al., (2009) on the same configuration revealed similar results which is in accordance with the above presented findings.

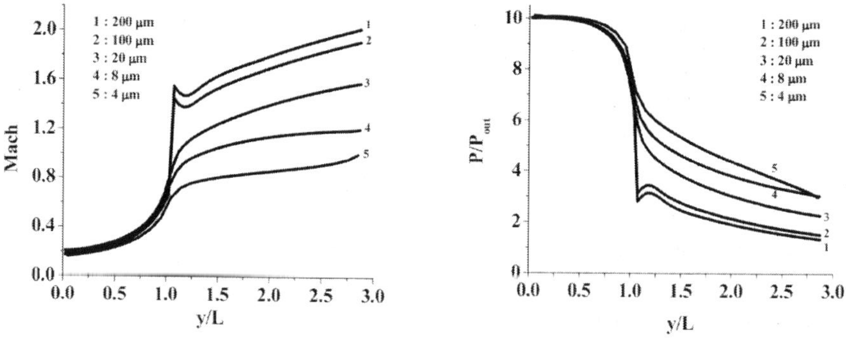

Figure 3.2.3: Comparison of centreline Mach distribution (left), centreline non-dimensional pressure distribution (right) case: 10 kPa with varying scale of the nozzle by AUSM+.

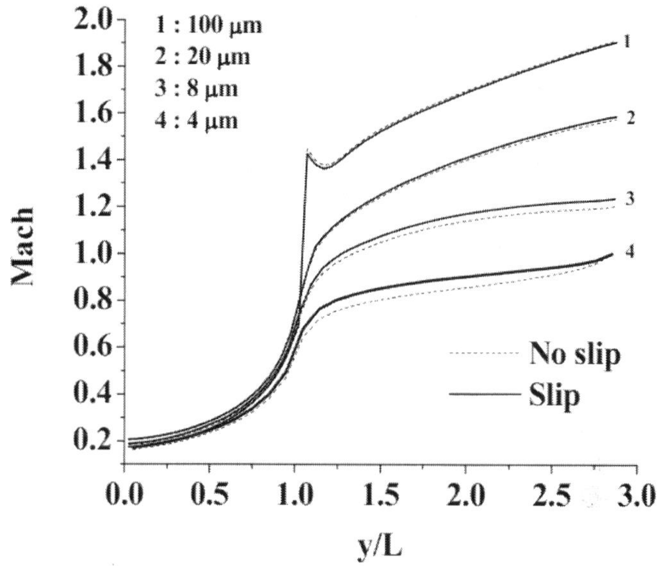

Figure 3.2.4: Comparison of slip/no-slip conditions, Case: 10 kPa by AUSM+.

Flow Separation in Microscale

In previous sub-sections, we have discussed thermal flow characteristics in straight microchannels and subsequently, the behavior of micronozzles with variable cross sectional areas. It can be noted that often bends or curvatures are unavoidable in microsystems. Studies of flow separation and related heat-loss in microscale are thus in need of more experimental measurements as well as numerical predictions to quantify the physics of the phenomena. Studies of thermal microflow with bends, cavities and separation flows over backward-facing steps as well as in micro-mixing systems are important for microfluidic devices. Literatures regarding these areas are inadequate till now. Microchannels with three different bend configurations (integrated with pressure sensors) have been fabricated and characterized by Lee et al., (2001). Their measurements of both mass-flow rate and pressure reveal that secondary flows can exist in microchannels with miter-bends, whereas fully attached flow occurs for curved and double turn bends. Yu et al., (2005) studied experimentally several microchannels (of 1-2 μm height) with cavities. They concluded that fully attached flow occurs for low Reynolds number similar to microchannel flows with fewer cavities. Rawool et al.,

(2005) studied numerically flow through microchannels with obstructions placed along the channels walls. They reported that recirculation zones are created downstream of these obstructions and the skin friction increases nonlinearly with the increase in obstruction height and decreases nonlinearly with the increase in Reynolds number. Xue & Chen, (2003) conducted numerical simulations of backward-facing step flows for various Knudsen numbers from the slip to transition flow regimes. It has been shown that flow separation, recirculation, and reattachment disappear for Knudsen number greater than 0.1. The movements of molecule with mean free paths larger than the backward-facing step dominate the flow behavior.

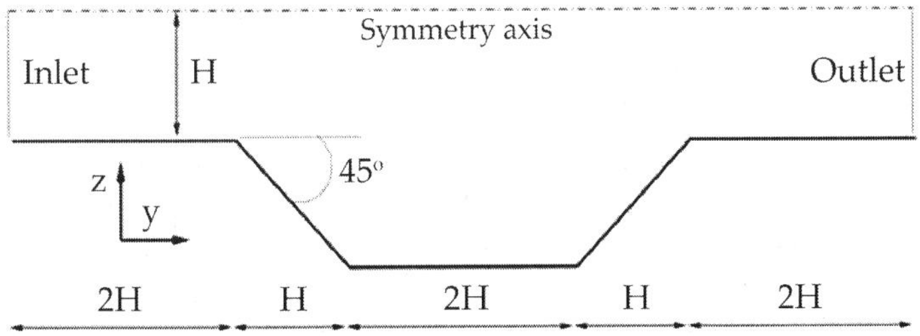

Figure 3.3.1: Schematic of flow geometry for flow-separation at microscale.

Chaudhuri et al., (2008) studied numerically the pressure driven flow through trapezoidal cavity configuration and confirmed the effect of miniaturization on gas flow-separation. Figure 3.3.1 shows the schematic of the flow geometry taken for this study (pressure driven air flow with ratio of inlet to outlet pressure equals to 1.26). Results for three test cases (case 1: H=5 μm, case 2: H=0.5 μm and case 3: H=50 μm) have been presented for demonstration. It can be seen, from Figure 3.3.2, that under similar conditions, flow separation does not occur for case 2 (highest Knudsen number) and flow remains fully attached. On the other hand, flow-separation appears most pronounced for case 3 with largest dimension while case 1 shows intermediate behavior. As expected, effect of slip conditions has been found negligible for case 3.

REACTIVE FLOW IN MICROCHANNELS

Microcombustors, microburners and microactuators, for instance, are the critical power components in MEMS devices. Many experimental techniques, developed for macroscale combustion devices, cannot be directly applied to study microcombustion due to the miniaturization effect. Micropower systems for application in portable electronics, microsensors, and other platforms are often the limiting components in terms of both size and performance. Many micropower systems require a high-temperature reservoir in the form of a combustor. Usually microcombustors can be categorized as (1) homogeneous gas- phase combustors and (2) heterogeneous catalytic combustors. Gas-phase microcombustors are limited by residence time constraints that can be quantified in terms of the Damkohler number. Performance of catalytic microcombustors is typically limited by diffusion of fuel species to the active surface as governed by the Peclet number. Miniaturization of the conventional reactor incurs high thermal losses and radical depletion on the wall which significantly affect the combustion behavior. Two of the most prominent questions are related to the lower limit of the combustor size and the combustion efficiency. Flow distributions of the reaction mixture are also very important factors for microreactor performance. Depending on geometry, composition, and flow rate, methane/air flames are typically quenched for dimensions less than 1-2 mm (Davy, 1817; Maekawa, 1975; Ono & Wakuri, 1977; Fukuda et al., 1981; Lewis & von Elbe, 1987; Linan & Williams, 1993). Although flame propagation at microscale is feasible, the interplay of kinetics and transport in flame stability and combustion characteristics of these systems are not yet well understood (Norton & Vlachos, 2003). In microscale, combustion can be advantageous due to possible faster ignition leading to reduced NOx production.

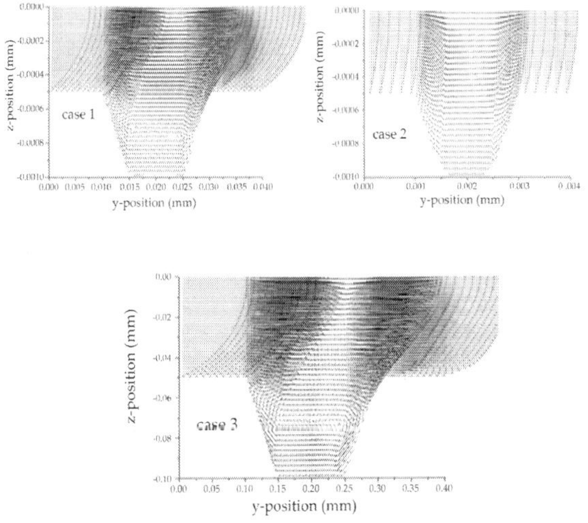

Figure 3.3.2: Velocity vector plot for different test cases.

Size constraint enhances the difficulty of systematic experimental investigations in microreactors. Undoubtedly, numerical simulations are indispensable for studying flow in microcombustors. Lee & Kwon, (2002) developed a theoretical model to simulate the flame propagation with non-negligible heat-loss to the wall in a closed vessel (chamber height ranging from 0.61 to 3.0 mm). Preheating or higher initial pressure of the combustible mixture of gases inside the combustor were found to be capable of increasing the combustion efficiency as well as the thermal efficiency of a power device. Many studies on the characterisation of microflames and stability flame regimes have been reported in the literature (Norton & Vlachos, 2003; Norton & Vlachos, 2004; Miesse et al., 2004; Li et al., 2005a; Yuasa et al., 2005; Shaurya et al. 2007). Federici & Vlachos, (2008) studied the stability of a premixed C3H8/air microflame in a heat recirculation reactor (HRR) and compared it with a single-channel microcombustor. In their scale analysis, Li et al., (2004) reported a relation between the hydrogen flame-temperature in a laminar microcombustor (radii of 3, 4 and 6 mm) and the external wall surface temperature where detailed reaction mechanisms were involved. Leach et al. (2006) presented an analytical model for the prediction of heat-exchange effects within the structure of a micro-channel combustor, and heat-loss from the structure to the environment. Yang et al., (2002) tested different cylindrical microcombustors (2-3 mm diameter). The designed combustors of 2-3 mm diameter with backward facing step were found to be very effective in controlling the position of the flame and widening the range of operation based on the inlet flow velocity as well as H2/air ratio. Later, Li et al., (2005b) investigated the effects of step height and external wall temperature on the structure of hydrogen-air flame. Numerical results showed that the external wall temperature increases drastically with decreasing the step height keeping unchanged the axial flame temperature.

Understanding the behavior of flame propagation with deflagration-to-detonation transition (DDT) in microchannels is typically important for the design of micropropulsion and micro power systems. Ott et al., (2003) predicted mechanism of laminar flame acceleration of air- acetylene system in narrow channel. Gamezo & Oran, (2006) extended the work of flame acceleration to micropropulsion application. Chaudhuri et al., (2007d) studied the phenomenon of flame acceleration of premixed combustible air-acetylene mixture towards the open end of an adiabatic microchannel by solving Navier-Stokes (NS) system of equations with single-step chemistry model. The length (L: y axis) and height (H: z axis) of the 2-D microchannel are 1cm and 600 μm respectively with left-end closed and right-end open. The flame has been initiated by a planar discontinuity in temperature and density located at 366 μm from the closed end. Burned and unburned gases are specified on the left and the right of the discontinuity respectively. The chemical reactions of air-acetylene mixture are considered using a single-step first-order kinetic model similar to that of Ott et al., (2003); Gamezo & Oran, (2006). The initial conditions and other physical and chemical parameters taken are similar to those presented in (Ott et al., 2003; Gamezo & Oran, 2006). Owing to the symmetry of the physical domain, only half of the domain has been simulated.

Symmetry and adiabatic wall conditions are applied accordingly. The position of the flame is defined as the location where the flame-front reaches the temperature of 1000K. Figure 4.1 depicts flame positions near the wall as well as at the centreline. It is evident from Figure 4.2 that the centerline flame velocity increases with time while reaching the end of the microchannel. Also, as shown in Figure 4.3, the transverse velocity component increases at the end of the microchannel as the flow evolves. The maximum outflow velocity reaches $\approx$ 138 m/s. Figure 4.4 shows the product mass fraction and streamwise velocity contours respectively at 0.000195s. The boundary layer growth and burning of boundary layer material clearly reveal the phenomenon of flame acceleration in microchannel. Gamezo & Oran, (2006) concluded that the maximum flame acceleration occurs when the channel height (640 μm) is about five times larger than the reaction zone of a laminar flame. Flame acceleration in microchannel close to this aspect ratio has been studied. The developed solver is capable of predicting the mechanism of flame acceleration, flame propagation and maximum outflow velocities. Results are in good agreement with the conclusion drawn by Gamezo & Oran, (2006) related to the maximum flame acceleration configuration of the adiabatic microchannel

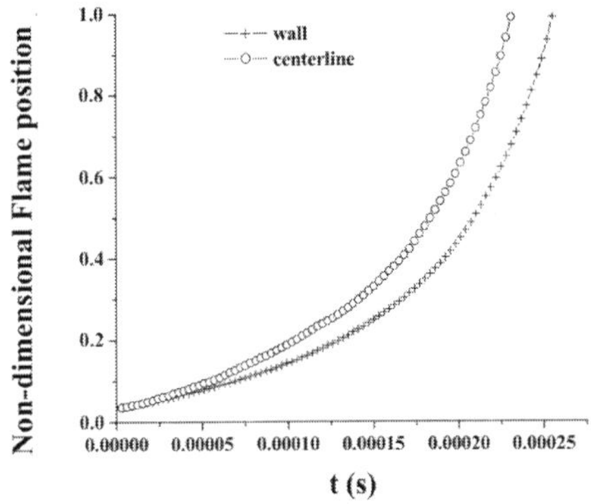

Figure 4.1: Time evolution of flame position.

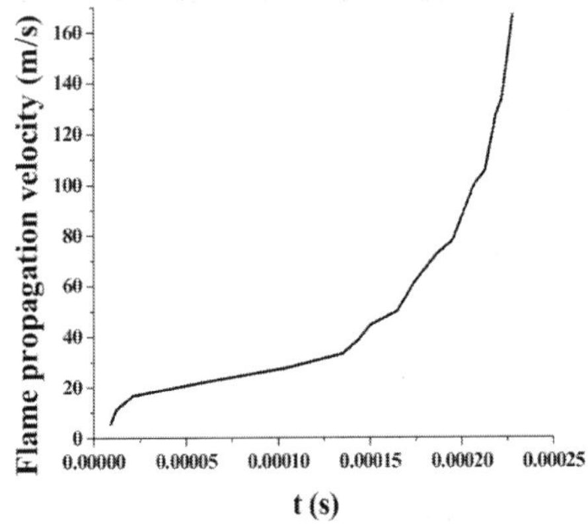

Figure 4.2: Flame propagation velocity.

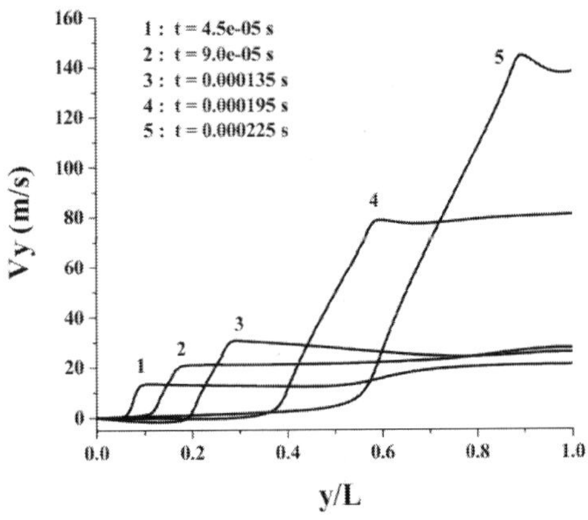

Figure 4.3: Time snaps of the centerline streamwise velocity component.

A detailed review on the application of micro-structured reactors for heterogeneous catalyzed gaseous phase reactions has been presented by Kolb & Hessel, (2004). Lee et al., (2002) proposed the concept of a MEMS scale reciprocating device powered by burnt gas and fabricated its prototype to investigate the applicability of the concept in a microscale power generator. Maruta et al., (2002) studied the extinction limits of self-sustaining methane-fueled combustion on platinum catalyst in non-adiabatic microscale flow reactor channels. Boyarko et al., (2005) experimentally studied fuel rich gaseous hydrogen and oxygen on catalytically active surfaces within 0.4 and 0.8 mm internal diameter microtubes. Good agreement with a plug flow model prediction of detailed gas-phase chemistry and surface kinetics has been reported. Schwarz et al., (2009) showed that microstructured reactors are well suited for application in strongly exothermic heterogeneous catalyzed gaseous phase reactions since they involve isothermal reactions over a wide range of concentrations and temperatures. Xu & Ju, (2004) developed a slip model for species to consider the rarefied gas effect at the boundary in microscale reactors and nanoscale structures. However, no further supporting studies are available for this slip model.

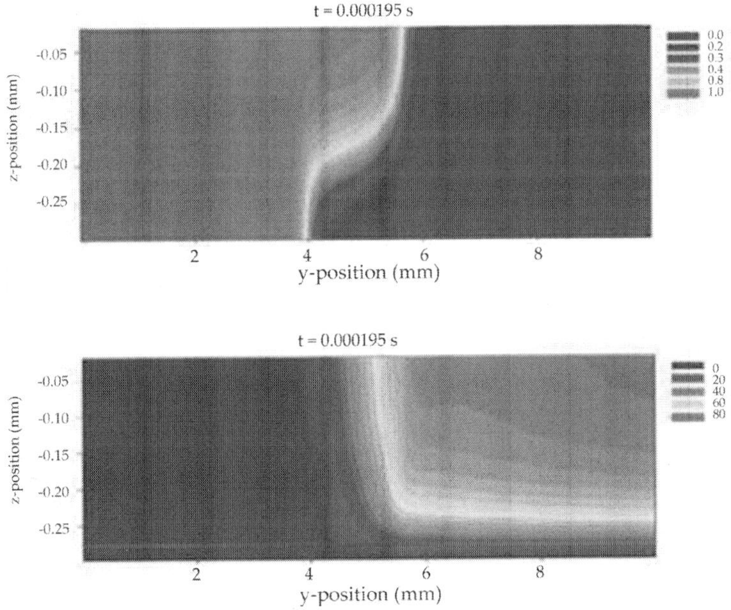

Figure 4.4: Product mass fraction contour (top) and velocity contour (bottom) at t: 0.000195 sec.

Apart from the studies of flame characteristics in microdimensions, viscous and diffusion effects on the propagation of supersonic combustion waves in microchannels have yet to be studied and understood. Kivotides, (2007) analysed the propagation of detonation waves in channels of height 10 !lm to illustrate the viscous effects on detonation characteristics.

Motivated by our previous studies of strong heat release in high-speed microchannel flows, the possibilities of reacting flows at microscale have been further investigated (Chaudhuri et al., 2007c). For this purpose, a stoichiometric air-acetylene mixture is considered at hypersonic speed (M=8) in three different microchannels having 10 mm length and different heights (60 !lm, 80 !lm 200 !lm for cases 1, 2 and 3, respectively). The inlet velocity for each case is taken as uniform with a constant temperature of 300 K and atmospheric pressure. All variables are obtained as a part of the solution at the outlet boundary and all variables are specified at the inlet boundary for each test case. Inlet conditions are assigned as the initial conditions for all variables throughout the computational domain. The chemical reactions of air-acetylene mixture are approximated as single-step first-order simple kinetics (similar to flame acceleration study). The effects of channel aspect ratio and wall temperature have been carefully investigated. Figure 4.5 illustrates the mass fraction of the combustible mixture, and temperature contours for case 1. It is clear from Figure 4.5 that the air-acetylene mixture gets completely burned inside the channel. The reaction zone initiating from the wall region meets at the centerline forming a bent structure of discontinuity. The double hill formation due to heat released by combustion in the burned region of the temperature contour is also evident from Figure 4.5. The high wall-friction associated with the microscale flow brings about the initiation of combustion near the wall region which, in turn, produces the complex discontinuity of Mach and temperature contours. The contour plots of the individual variables for case 2 & 3 are shown in Figs. 4.6 & 4.7. The height-to-length ratios (H/L) of the microchannel are gradually increased. It is clearly evident that with the increase in H/L, the zone of burned gas moves towards the right hand side (RHS) of the microchannel. For highest H/L, the reacted burned gas zone remains detached near the centerline. It can be clearly observed, from Figure 4.8, that the point of initiation of combustion from the wall and corresponding centerline distribution of mass fraction of combustible mixture gradually shifts towards the RHS of the microchannel from case 1 to case 3. The rise in temperature near the wall due to wall friction initiates the chemical reaction of the combustible mixture close to the wall region and the reacted zone reaches the centerline for smaller height-to-length ratio of the microchannels. The wall temperature plays an important role on self-sustainment or reignition of the microcombustion in hypersonic flows. Since the assumption of single-step simple chemistry model is a major approximation of the simulation, the present results can be seen as a preliminary prediction for understanding the general trends of microcombustion phenomena.

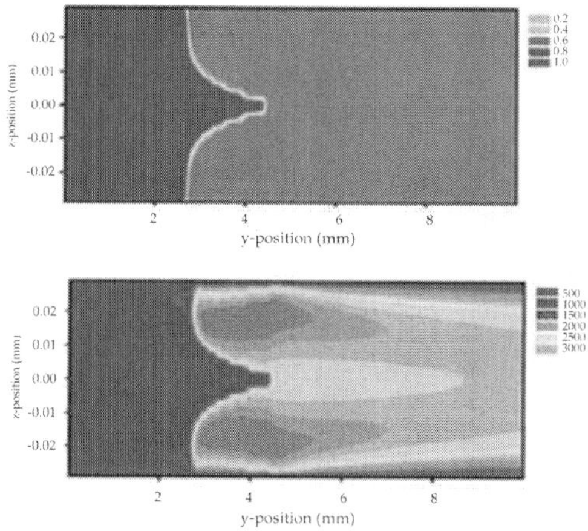

Figure 4.5: Contours of fuel-mixture mass fraction (top) and temperature (bottom) for case 1.

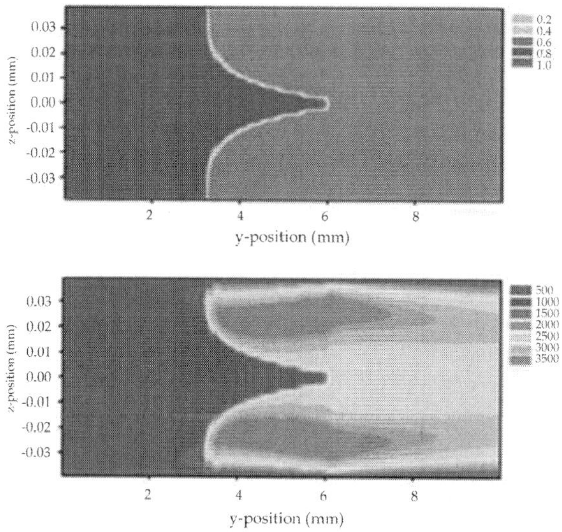

Figure 4.6: Contours of fuel-mixture mass fraction (top) and temperature (bottom) for case 2.

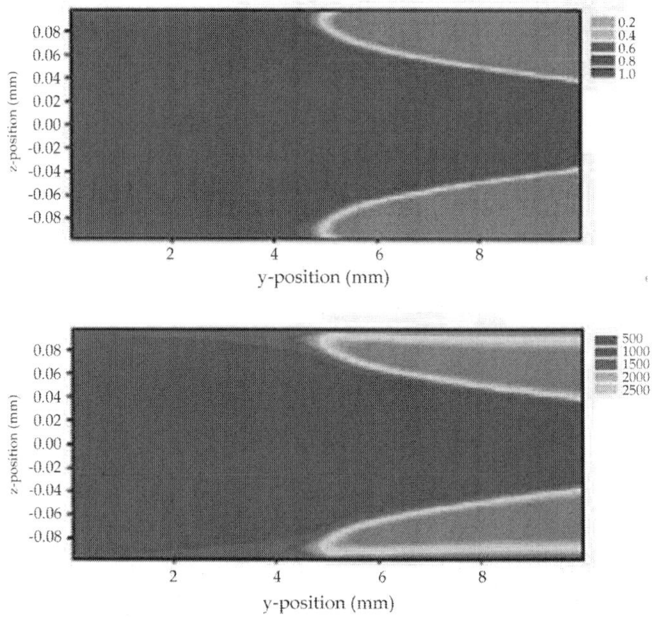

Figure 4.7: Contours of fuel-mixture mass fraction (top) and temperature (bottom) for case 3.

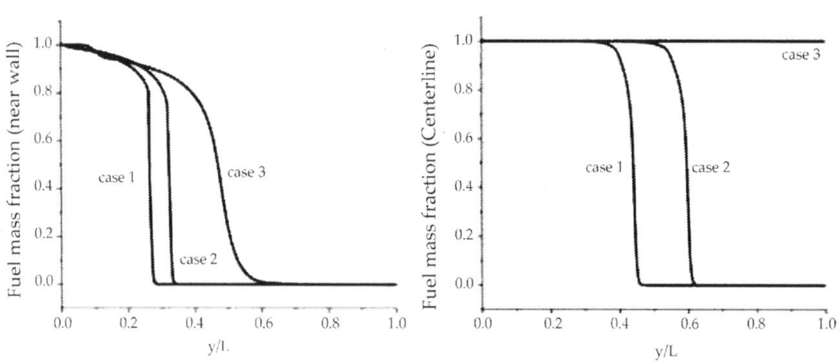

Figure 4.8: Fuel mass fraction distribution near wall (left) and centerline.

SHOCK-WAVES IN MICROSCALES

Thermal losses severely degrade the efficiency of miniaturized devices like internal combustion engines and gas turbines. In microscale devices shock waves can be exploited to produce shock-assisted (or shock-induced) combustion (Brouillette 2003). The behavior of conventional shock tubes at low pressure deviates from the ideal 1D gas dynamics theory. The properties of shock waves for low initial pressure were first examined by Duff, (1959). In general, shock tube flows are sensitive to boundary layer effects because they lead to flow non-uniformities, shock-wave attenuation and reduction in test time. Mirels, (1957) presented a methodology to compute the attenuation of shock waves by the method of characteristics. Later, Mirels, (1963, 1966) developed a theoretical model incorporating the influence of boundary layer on shock wave propagation. In microdimensions the shock tube behavior also suffers from strong deviation from the ideal case. Due to recent progress in the development of microdevices, several authors studied the characteristics of shock-wave propagation in mini/microchannels (Sun et al., 2001; Brouillette, 2003; Udagawa et al., 2007; Zeitoun and Burtschell, 2006; Parisse et al., 2008). Brouillette, (2003) proposed a 1D model to account for the scaling effects via molecular diffusion phenomena to analyse the generation and the propagation of shock waves. Zeitoun & Burtschel, (2006) numerically showed that the attenuation of the shock wave intensity is stronger for lower initial pressure or/and for smaller tube diameter. The effects of slip boundary conditions were found to be important. Zeitoun et al., (2007) also compared the prediction of shock wave propagation in the microtubes and microchannels using three different numerical approaches, namely DSMC method, BGK model and Navier-Stokes system of equations with slip models. In recent works of Mirshekari & Brouillette, (2009), a combined approach using 1D model and numerical simulations has been proposed to predict the performance of a microscale shock tubes with 3D effects incorporating source terms in conservation equations. It can be realized that at lower pressure ratios the shock wave tends to become nonexistent in microscale. In terms of applications, microscale shock tube material with lower thermal conductivity is recommended by Mirshekari & Brouillette, (2009) to achieve desired high temperature-jump. We carried out three demonstratives test-cases to show the effect of lowering the initial pressure (P1) of driven-gas (argon) for a 1mm (L) by 10 μm (H) channel(case 1: 1atm, case 2: 0.5atm and case 3: 0.1atm). The pressure of the driver-gas (oxygen), P4, has been calculated from the theoretical value which gives the pressure ratio (P_4/P_1) as function of the incident Mach number (M_s).

$$\frac{P_4}{P_1} = \frac{1 + \frac{2\gamma_1}{\gamma_1 + 1}\left(M_s^2 - 1\right)}{\left[1 - \frac{\gamma_4 - 1}{\gamma_4 + 1}\left(\frac{a_1}{a_4}\right)\left(M_s - \frac{1}{M_s}\right)\right]^{\frac{2\gamma_4}{\gamma_4 - 1}}} \tag{7}$$

where a_1 and a_4 are the speed of sound for driven-gas and driver-gas respectively.

M_s has been taken as 1.6 for the three test-cases. Owing to the symmetry of the problem only half of the physical domain has been simulated considering symmetry boundary conditions (on the top) and first-order slip boundary conditions (on the bottom). Based on Knudsen number criteria and previous studies (Zeitoun & Burtschel, 2006), the effect of thermal creep was neglected. Isothermal wall conditions (with a fixed wall temperature 300K) are imposed for all the considered cases. Unsteady computations have been carried out and the simulations have been stopped before the shock wave and the expansion fan reach the right and left boundaries of the computational domain.

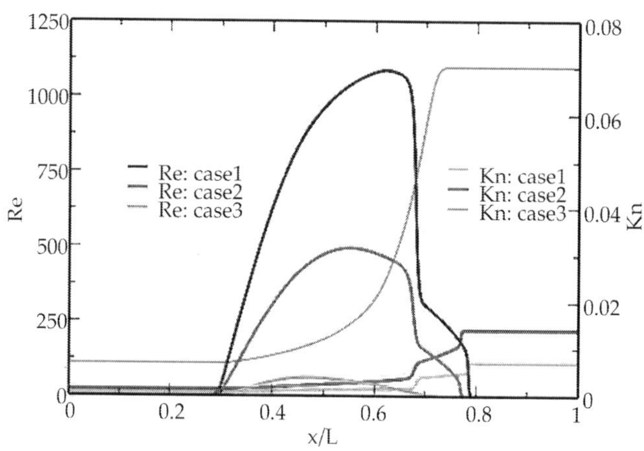

Figure 5.1: Comparison of Reynolds number and Knudsen number.

Reynolds and Knudsen numbers, based on centerline flow-field properties, have been shown in Figure 5.1. It is evident that Knudsen number is higher for driven-gas section having lower pressure, and Knudsen number is progressively higher from case 1 to case 3. Presence of dominant viscous effects can also be realized from lowest Reynolds number for case 3.

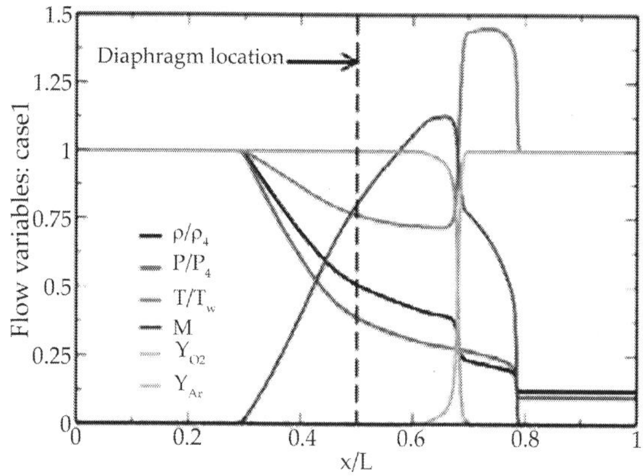

Figure 5.2: Flow variables for case 1.

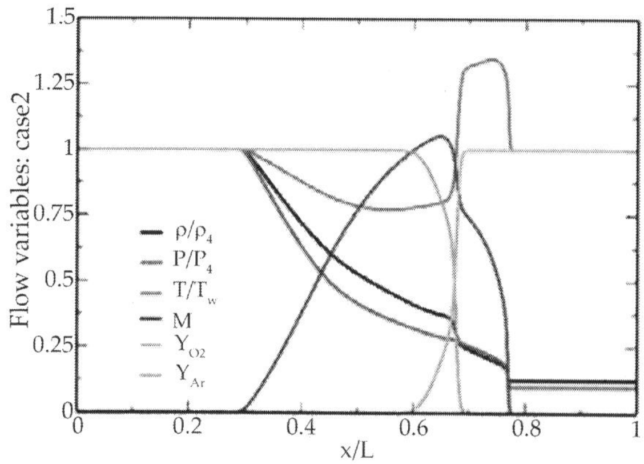

Figure 5.3: Flow variables for case 2.

Figs. 5.2-5.4 show the flow field parameters at non-dimensional time $\tau = ta_1/H = 20$. It can be seen that the strength of the shock gradually becomes weaker in cases from 1 to 3. The contact discontinuity location can be visualized from species mass fraction profile for each test case. Comparisons of dimensionless streamwise velocities (here, we used the speed of sound a_1 as a reference velocity) have been illustrated in Figure 5.5. The attenuation and retardation of shock propagation are clearly visible from this figure. Indeed for case 3, shock wave became almost nonexistent compared to the other two cases.

Finally, the shock location has been recorded and shown in Figure 5.6 for all cases. The ideal curve corresponds to the theoretical $M_s = 1.6$. It is evident that damping of shock propagation increases with the increase of Knudsen number associated with lower initial driven-gas pressure

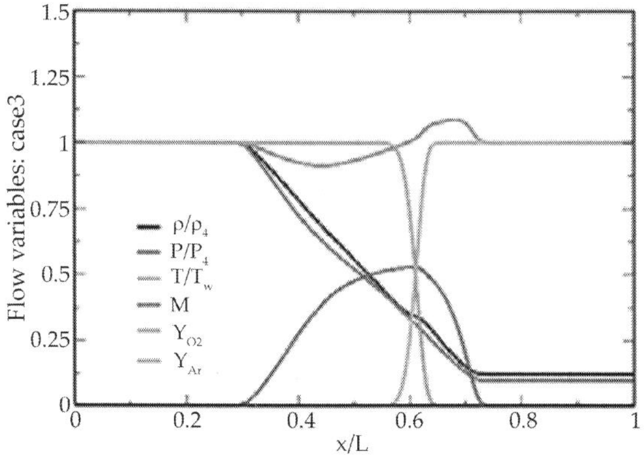

Figure 5.4: Flow variables for case 3.

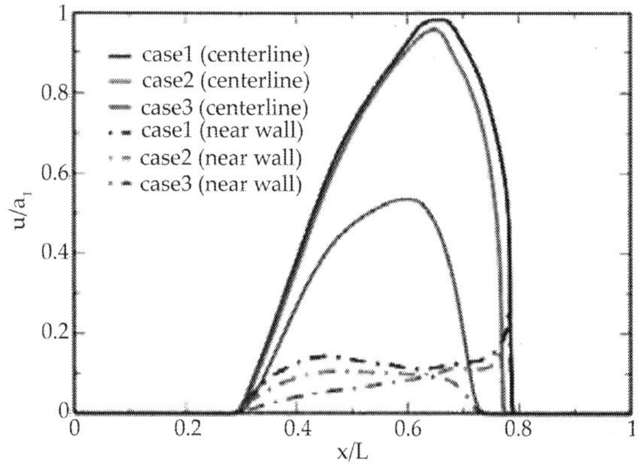

Figure 5.5: Comparison of streamwise velocities.

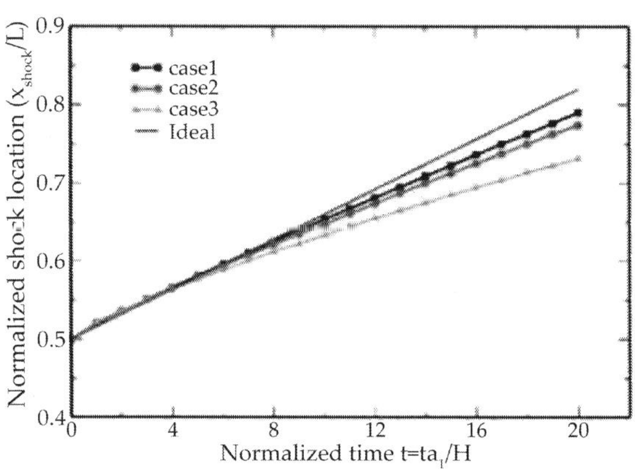

Figure 5.6: Evolution of shock-position.

CONCLUSIONS

In this chapter, we have provided a brief literature survey on compressible, reacting and non-reacting microscale gas flows including some of our recent numerical findings. The applicability of Navier-Stokes solutions with wall-slip models proved to be very cost- effective numerical approach for successful prediction of gas flows and thermal characteristics in microdevices. In particular, the validation of implemented first-order slip model has been achieved in solving various relevant benchmark problems. First, partially heated microchannel, resembling the classical Graetz problem, has been investigated with respect to the effect of Knudsen number as well as gas species. Then, high-speed supersonic and hypersonic flows with heat transfer in microchannel have been investigated. It has been found that the wall temperature jump is higher for higher inlet Mach numbers. Differences between conventional and microscale nozzle flows have been illustrated through different numerical simulations. Specifically, shock-free supersonic flow has been observed in micronozzles, contrary to conventional full-scale nozzles, and boundary-layer separation phenomenon has been found to be less pronounced in microscales. For the reacting cases, a specific solver has been developed for predicting the mechanism of flame acceleration and microcombustion in high-speed premixed stoichiometric air-acetylene mixture inside microchannels. The complex interactions among hydrodynamic process, shock structures and combustion have been predicted using a simplified single-step chemistry model. It can be concluded that microcombustion in high-speed flows can be achieved in suitably preheated stoichiometric premixed fuel-oxidizer. Furthermore, thermal loss is an important factor in high-speed microcombustion. In addition, in order to get better insights into the physics of shock-wave phenomena, attenuation of shock waves in microdimensions through both viscous and thermal effects has been reported.

One can notice that theoretical knowledge is presently more advanced for gas-flows than for liquid/two-phase flows in microchannels. However, there exists the need for precise experimental data both for steady and unsteady reacting and non-reacting microscale gas flows to definitely validate the choice of the best boundary model in the slip flow regime. Application of high-order slip models needs further intensive numerical investigations. Determination of the appropriate values of the accommodation coefficients also remains an open issue. Theoretical investigations relative to unsteady or thermally driven microflows would also need to be supported by smart experiments. Although the present two- dimensional studies reveal many important features of microscale flows and gaseous reactions, three-dimensional studies are required to get better insight into the physics of microflows. Incorporation of heat-loss mechanism through solid wall and detailed multi- step chemical reaction models, in case of reacting simulations, are also crucial aspects for more realistic and high-fidelity numerical predictions.

REFERENCES

1. Alexeenko, A.A.; Levin, D. A.; Gimelshein, S. F.; Collins, R. J. & Markelov, G. N. (2002). Numerical simulation of high-temperature gas flows in a millimeter-scale thruster. *J. of Thermophysics & Heat Transfer*, 16, 1, 10-16
2. Arkilic, E.B.; Breuer, K.S. & Schmidt, M.A. (1994). Gaseous flow in microchannels. FED-Vol. 197, *Application of Microfabrication to Fluid Mechanics, ASME*, 57-66
3. Arkilic, E. B.; Breuer, K. S. & Schmidt, M. A. (1997a). Gaseous slip flow in long microchannels. *J. Microelectromech. Syst.*, 6,167-178
4. Arkilic, E.B.; Schmidt, M.A. & Breuer, K.S. (1997b). TMAC measurement in silicon micromachined channels. *Rarefied Gas Dynamics*, 20
5. Arkilic, E B.; Breuer, K. S. & Schmidt, M. A. (2001). Mass flow and tangential momentum accommodation in silicon micromachined channels. *J. Fluid Mech.*, 437, 29-43
6. Bayt, R. L.; Ayon, A.A. & Breuer, K. S. (1997). A performance evaluation of MEMS based micronozzle. *AIAA* 97-3169
7. Bayt, R. L. & Breuer, K. S. (1998). Viscous effects in supersonic MEMS-fabricated micronozzles. *Proceedings of 3rd ASME Microfluids Symposium*. Anaheim, CA
8. Bayt, R. L. (1999). Fabrication and testing of a MEMS based micropropulsion systems. *PhD thesis* MIT, Cambridge, MA
9. Beskok, A.; Trimmer, W. & Karniadakis, G. E. (1996). Rarefaction and compressibility effects in gas microflows, *J. Fluids Eng.*, 118, 3, 448-456
10. Boyarko, G.A.; Sung, C. J. & Schneider, S. J. (2005). Catalyzed combustion of hydrogen-oxygen in platinum tubes for micro-propulsion applications. *Proceedings of the Combustion Institute*, 30, 2481-2488
11. Brouillette, M. (2003). Shock waves at microscales. *Shock Waves*, 13, 3-12
12. Celata, G. P. ; Cumo, M.; McPhail, S. J.; Tesfagabir, L. & Zummo, G. (2007). Experimental study on compressible flow in microtubes. *Int. J. Heat & Fluid Flow*, 28, 28-36

13. Chaudhuri, A.; Guha, C. & Dutta, T. K. (2006) Numerical study of micro-scale gas flow using finite volume method, *Journal of Physics Conference series*, 34, 291-297
14. Chaudhuri, A.; Guha, C. & Dutta, T. K. (2007a). Finite volume simulation of supersonic to hypersonic gas flow and heat transfer through microchannels, *Chem. Eng. Technol.,* 30,1, 41-45
15. Chaudhuri, A.; Guha, C. & Dutta, T. K. (2007b). Numerical study of fluid flow and heat transfer in partially heated microchannels using explicit finite volume method, *Chem. Eng. Technol.,* 30, 4, 425-430
16. Chaudhuri, A.; Guha, C. & Dutta, T. K. (2007c). Finite volume simulation of high speed combustion of premixed air-acetylene mixtures in a microchannel, *Chem. Eng. Technol.,* 30, 5, 615-620
17. Chaudhuri, A.; Guha, C. & Dutta, T. K. (2007d). Numerical study of flame acceleration in microchannel, *Proceedings of NSTI, Nanotech 2007, Santa Clara Convention Center*, Santa Clara, California, Vol. 3, 149-152, May 20-24, 2007
18. Chaudhuri, A.; Guha, C. & Dutta, T. K. (2008). Numerical study of microscale gas flow- separation using explicit finite volume method, *Proceedings of World Academy of Sciences, Engineering and Technology*, Vol. 36, 338-341, ISSN 2070-3740, December 2008
19. Chen, C. S. (2004). Numerical method for predicting three-dimensional steady compressible flow in long microchannels. *J. Micromech. Microeng.*, 14, 1091-1100
20. Chen, K.; Winter, M. & Huang, R. F. (2005). Supersonic flow in miniature nozzles of planar configuration, *J. Micromech. Microeng.* 15, 1736-1744
21. Choi, S.B.; Barron, R.F. & Warrington, R.O. (1991). Fluid flow and heat transfer in microtubes. In Micro mechanical Sensors, Actuators, and Systems, *ASME Winter Annual meeting*, 123-133, Atlanta, GA
22. Colin, S.; Lalonde, P. & Caen, R. (2004). Validation of a second-order slip flow model in rectangular microchannels. *Heat Transfer Eng.*, 25, 23-30
23. Colin, S. (2005). Rarefaction and compressibility effects on steady and transient gas flows in microchannels. *Microfluid Nanofluid*, 1, 268-279
24. Davy, H. (1817). Some researches on flame. *Transactions of the Royal Society of London*, 107, 45-76

25. Dongari, N.; Agrawal, A. & Agrawal, A . (2007). Analytical solution of gaseous slip flow in long microchannels. *Int. J. Heat & Mass Transfer*, 50, 3411-3421
26. Duff, R.E. (1959). Shock-tube performance at low initial pressure. *Phys. Fluids,* 2(2), 207-216 Ewart, T.; Perrier, P.; Graur, I. & Me'olans, J. G. (2007). Tangential momemtum accommodation in microtube. *Microfluid Nanofluid*, 3, 689-695
27. Federici, J.A. & Vlachos, D.G. (2008). A computational fluid dynamics study of propane/air microflame stability in a heat recirculation reactor. *Combustion and Flame,* 153, 258- 269
28. Fukuda, M.; Koji, K. & Sakamoto, M. (1981). On quenching distance of mixtures of methane and hydrogen with air. *Bulletin of the JSME*, 24(193), 1192-1197
29. Gad-el-Hak, M. (1999). The fluid mechanics of microdevices-The freeman scholar lecture. *J. Fluids Eng.*, 121, 5-33
30. Gamezo, V. N. & Oran, E. S. (2006). Flame acceleration in narrow channels: Application for micropropulsion in low gravity environment, *AIAA J.*, 44, 329-336
31. Hammel, J. R. (2002). Development of an unstructured 3D DSMC/ Particle-in-cell code and the simulation of microthruster flows. *MS Thesis,* Mech. Engg. Worcester Polytechnic Institute
32. Hao, P. F.; Ding, Y.; Yao, Z.; He, F. & Zhu, K. (2005). Size effect on gas flow in micro nozzles. *J. Micromech. Microeng.*, 15, 2069-2073
33. Harley, J.C.; Huang, Y.; Bau, H.H. & Zemel, J.N. (1995). Gas flow in microchannels. *J. Fluid Mech.*, 284, 257-274
34. Jain, V. & Lin, C. X. (2006). Numerical modeling of three-dimensional compressible gas flow in microchannels. *J. Micromech. Microeng.*, 16, 292-302
35. Janson, S. W.; Helvajian, H. & Breuer, K. (1999). MEMS, microengineering and aerospace systems. *AIAA* 99-3802
36. Jiang, G. & Shu, C. W. (1996). Efficient implementation of weighted ENO schemes, *J. Comp. Phys.*, 126, 202-228
37. Karniadakis, G. E. & Beskok, A. (2002). *Microflows: fundamentals and simulation.* Springer, Berlin Heidelberg New York
38. Kivotides, D. (2007). Viscous microdetonation physics. *Physics Letters A*, 363, 458-467

39. Kolb, G. & Hessel, V. (2004). Micro-structured reactors for gas phase reactions. *Chemical Engineering Journal*, 98, 1-38
40. Kujawa, J.; Hitt, D.L. & Cretu, G. (2003). Numerical simulation of supersonic flow in MEMS nozzle geometries with heat loss, *AIAA Paper* 2003-3585
41. Le, M. & Hassan, I. (2006). Simulation of heat transfer in high speed microflows, *App. Thermal Eng*., 26, 2035-2044
42. Leach, T. T.; Cadou, C. P. & Jackson, G. S. (2006) Effect of structural conduction and heat loss on combustion in micro-channels. *Combustion Theory and Modelling*, 10, 1, 85-103 Lee, S. Y. K.; Wong, M. & Zohar, Y. (2001). Gas flow in microchannels with bends. *J. Micromech. Microeng*., 11, 635-644
43. Lee, D. H. & Kwon, S. (2002). Heat transfer and quenching analysis of combustion in a micro combustion vessel. *J. Micromech. Microeng*, 12, 670-676
44. Lee, D. H.; Park, D. E.; Yoon, J. B.; Kwon, S. & Yoon, E. (2002). Fabrication and test of a MEMS combustor and reciprocating device. *J. Micromech. Microeng*., 12, 26-34 Lewis Jr., D. H. (2000). Digital micropropulsion. *Sensors and Actuators*, 80, 143-154 Lewis, B. & von Elbe, G. (1987). *Combustion, flames and explosions of gases*. Orlando, FL: Academic Press.
45. Li, Z. W.; Chou, S. K.; Shu, C.; Yang, W. M. & Xue, H. (2004). Predicting the temperature of a premixed flame in a microcombustor, *Journal of Applied Physics*, 96, 3524-3530
46. Li, Z.W.; Chou, S.K.; Shu, C.; Xue, H. & Yang, W.M. (2005a). Characteristics of premixed flame in microcombustors with different diameters. *Applied Thermal Engineering*, 25, 271-281
47. Li, Z. W.; Chou, S. K.; Shu, C. & Yang, W. M. (2005b). Effects of step height on wall temperature of a microcombustor. *J. Micromech. Microeng*., 15, 207-212
48. Lin, C. X. & Gadepalli, V. V. V. (2008). A computational study of gas flow in a De-Laval micronozzle at different throat diameters. *Int. J. Numer. Meth. Fluids,* DOI: 10.1002/fld.1868
49. Linan, A. & Williams, F. A. (1993). *Fundamental aspects of combustion*. New York: Oxford University Press.
50. Liou, M. S. & Steffen Jr. C. J. (1993). A new flux splitting method. *J. of Computational Physics*, 107, 23-39

51. Liou, W. W. & Fang, L. Y. (2001). Heat transfer in microchannel devices using DSMC. *J. Microelectromech. Syst.*, 2001, 10, 274-279

52. Liu, J.; Tai, Y. C. & Ho, C. M. (1995). MEMS for pressure distribution studies of gaseous flows in microchannels. *In: IEEE international conference on micro electro mechanical systems.* Amsterdam, Netherlands, 209-215

53. Liu, M.; Zhang, X.; Zhang, G. & Chen, Y. (2006). Study on micronozzle flow and propulsion performance using DSMC and continuum methods. *Acta Mechanica Sinica*, 22, 409- 416

54. Maekawa, M. (1975). Flame quenching by rectangular channels as a function of channel length for methane-air mixture. *Combustion Science and Technology*, 11, 141-145.

55. Maruta, K.; Takeda, K.; Ahn, J.; Borer, K.; Sitzki, L.; Ronney, P. D. & Deutschmann, O. (2002). Extinction Limits of Catalytic Combustion in Microchannels. *Proceedings of the Combustion Institute*, 29, 957-963

56. Maurer, J.; Tabeling, P.; Joseph, P. & Willaime, H. (2003). Second-order slip laws in microchannels for helium and nitrogen. *Phys Fluids*, 15, 2613-2621

57. Miesse, C. M.; Masel, R. I.; Jensen, C. D.; Shannon, M. A & Short, M. (2004). Submillimeter- Scale Combustion. *AIChE Journal*, 50, 12, 3206-3214

58. Mirels, H. (1957). Attenuation in a shock tube due to unsteady-boundary -layer action. *Technical Report TN 1333*, National Advisory Committee for Aeronautics Mirels, H. (1963) Test time in low-pressure shock tubes. *Phys. of Fluids*, 6, 9, 1201-1214

59. Mirels, H. (1966). Correlation formulas for laminar shock tube boundary layer. *Phys. Of Fluids*, 9, 7, 1265-1272

60. Mirshekari, G. & Brouillette, M. (2009). One-dimensional model for microscale shock tube flow. *Shock Waves*, 19, 25-38.

61. Morffiigo, J. A.; Quesada, J. H. & Requena, F. C. (2007). Slip-model performance for underexpanded micro-scale rocket nozzle flows. *J. of Thermal Science*, 16, 223-230

62. Norton, D. G.; Vlachos, D. G. (2003). Combustion characteristics and flame stability at the microscale: a CFD study of premixed methane/air mixtures. *Chemical Engineering Science*, 58, 4871 – 4882 Norton,

D. G.; Vlachos, D. G. (2004). A CFD study of propane/air microflame stability. *Combustion and Flame*, 138, 97-107

63. Ott, J. D.; Oran, E. S. & Anderson, J. D. (2003). A mechanism for flame acceleration in narrow tubes, *AIAA J.*, 41, 1391-1396
64. Papautsky, I.; Ameel, T. & Frazier, A. B. (2001). A review of laminar single-phase flow in microchannels. *Proceedings of ASME Int. Mech. Eng. Cong. & Expo*. New York Parisse, J. D.; Giordano, J.; Perrier, P.; Burtschell, Y. & Graur, I. A. (2008). Numerical investigation of micro shock waves generation. *Microfluid Nanofluid*, DOI 10.1007/s10404-008-0336-y
65. Pfahler, J.; Harley, J.; Bau, H. & Zemel, J. N. (1990). Gas and liquid transport in small channels. *Microstruct Sensors Actuators ASME*, 19, 149-157
66. Pong, K.C.; Ho, C.M. Liu J. & Tai, Y.C. (1994). Non-linear pressure distribution in uniform microchannels. *In Application of Microfabrication to Fluid Mechanics, ASME Winter Annual meeting*, 51-56, Chicago, Ill
67. Porodnov, B.T.; Suetin, P.E. ; Borisov, S.F. & Akinshin, V.D. (1974). Experimental investigation of rarefied gas flow in different channels. *J. of Fluid Mech.*, 64, 417-437
68. Raju, R.; Pandey, B.P. & Roy, S. (2002). Finite element model of fluid flow inside a micro- thruster, *Nano tech 2002-"At the edge of revolution", AIAA-2002-5733*, Houston, Texas
69. Raju, R. & Roy, S. (2005). Hydrodynamic study of high speed flow and heat transfer through a microchannel, *J. Thermophys. Heat Transfer*, 19, 106-113
70. Rawool, A. S.; Mitra, S. K. & Kandlikar, S. G. (2005). Numerical simulation of flow through microchannels with designed roughness. *Microfluid Nanofluid*, DOI 10.1007/s10404- 005-0064-5
71. Reed, B. (2003). Decomposing solid micropropulsion nozzle performance issues. *AIAA 2003- 672*
72. Rossi, C.; Rouhani, M. D. & Daniel, E. (2000). Prediction of the performance of a Si- micromachined microthruster by computing the subsonic gas flow inside the thruster, *Sensors and Actuators, Physical A*, 87, 96-104

73. Rossi, C.; Conto, T. D.; Esteve, D. & Larangot, B. (2001) . Design, fabrication and modelling of MEMS-based microthrusters for space Application. *Smart Mater. Struct.* 10 1156- 1162
74. Rostami, A. A.; Mujumdar, A. S. & Saniei, N. (2002). Flow and heat transfer for gas flowing in microchannel: a review, *Heat and Mass Transfer*, 38, 359-367
75. San, O.; Bayraktar, I. & Bayraktar, T. (2009). Size and expansion ratio analysis of micro nozzle gas flow. *doi:10.1016/j. icheatmasstransfer.2009.01.021*
76. Schwarz, O.; Duong, P.Q.; Schafer, G. & Schomacker, R. (2009). Development of a microstructured reactor for heterogeneously catalyzed gas phase reactions: Part II.
77. Reactor characterization and kinetic investigations *Chemical Engineering Journal*, 145, 429-435
78. Shaurya, S.; Armijo, A. D.; Masel, R. I. & Shannon, (2007). M. A. Flame dynamics in sub- millimetre combustors. *Int. J. Alternative Propulsion*, 1, 2/3, 325-338
79. Shih, J.C.; Ho, C.; Liu, J. & Tai, Y. (1996). Monatomic and polyatomic gas flow through uniform microchannels. *In Application of Microfabrication to Fluid Mechanics, ASME Winter Annual meeting*, 197-203, Atlanta, GA
80. Sreekanth, A. K. (1969). Slip flow through long circular tubes. *In: Trilling L, Wachman HY (eds) Proceedings of the 6th international symposium on rarefied gas dynamics.* Academic Press, New York, 667-680
81. Sun, M.; Ogawa, T. & Takayama, K. (2001). Shock propagation in narrow channels. In: Lu FK (ed) *Proceedings of 24th international symposium on shock waves*, 1320-1327
82. Udagawa, S.; Maeno, K.; Golubeva, I. & Garen, W. (2007). Interferometric signal measurement of shock waves and contact surfaces in small scale shock tube. *Proccedings of 26th international symposium on shock waves*, 2060
83. Wu, P. & Little, W.A. (1984). Measurement of the heat transfer characteristics of gas flow in fine channel heat exchanger used for microminiature refrigerators. *Cryogenics*, 24, 8, 15-420

84. Xu, B. & Ju, Y. (2004). Numerical modeling of the heterogeneous combustion in a microscale chemical reactor. *AIAA 2004-304*, 42nd AIAA Aerospace Sciences Meeting and Exhibit, Reno Hilton, Reno, Nevada

85. Xu, J. & Zhao, C. (2007). Two-dimensional numerical simulations of shock waves in micro convergent-divergent nozzles. *Int. J. of Heat & Mass Transfer*, 50, 2434-2438

86. Xue, H. & Chen, S. (2003). DSMC simulation of microscale backward-facing step flow. *Microscale Thermophysical Engineering*, 7, 69-86

87. Yan, F. & Farouk, B. (2002).Computation of fluid flow and heat transfer in ducts using the direct solution Monte Carlo method, *J. of Heat Transfer,* 124, 609-616

88. Yang, W.M.; Chou, S.K.; Shu, C.; Li, Z.W. & Xue, H. (2002). Combustion in micro-cylindrical combustors with and without a backward facing step. *Applied Thermal Engineering*, 22, 1777-1787

89. Yu, Z. T. F; Lee, Y; Wong, M. & Zohar, Y. (2005). Fluid flows in microchannels with cavities. *J. of Microelectromechanical Systems*, 14, 6, 1386 - 1398

90. Yuasa, S.; Oshimi, K.; Nose, H. & Tennichi, Y. (2005). Concept and combustion characteristics of ultra-micro combustors with premixed flame. *Proceedings of the Combustion Institute*, 30, 2455-2462

91. Zeitoun, D.E., Burtschel, Y. (2006). Navier-stokes computations in micro shock tubes. *Shock Waves,* 15, 241-246

92. Zeitoun, D. E.; Burtschell, Y. ; Graur, I.A. ; Ivanov, M. S.; Kudryavtsev, A. N. & Bondar, Y. A. (2007). Shock waves at microscalesn. *Proceedings of West-east high speed flow field Conference*, 19-22,Moscow, Russia

Chapter

RECENT NUMERICAL METHODS IN ELECTROCARDIOLOGY

12

Youssef Belhamadia

University of Alberta, Campus Saint-Jean Edmonton, Alberta, Canada

INTRODUCTION

Heart diseases are the leading cause of death in the world. Many questions have not yet been answered regarding the electrical waves propagation in cardiac tissue, and the mechanism of ventricular fibrillation that is produced by one or many spiral propagation waves of the excitation cardiac wall. Numerical modeling can play a crucial role and provides the necessary tools to answer some of these questions. However, the mathematical models, which give the best reflection of electrophysiological waves in cardiac tissue, are extremely complicated and present a significant computational challenges.

Citation: Youssef Belhamadia, "Recent Numerical Methods in Electrocardiology", IntechOpen New Developments in Biomedical Engineering, 2010, DOI: 10.5772/7619.

The bidomain model is considered as the mathematical equations that have been used for simulating cardiac electrophysiological waves for many years (see Sundnes (2002), and Pierre (2006) and the reference therein). This model represents the cardiac tissue at a macroscopic scale by relating the transmembrane potential, the extracellular potential, and the ionic currents. The biodomain model consists of a system of two nonlinear partial differential equations coupled to a system of ordinary differential equations. From the numerical point of view, the model is computationally very expensive. The major difficulties are due to the computational grids size that must be very fine to get a realistic simulation of cardiac tissue. Indeed, the action potential is a wave with sharp depolarization and repolarization fronts and this wave travels across the whole computational domain calling for a very fine uniform mesh.

One popular way of reducing the computational challenges of the bidomain model is the use of the monodomain model. This model considers a single nonlinear partial differential equation coupled with the same system of ordinary differential equations for the ionic currents. Although, it has been reported that the CPU requirements are reduced when simplifying the bidomain model to a monodomain model (see Sundnes et al. (2006)), both models still encounter computational difficulties because of the need for fine meshes and small time-steps.

Many methods have been introduced in the literature to overcome these difficulties. The operator splitting is usually performed to separate the large non-linear system of ODEs and thus introduces subproblems easier to solve. A first-order (Godunov method) and a second-order (Strang method) accurate splitting technique can be employed. For more details the reader is referred to Sundnes et al. (2005), Lines, Buist, Grottum, Pullan, Sundnes & Tveito (2003); Lines, Grottum & Tveito (2003), and Weber Dos Santos et al. (2003)). To reduce the computational time at each time step, parallel computing techniques are used (see Colli Franzone & Pavarino (2004), Karpoukhin et al. (1995) and Weber dos Santos et al. (2004)). Several timestepping strategies have also been used, fully implicit (Bourgault et al. (2003), and Murillo & Cai (2004)), and semi-implicit (Franzone & Pavarino (2004), Ethier & Bourgault (2008)).

Recently, mesh adaptation methods have been introduced to reduce the size of the spatial mesh as well as the computational time. This method consists in locating finer mesh cells near the depolarisation-repolarization front position while a coarser mesh is used away from the front. In the context of isotropic unstructured meshes, the reader is referred to Cherry et al. (2003), Colli Franzone et al. (2006) and Trangenstein & Kim (2004) for more details. However, for two and three dimensional anisotropic mesh adaptation, where mesh cells are elongated along a specified direction, the reader is referred to Belhamadia (2008a;b); Belhamadia et al. (2009).

The scope of this book chapter is to present the recent adaptive technique introduced in Belhamadia (2008a;b) for simulating the two-dimensional cardiac electrical activity. The method proposed reduces greatly the size of the spatial mesh as well as the computational time. Also, an accurate prediction of the depolarization and repolarization fronts is obtained showing the advantages of the proposed method.

This work is organized as follows. Section 2 presents a brief description of the bidomain and monodomain models with Aliev-Panfilov ion kinetics. Also, the finite element discretization for these models are presented. Section 3 is devoted to a description of the time-dependent adaptive strategy while the last section presents two-dimensional numerical results representing the re-entrant waves.

MATHEMATICAL MODELS

The bidomain and modomain models will be now presented. The first model consists of a nonlinear partial differential equation for the transmembrane potential V_m coupled with an elliptic one for the extracellular potential φ_e, as well as an ordinary differential equation, for at least one variable, representing the ionic currents. This system of equations takes the following form:

$$\begin{cases} \frac{\partial V_m}{\partial t} - \nabla \cdot (G_i \nabla V_m) = \nabla \cdot (G_i \nabla \phi_e) + I_{ion}(V_m, W) \\ \nabla \cdot ((G_i + G_e)\nabla \phi_e) = -\nabla \cdot (G_i \nabla V_m) \\ \frac{\partial W}{\partial t} = g(V_m, W), \end{cases} \quad (1)$$

where G_i and G_e are the symmetric intra- and extra-cellular conductivity tensors. The definition of thc functions $I_{ion}(V_m,w)$ and $g(V_m,w)$ depends on the ionic model. Modern cardiac ionic models include generally a set of 10 to 60 ordinary differential equations. However, in this work the Aliev-Panfilov model (see Aliev & Panfilov (1996)) is presented which consists of the following equations:

$$I_{ion} = kV_m(V_m - a)(1 - V_m) - V_m W,$$

$$g(V_m, W) = \left(\epsilon + \frac{\mu_1 W}{\mu_2 + V_m}\right)(-W - kV_m(V_m - a - 1)).$$

If we assume equal anisotropy ratio of the intra- and extra-cellular media, it is well known that the bidomain equations can be reduced to the monodomain model. The resulting system consists of one nonlinear partial differential equation for the transmembrane potential V_m coupled with an ordinary differential equation for the ionic currents. The monodomain equations using Aliev-Panfilov model take the following form:

$$\begin{cases} \dfrac{\partial V_m}{\partial t} - \nabla \cdot (G \nabla V_m) = I_{ion}(V_m, W) \\ \dfrac{\partial W}{\partial t} = g(V_m, W), \end{cases} \tag{2}$$

Several time derivative discretization have been introduced for the bidomain model (see Ethier & Bourgault (2008), and Keener & Bogar (1998)). Also, the reader is referred to Belhamadia (2008b) for more discussion about different time schemes and their impact on twodimensional mesh adaptation. In this work, a fully implicit backward second order scheme (Gear) is employed as time discretization. Starting from V_m^{n-1} and W^{n-1} at time t^{n-1} and from V_m^n and W^n at time t^n, Gear scheme gives:

$$\frac{\partial V_m}{\partial t}(t^{(n+1)}) \simeq \frac{3V_m^{(n+1)} - 4V_m^{(n)} + V_m^{(n-1)}}{2\Delta t},$$

and

$$\frac{\partial W}{\partial t}(t^{(n+1)}) \simeq \frac{3W^{(n+1)} - 4W^{(n)} + W^{(n-1)}}{2\Delta t}.$$

The variational formulation of the system of nonlinear equation (1) is straightforward and obtained by multiplying this system by test functions $(\psi_v, \psi_\varphi, \psi_w)$ such that:

$$\left\{\begin{array}{l}\displaystyle\int_{\Omega}\frac{3V_m^{(n+1)}-4V_m^{(n)}+V_m^{(n-1)}}{2\Delta t}\psi_v\,d\Omega+\int_{\Omega}G_i\nabla V_m^{(n+1)}\cdot\nabla\psi_v\,d\Omega\\ \displaystyle+\int_{\Omega}G_i\nabla\phi_e^{(n+1)}\cdot\nabla\psi_v\,d\Omega=\int_{\Omega}I_{ion}(V_m^{(n+1)},W^{(n+1)})\psi_v\,d\Omega\\ \displaystyle-\int_{\Omega}(G_i+G_e)\nabla\phi_e^{(n+1)}\cdot\nabla\psi_\phi\,d\Omega=\int_{\Omega}G_e\nabla V_m^{(n+1)}\cdot\nabla\psi_\phi\,d\Omega\\ \displaystyle\int_{\Omega}\frac{3W^{(n+1)}-4W^{(n)}+W^{(n-1)}}{2\Delta t}\psi_w\,d\Omega=\int_{\Omega}g(V_m^{(n+1)},W^{(n+1)})\,\psi_w\,d\Omega.\end{array}\right. \tag{3}$$

Similarly, the variational formulation of the system of nonlinear equation (2) takes the following form:

$$\left\{\begin{array}{l}\displaystyle\int_{\Omega}\frac{3V_m^{(n+1)}-4V_m^{(n)}+V_m^{(n-1)}}{2\Delta t}\psi_v\,d\Omega+\int_{\Omega}G\nabla V_m^{(n+1)}\cdot\nabla\psi_v\,d\Omega\\ \displaystyle=\int_{\Omega}I_{ion}(V_m^{(n+1)},W^{(n+1)})\psi_v\,d\Omega\\ \displaystyle\int_{\Omega}\frac{3W^{(n+1)}-4W^{(n)}+W^{(n-1)}}{2\Delta t}\psi_w\,d\Omega=\int_{\Omega}g(V_m^{(n+1)},W^{(n+1)})\,\psi_w\,d\Omega.\end{array}\right. \tag{4}$$

In all numerical simulations, a quadratic (P_2) for spatial discretization and Newton's method are employed to solve the non linear system above at each time step. Linear system resulting from Newton's method is solved by iterative methods, an incomplete LU decomposition (ILU) GMRES solver Saad (1996) from the PETSc library Balay et al. (2003).

ADAPTIVE METHOD

As already mentioned, the accurate prediction of the depolarization-repolarization fronts in cardiac tissue is crucial. It is well known that a typical simulation of time-dependent cardiac electrophysiological waves using the whole heart may require about 10^7 grid points (see Cherry et al. (2003) and Ying (2005)), which leads to numerical challenges beyond the limit of the existing computational resources. To partially avoid these challenges, the mesh has to be adapted at each time step near the depolarization-repolarization fronts while coarser mesh are sufficient away from these fronts. This can be done with appropriate mesh adaptation techniques. In the context of the electrical wave of the heart, two different methods for estimating the error, depending on the dimension of the problem, have been introduced. A hierarchical error estimator described in Belhamadia (2008b) was used for a two-dimensional case, and an error estimator based on a definition of edge length using a solution dependent metric described in Belhamadia et al. (2009) was used for a three-dimensional case.

A brief description of adaptive methods for time dependent problems will now be presented. Only the case of the monodomain model will be presented and similar strategy can be presented for the bidomain model. The objective of this method is to build at each time step t^n a fine mesh in all regions where the variables V_m and W evolve (V_m, W and φ_e in case of the bidomain model) and a coarse mesh in these other regions. Therefore, an accurate solution is obtained and the total number of elements is greatly reduced at each time step. The overall adaptive strategy is the following:

1. Start from the solutions $V_m^{(n-1)}, V_m^{(n)}, W^{(n-1)}$ and $W^{(n)}$ and a mesh $M^{(n)}$ at time $t^{(n)}$;
2. Solve the system (2) on mesh $M^{(n)}$ to obtain a first approximation of the solutions (denoted $\tilde{V_m}^{(n+1)}$ and $\tilde{W}^{(n+1)}$) at time $t^{(n+1)}$;
3. Adapt the mesh on the two expressions $\dfrac{\tilde{V_m}^{(n+1)} + V_m^{(n)} + V_m^{(n-1)}}{3}$ and $\dfrac{\tilde{W}^{(n+1)} + W^{(n)} + W^{(n-1)}}{3}$ 3 to obtain a new mesh $M^{(n+1)}$;
4. Reinterpolate $V_m^{(n-1)}, V_m^{(n)}, W^{(n-1)}$ and $W^{(n)}$ on mesh $M^{(n+1)}$;
5. Solve the system (2) on mesh $M^{(n+1)}$ for V_m^{n+1} and W^{n+1} .
6. Next time step: go to step 2.

NUMERICAL RESULTS

The mechanism of ventricular fibrillation is believed to be produced by one or many spiral propagation waves in the myocardium. The reader is referred to Biktashev et al. (1999), Jalife (2000), and Panfilov & Kerkhof (2004) and the reference therein for a complete discussion. From the numerical point of view, there are many strategies to initiate a spiral wave (see Bourgault et al. (2003), and Ethier & Bourgault (2008)). In this section, the performance of the adaptive method will be presented. A two-dimensional problem representing the re-entrant waves will be presented using the monodomain and bidomain model.

Monodomain Model

This section is devoted to a test case using the monodomain model. The computational domain is the square [0, 100] × [0, 100]. Homogeneous Neumann conditions are imposed on all sides, and the following parameters values have been used:

$k = 8$	$a = 0.15$
$\epsilon = 0.002$	$\mu_1 = 0.2$
$\mu_2 = 0.3$	$G = 1$
$\Delta t = 0.5$	

Figure 1 presents the transmembrane potential V_m and the recovery variable W at the center of the computational domain as a function of time. The numerical solutions are obtained using adapted meshes with only an average of 5900 triangular elements leading to 23000 dof since we use quadratic (P_2) for spatial discretization. The total number of elements is reduced due to the use of the anisotropic adapted meshes. The reader is referred to Belhamadia (2008b) for more details about quantitative results and comparisons between structured and adapted meshes.

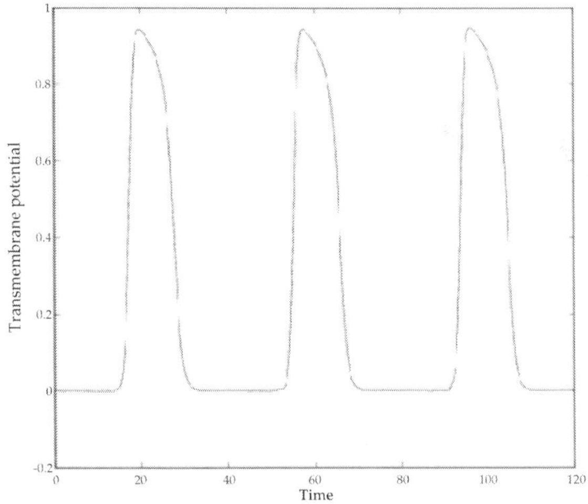

a) V_m vs time

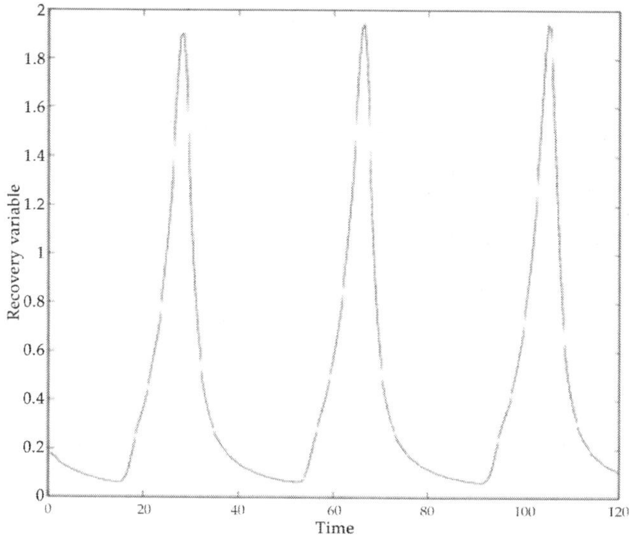

b) W vs time

Figure 1: Transmembrane potential and recovery variable at the point (50, 50, 50) as a function of time using adapted meshes.

Figure 2 a) b) shows the solutions V_m and W at time t = 8 t.u. while the adapted mesh at the same time is presented in figure 2 c). A close up view of the mesh on the interface is presented in figure 2 d). It is clearly shown that the mesh is refined only in the vicinity of the front position while keeping sufficient resolution in other regions. The gain in computational time is obvious using the adaptive method since the total number of elements is greatly reduced.

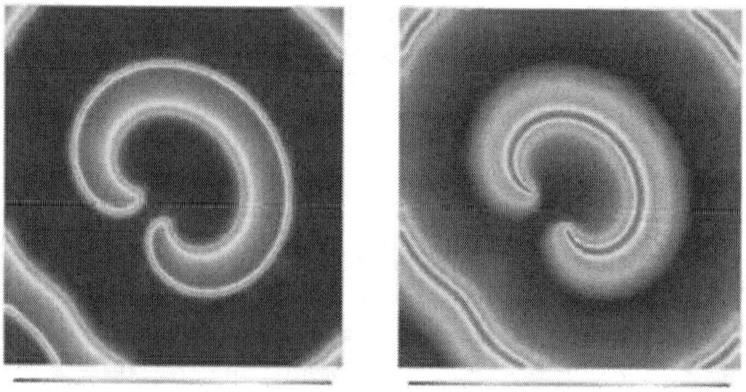

a) Transmembrane potential V_m b) Recovery variable W

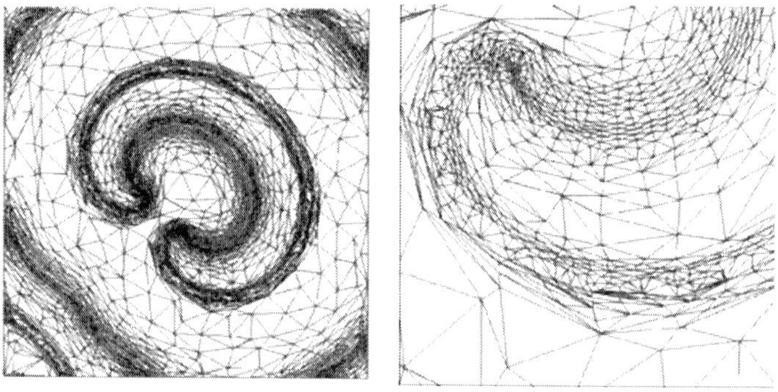

c) Adapted mesh d) Close up of the mesh

Figure 2: Numerical Solutions using adapted meshes: monodomain model case.

Bidomain Model

A test case using the bidomain model is now presented. The computational domain, the boundary conditions, and the physical parameters are the same as the previous section. However, the intra- and extra-cellular conductivity tensors are

$$G_i = \begin{pmatrix} 3 & 0 \\ 0 & 0.32 \end{pmatrix} \quad \text{and} \quad G_e = \begin{pmatrix} 2 & 0 \\ 0 & 1.37 \end{pmatrix}$$

As the previous section, the advantage of the adaptive method can be also illustrated in the case of unequal anisotropy ratios. The transmembrane potentials, and the recovery variable at the center of the computational domain as a function of time are similar to figure 1 and are not presented in this work to avoid a repetition. The numerical solutions are obtained using adapted meshes with only an average of 7100 triangular elements (29000 dof).

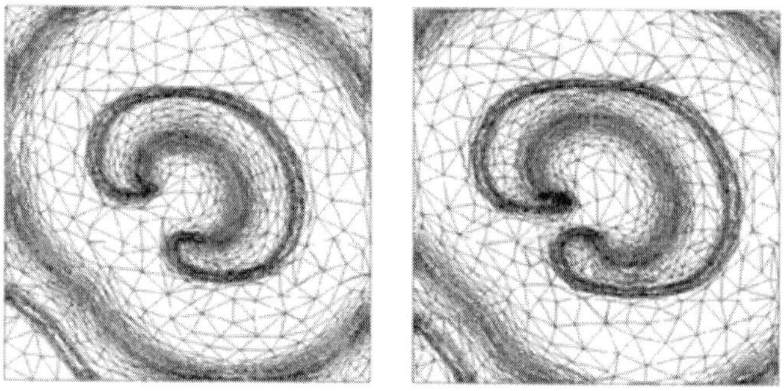

a) Adapted mesh at $t = 5$ t.u. b) Adapted mesh at $t = 10$ t.u.

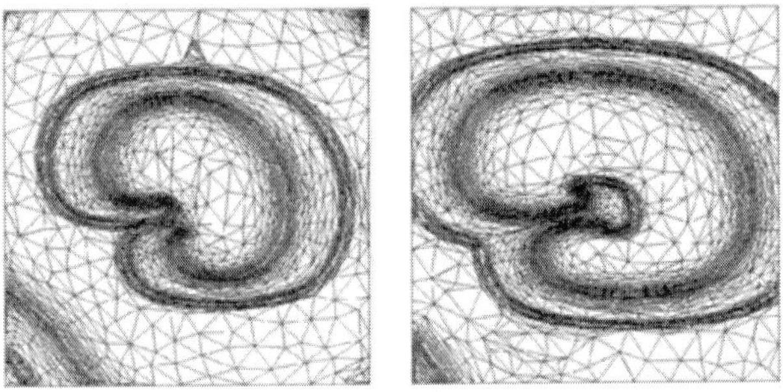

c) Adapted mesh at $t = 15 t.u.$ d) Adapted mesh at $t = 25$ t.u.

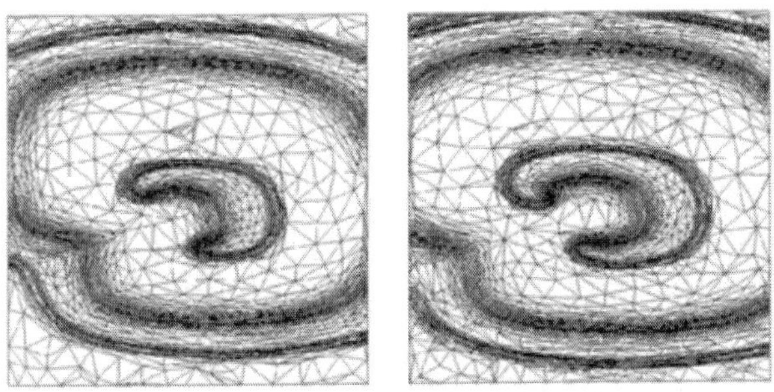

e) Adapted mesh at $t = 35$ t.u. f) Adapted mesh at $t = 40$ t.u.

Figure 3: Mesh evolution using the bidomain model.

Figure 3 illustrates the evolution of the adapted mesh. The front position is well captured and the solution seems uniformly accurate over time steps. Finally, the numerical solutions of the transmembrane potential, the extracellular potential and the recovery variable are shown in figure 4. As could be seen, the depolarization and repolarization fronts are smooth and well captured on the adapted anisotropic meshes.

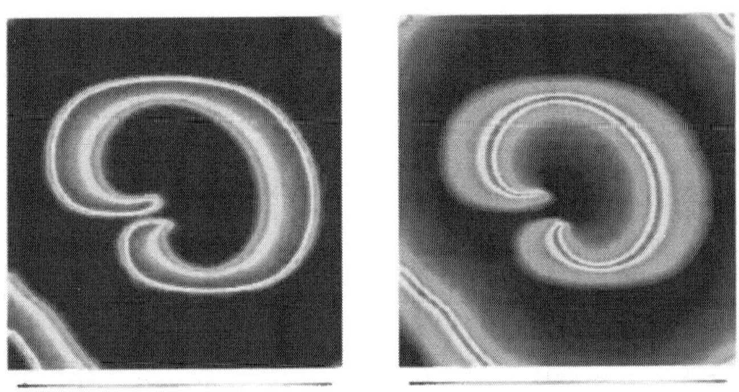

a) Transmembrane potential V_m b) Recovery variable W

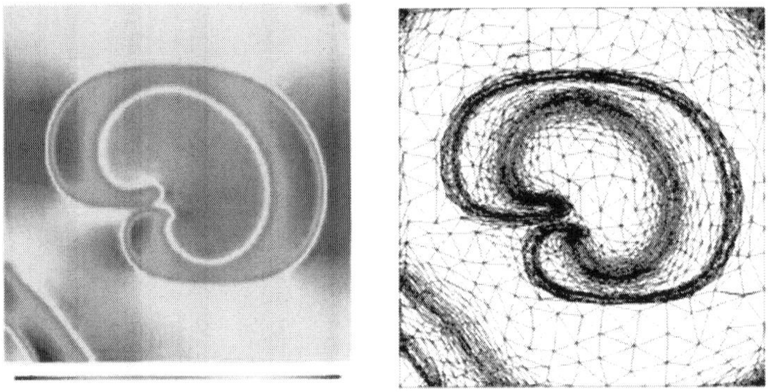

c) Extracellular potential V_m d) Adapted mesh at $t = 13$ t.u.

Figure 4: Numerical Solutions using adapted meshes: bidomain model case.

CONCLUSIONS

A recent numerical method for the transmembrane potential was presented. The accuracy of the method was obtained by using an anisotropic time-dependent adaptive method. A twodimensional problem representing the re-entrant waves was shown using the monodomain and bidomain model. Although only a two dimensional case was presented, this method is general and can be extended to three dimensional case. Results using realist heart geometry and with the monodomain model are recently presented in Belhamadia et al. (2009). The method proposed in this work uses two-variable ionic model. It will be interesting to see how the method performs with more complex ionic models.

ACKNOWLEDGMENTS

The authors acknowledge the financial support of NSERC.

REFERENCES

1. Aliev, R. & Panfilov, A. (1996). A Simple Two-Variable Model of Cardiac Excitation, Chaos, Solitons and Fractals 7(3): 293–301.
2. Balay, S., Buschelman, K., Eijkhout, V., Gropp, W., Kaushik, D., Knepley, M., McInnes, L. C., Smith, B. & Zhang, H. (2003). PETSc Users Manual, Technical Report ANL-95/11-Revision 2.1.6, Argonne National Laboratory, Argonne, Illinois. http://www.mcs.anl.gov/petsc/.
3. Belhamadia, Y. (2008a). An Efficient Computational Method for Simulation of Electrophysiological Waves, Conf Proc IEEE Eng Med Biol Soc. pp. 5922–5925.
4. Belhamadia, Y. (2008b). A Time-Dependent Adaptive Remeshing for Electrical Waves of the Heart, IEEE Transactions on Biomedical Engineering 55(2, Part-1): 443–452.
5. Belhamadia, Y., Fortin, A. & Bourgault, Y. (2009). Towards Accurate Numerical Method for Monodomain Models Using a Realistic Heart Geometry, Mathematical Biosciences . 220(2): 89-10.
6. Biktashev, V., Holden, A., Mironov, S., Pertsov, A. & Zaitsev, A. (1999). Three Dimensional Aspects of Re-Entry in Experimental and Numerical Models of Ventricular Fibrillation, Int. J. Bifurcation & Chaos 9(4): 694–704.
7. Bourgault, Y., Ethier, M. & LeBlanc, V. (2003). Simulation of Electrophysiological Waves With an Unstructured Finite Element Method, Mathematical Modelling and Numerical Analysis 37(4): 649–662.
8. Cherry, E., Greenside, H. & Henriquez, C. S. (2003). Efficient Simulation of Three-dimensional Anisotropic Cardiac Tissue Using an Adaptive Mesh Refinement Method, Chaos: An Interdisciplinary Journal of Nonlinear Science 13(3): 853–865.
9. Colli Franzone, P., Deuflhard, P., Erdmann, B., Lang, J. & Pavarino, L. F. (2006). Adaptivity in Space and Time for Reaction-Diffusion Systems in Electrocardiology, SIAM Journal on Scientific Computing 28(3): 942–962.
10. Colli Franzone, P. & Pavarino, L. F. (2004). A Parallel Solver for Reaction-Diffusion Systems in Computational Electrocardiology, Math. Models and Methods in Applied Sciences 14(6): 883–911.
11. Ethier, M. & Bourgault, Y. (2008). Semi-implicit time-discretization

schemes for the bidomain model, SIAM Journal of Numerical Analysis 46(5): 2443–2468.

12. Franzone, P. C. & Pavarino, L. F. (2004). A parallel solver for reaction-diffusion systems in computational electrocardiology, Mathematical Models and Methods in Applied Sciences 14(6): 883–912.
13. Jalife, J. (2000). Ventricular Fibrillation: Mechanisms of Initiation and Maintenance, Annual Review of Physiology 60: 25–50.
14. Karpoukhin, M., Kogan, B. & Karplus, J. W. (1995). The Application of a Massively Parallel Computer to the Simulation of Electrical Wave Propagation Phenomena in the Heart Muscle Using Simplified Models, HICSS 5: 112–122.
15. Keener, J. P. & Bogar, K. (1998). A numerical method for the solution of the bidomain equations in cardiac tissue, Chaos 8: 234–241.
16. Lines, G., Buist, M., Grottum, P., Pullan, A., Sundnes, J. & Tveito, A. (2003). Mathematical models and numerical methods for the forward problem in cardiac electrophysiology, Comput. Visual. Sc. 5: 215–239.
17. Lines, G., Grottum, P. & Tveito, A. (2003). Modeling the electrical activity of the heart: A bidomain model of the ventricles embedded in a torso, Comput. Visual. Sc. 5: 195–213.
18. Murillo, M. & Cai, X. (2004). A fully implicit parallel algorithm for simulating the non-linear electrical activity of the heart, Numerical Linear Algebra with Applications 11: 261–277.
19. URL: http://www3.interscience.wiley.com/cgi-bin/abstract/107632540/ABSTRACT Panfilov, A. V. & Kerkhof, P. (2004). Quantifying Ventricular Fibrillation: in Silico Research and Clinical Implications, IEEE Transactions on Biomedical Engineering 51(1): 195–196.
20. Pierre, C. (2006). Modlisation et Simulation de l'Activit lectrique du Coeur dans le Thorax, Analyse Numrique et Mthodes de Volumes Finis, PhD thesis, Universit de Nantes.
21. Saad, Y. (1996). Iterative Methods for Sparse Linear Systems, PWS Publishing Company.
22. Sundnes, J. (2002). Numerical Methods for Simulating the Electrical Activity of the Heart, PhD thesis, University of Oslo.
23. Sundnes, J., Lines, G. & Tveito, A. (2005). An Operator Splitting Method for Solving the Bidomain Equations Coupled to a Volume

Conductor Model for the Torso, Mathematical biosciences 194(2): 233–248.

24. Sundnes, J., Nielsen, B., Mardal, K., Cai, X., Lines, G. & A., T. (2006). On the Computational Complexity of the Bidomain and the Monodomain Models of Electrophysiology, Ann Biomed Eng 34(7): 1088–1097.
25. Trangenstein, J. & Kim, C. (2004). Operator Splitting and Adaptive Mesh Refinement for the Luo-Rudy I Model, Journal of Computational Physics 196(2): 645–679.
26. Weber Dos Santos, R., Plank, G., Bauer, S. & Vigmond, E. (2003). Preconditioning techniques for the bidomain equations, Proceedings of the 15th International Conference on Domain Decomposition Methods, Springer, Lecture Notes in Computational Science and Engineering (LNCSE)., pp. 571–580. URL: http://www.ddm.org/DD15/proceedings-dd15.html
27. Weber dos Santos, R., Plank, G., Bauer, S. & Vigmond, E. (2004). Parallel Multigrid Preconditioner for the Cardiac Bidomain Model, IEEE Trans. Biomed. Eng. 51(11): 1960–1968.
28. Ying, W. (2005). A Multilevel Adaptive Approach for Computational Cardiology, PhD thesis, Duke University, Durham, USA.

Chapter

13

NUMERICAL SIMULATION OF THE BLOOD FLOW THROUGH A BRAIN VASCULAR ANEURYSM WITH AN ARTIFICIAL STENT USING THE SPH METHOD

Leonardo Di G. Sigalotti[1], Jaime Klapp[2], Karla Pedroza[3], Edgar Nathal[4], Carlos E. Alvarado-Rodríguez[5]

[1]Area of Physics of Irreversible Processes, Department of Basic Sciences, Autonomous Metropolitan University-Azcapotzalco (UAM-A), Mexico City, Mexico.
[2]Physics Department, National Institute of Nuclear Research (ININ), Ocoyoacac, Mexico.
[3]Los Cabos H+ Hospital, San José del Cabo, Mexico.
[4]National Institute of Neurology and Neurosurgery "Manuel Velasco Suárez", Mexico City, Mexico.
[5]Departament of Chemical Engineering, DCNyE, University of Guanajuato, Guanajuato, Mexico

Citation: Sigalotti, L. , Klapp, J. , Pedroza, K. , Nathal, E. and Alvarado-Rodríguez, C. (2018), "Numerical Simulation of the Blood Flow through a Brain Vascular Aneurysm with an Artificial Stent Using the SPH Method". Engineering, 10, 891-912. doi: 10.4236/eng.2018.1012062.

ABSTRACT

We present numerical simulations of blood flow through a brain vascular aneurysm with an artificial stent using Smoothed Particle Hydrodynamics (SPH). The aim of this work is to analyze how the flow into an aneurysm changes using different stent configurations. The initial conditions for the simulations were constructed from angiographic images of a real patient with an aneurysm. The wall shear stresses, pressure and highest velocity within the artery, and other particular quantities are calculated which are of medical specific interest. The numerical simulations of the cerebral circulation help doctors to determine if the patient's own vascular anatomy has the conditions to allow arterial stenting by endovascular method before the surgery or even evaluate the effect of different stent structure and materials. The results show that the flow downstream the aneurysm is highly modified by the stent configuration and that the best choice for reducing the flow in the aneurysm is to use a completely extended Endeavor stent.

Keywords: Brain Vascular Flow, Aneurysms, Blood Flow, Particle Methods, Numerical Modeling

INTRODUCTION

Brain vascular pathologies are difficult to treat and have a broad field of research. A brain aneurysm is a vessel dilation caused by a weakness of the artery's wall, with risk of rupture or cerebral compression when reaching a big size. But even with the best technology many patients have lost the opportunity to cure because of a special manifestation of the disease. In these cases, simulation of the angioarchitecture is crucial for planning the treatment. The general incidence of bleeding related to an aneurysm rupture is 9.1/100,000 persons/year, but there are countries with two times more risk of bleeding like Japan with 22.7/100,000 persons/year [1] .

There are different shapes of aneurysms, but in general they can be classified as saccular and fusiform. The first one is the most common form in 90% of cases. Several studies have found that the localization, size and shape are important factors of rupture, a trial called ISUIA (International Study of Unruptured Intracranial Aneurysms). A critical review of the ISUIA found that aneurysms greater than 10 mm and located at posterior circulation have more risk of rupture [2] as well as a long perpendicular height (the measurement of the aneurysm height at a perpendicular direction from the center of the aneurysm neck to the aneurysm dome) [3] [4] . The standard treatment of brain aneurysm ruptured or unruptured are open surgery with clipping or revascularization and endovascular therapy that can be with coiling or stenting [5] [6] [7] . Most of the aneurysms that can be treated by endovascular methods are candidates to embolization with coils which are small pieces of platine that fill the aneurysm cavity avoiding the blood stress on the aneurysm wall. However, not all the aneurysms can be treated in that way. For instance, this is the case of fusiform aneurysms, of saccular aneurysms with wide neck and giant aneurysms that are very big to be filled by coils, which must be treated with surgery or with a special device called flow diverter, stents and low profile endovascular devices [8] [9] [10] . This kind of devices are like expandable cylinders put by endovascular access that cover the inner membrane of the vessels avoiding or lowering the flow of blood within the aneurysm that could eventually lead to thrombosis. The stents are built with nitinol, platinum-tungsten and cobalt, and chromium with diameters of 2.5 to 6 mm and lengths between 14 and 40 mm. Depending upon the reduction of the vessel diameter, the flow and the porosity-dependent pressure, they can be classified as laser?cut or braided devices. The stents are composed of a mesh that is organized following an architecture of open-cells or closed cell [11] (see Figure 1).

There are actually a lot of questions about the best materials to employ and the design of the endovascular devices for complex aneurysm treatment. Further developments should focus on improving the deployment and the best stent for every patient. On the other hand, the brain vascular computer simulations must reproduce the vascular tree or at least some of its parts in order to estimate the response to the treatment prior to the accomplishment of a modification of the cerebral vascular anatomy.

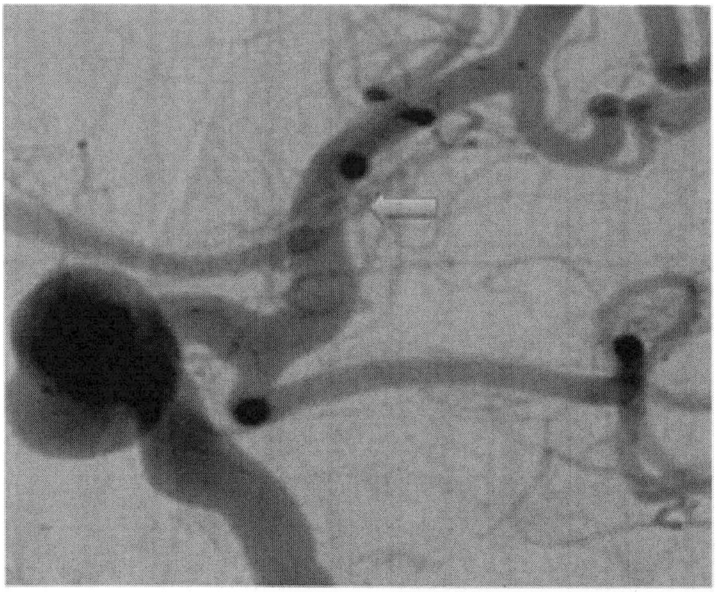

Figure 1: Stent image as seen in an angiogram. The blue arrow points to the position of the stent.

Virtual stenting may simulate the cerebral blood flow change in a vessel with aneurysmal pathology in a specific patient using the hemodynamic modification analysis with a specific device [12] as well as finite element methods [13] . In order to enhance prediction, the effects of blood viscosity, soft materials, thickness and the architecture of the complex network even in not completely expanded stents need be simulated. In this paper, the numerical simulations were performed using the method of Smoothed Particle Hydrodynamics (SPH). SPH is a particle-based, Lagrangian method used for simulating the mechanics of continuous media [14] . The basic idea behind SPH is very simple. It starts by reconstructing a continuous field at any point in space and time from a cloud of discrete particles, which are the material property carriers. This is accomplished by a three-dimensional interpolation of the scattered data by means of a kernel function [14] . Medical applications using the SPH method have started to appear very recently. For example, interactive blood simulations for virtual surgery using the SPH method have been performed by Muller et al. [15] , while SPH of left ventricular blood flow has been reported by Caballero et al. [16] . Although, computational hemodynamic analyses of endovascular devices have been performed using conventional methods [17] [18] [19] , so far no simulations with SPH exist in the literature. The initial conditions for the

SPH simulations are constructed using angiographic images of a real patient with an aneurysm. While this is the gold standard test for obtaining the best quality in angioarchitecture, such images can also be obtained from a CT scan or a MRI with 3D reconstruction [20] [21] , as shown in Figure 2.

The paper is organized as follows. A description of the numerical models and parameters are given in Section 2. Section 3 describes the main results, and Section 4 contains the relevant conclusions.

NUMERICAL MODELS AND PARAMETERS

The numerical simulation of blood flow through a brain vascular aneurysm with an artificial stent was performed using the SPH method [14] [22] .

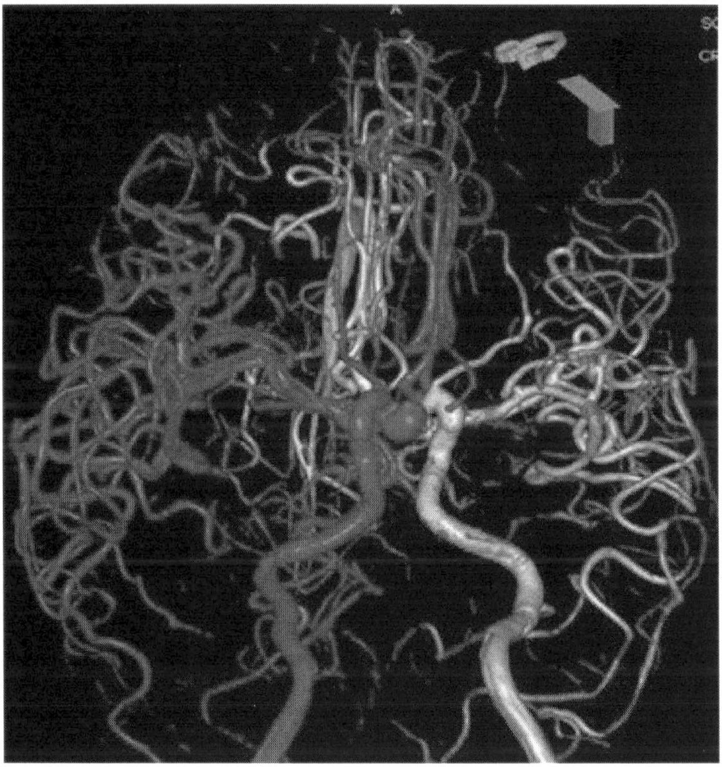

Figure 2: Patient angiography with 3D reconstruction.

The code solves the conservation laws of mass and momentum for a viscous fluid with prescribed boundary conditions. The discretization of the governing equations is performed over a cloud of scattered particles, which are the carriers of the fluid properties. For more details on the method, the reader is referred to Monaghan [14] and Liu and Liu [22] . The aim of this work is to analyze the hemodynamic flow as it goes into the aneurysm using different stent configurations from which it is possible to calculate the wall shear stress tensor, the pressure and the highest velocity within the artery as well as other particular quantities which may have a specific interest for medical evaluation.

The numerical simulations were performed using a modified version of the DualSPHysics code, which is described in detail by Crespo et al. [23] . One of the main advantages of the DualSPHysics code is its parallel structure. The code is written in the C++ language using the Open Multi-processing application that allows performing calculations in parallel using different numbers of cores according to the computer equipment used. In addition, there is a version written in CUDA (Compute Unified Device Architecture) that allows the use of GPU processors (Graphics Processor Unit).

Artery, Aneurysm and Stent Geometry

The geometry of the aneurysm was taken from the angiographic image of a real patient as illustrated in the Figure 3 from three different perspectives, front (YZ plane), side (XZ plane) and top (XY plane). A virtual file with this geometry was produced from the results of the angiography and exported to the SPH code to create the boundary conditions of the artery and aneurysm. Figure 4 shows the distribution of boundary SPH particles (right) employed to reconstruct the true artery and aneurysm geometries (left).

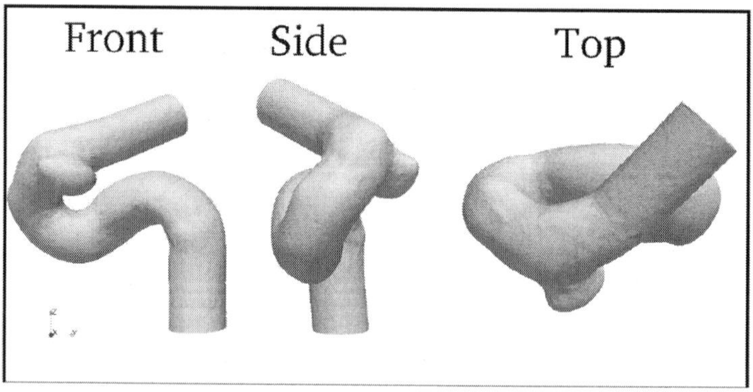

Figure 3: The aneurysm of a patient taken from an angiographic image. The image is shown from three different perspectives: front (YZ plane), side (XZ plane) and top (XY plane).

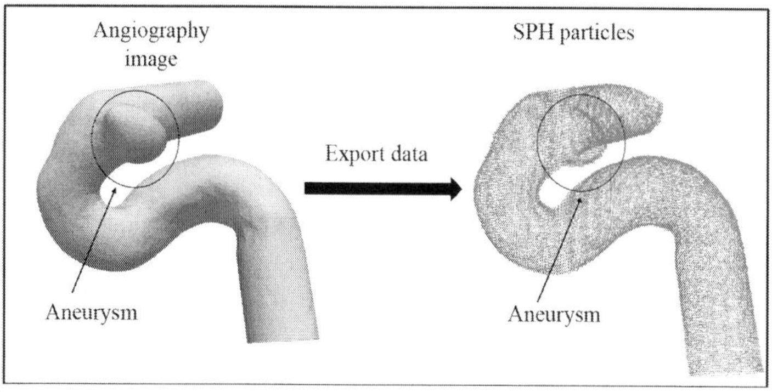

Figure 4: SPH particle reconstruction of the angiographic image of an aneurysm. The boundary of the artery is made up of SPH particles.

In the present simulations, the assumption is made that both the artery and the aneurysm are rigid bodies.

Two generic stent geometries were considered to analyze the flow behavior into the aneurysm when one stent is introduced into the artery. The first stent used in this study is equivalent to a second generation Endeavor stent for which the strut wall thickness is 0.09 mm. The geometry of the stent is shown in the left frame of Figure 5. The second stent was constructed by taking an octagonal geometry as the basis of the mesh. The geometry of the octagonal stent is depicted in the right frame of Figure 5. The stent model geometries were reproduced with the aid of the software Blender v2.67 and the stent insertion into the aneurysm was made using tools of the software Blender, which creates a file that is finally imported into the DualSPHysics code to reproduce numerically the geometry of both the stent and the artery using SPH particles.

The geometry from the angiographic image was used as the basis to model both stents. Four different cases of study are considered. In the first case, the Endeavor stent is extended completely into the artery, so that it almost fits the wall boundaries of the artery as shown in the left frame of Figure 6. The second case deals with the Endeavor stent not being completely extended and so it is slightly separated from the wall of the artery as shown in the right frame of Figure 6.

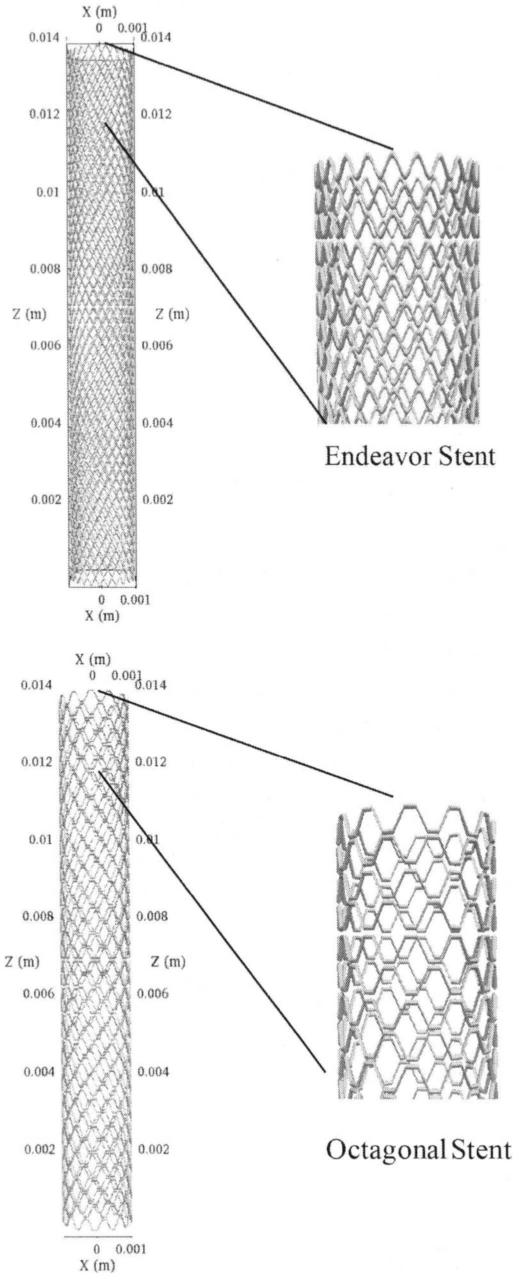

Figure 5: Geometry and dimensions of the Endeavor (left) and octagonal (right) stents as used in the numerical simulations.

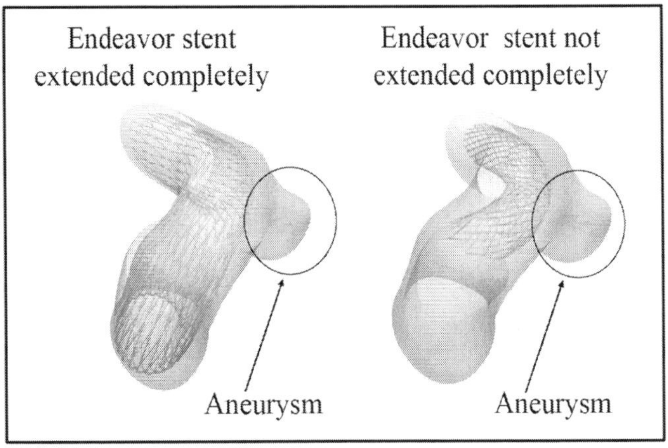

Figure 6: Completely (left) and partially extended (right) Endeavor stents within the artery as employed in the numerical simulations.

The third and fourth cases of study maps the geometry obtained from the angiographic image with an octagonal stent into the artery. In the third case, the octagonal stent is again extended completely so that the mesh fits the inner wall of the artery, while in the fourth case the stent is not extended completely and so it looks separated from the wall of the artery. These geometries are both displayed in Figure 7.

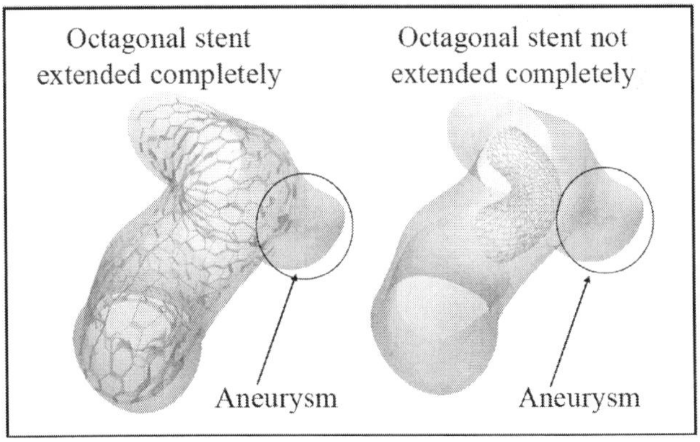

Figure 7: Completely (left) and partially extended (right) octagonal stents into the artery as employed in the numerical simulations.

Parameters

The parameters used in the numerical simulations are listed in Table 1.

The reference density corresponds to the average value of whole blood for a human, while v = 3.36 mPa, s is the blood dynamical viscosity at 27°C. The background pressure of 7500 Pa corresponds to the normal diastolic arterial pressure. The smoothing length is the width of the interpolating kernel function used in SPH and determines the spatial resolution of the calculation. For about half million particles a value of 1.732 mm is enough for resolving small features of the flow and representing the arterial irregularities with sufficiently good accuracy. The time duration of a simulation is typically 6.0 seconds. At the initial time (t = 0 s), the blood within the aneurysm is assumed to be at rest (i.e., v = 0).

Boundary Conditions

No slip boundary conditions are used at the solid wall of the artery, using the dynamic particle method implemented in the DualSPHysics code [23] . At the inlet of the aneurysm the flow velocity is set by a sequential flow rate characterizing a systole and diastole pulsatile flow wave in the cardiac cycle, the two mean phases of each heartbeat. The flow rate at the inlet corresponding to this pulsatile blood flow is shown in the right frame of Figure 8. Non-reflecting outlet boundary conditions, as were recently developed for pipe flow simulations by Alvarado-Rodríguez et al. [24] , were implemented at the outlet of the artery. This outflow boundary conditions are implemented using an outflow zone downstream of the outlet, where particles are moved using an outgoing wave equation for the velocity field so that feedback noise from the outlet boundary is greatly reduced. The inlet and outlet flow geometry is depicted in the left frame of Figure 8.

RESULTS

The results of the numerical simulations are divided into three parts. In the first part, we describe the results in terms of the blood flow rate at the inlet and outlet planes, the velocity at the center of the aneurysm and the velocity field through the artery for each case of study.

Table 1: Parameters used for the numerical simulations

Parameter	Value
Reference density (ρ_0)	1056 kg · m^{-3}
Dynamical viscosity (v)	3.36 mPa · s
Speed of sound (c)	40 m · s^{-1}
Background pressure (p)	7500 Pa
Smoothing length (h)	1.732×10^{-3} m
Gravity (g)	−9.8 m · s^{-2}
CFL number	0.2
Total number of particles	493,904

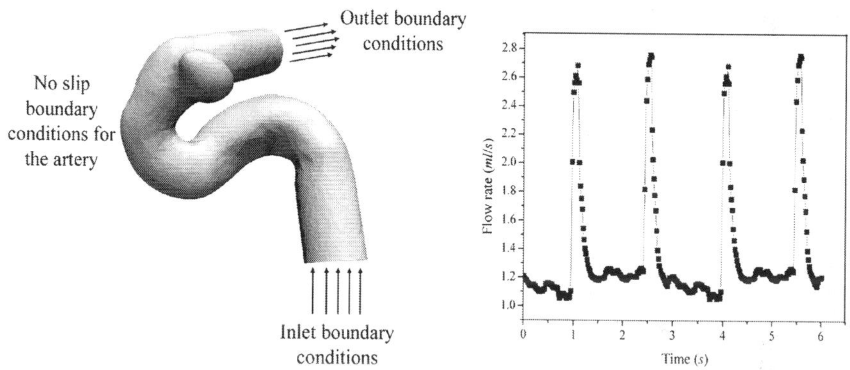

Figure 8: Boundary conditions. The inlet flow rate shown in the right frame was used as the inlet boundary conditions, while at the outlet of the artery fully non-reflecting boundary conditions are employed.

In the second and third parts, results are given for the pressure and the wall shear stress tensor, respectively. All the numerical results were obtained using the tools of DualSPHysics and the free software ParaView 5.1.2 for large data visualization.

Velocity and Flow Rates

The inlet flow rate profile is the same for all cases of study. However, the flow rate profile at the center of the artery before the stent zone is modified due to the presence of the stent. The flow rates are very similar to the inlet profile for all cases except for the case with the octagonal stent when it is extended completely. This is due to the geometry of the stent and its thickness. The thickness of the stent canalizes the flow in the center of the artery by a reduction of the transversal area, thereby increasing the flow velocity. The flow rates for each case of study are seen to differ at the zone just before the stent as shown in Figure 9. The flow rate was measured at an inner station of the artery as shown by the black point of coordinates (0, 0, 0.03).

At the outlet of the artery, the flow is also seen to be greatly affected by the stent configuration.

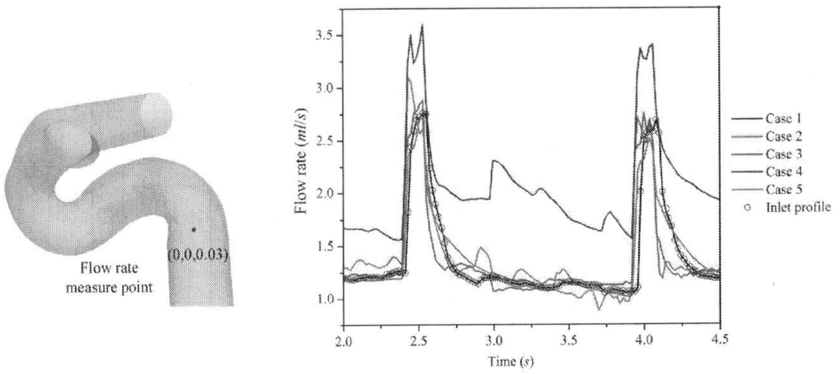

Figure 9: Flow rate profile within the artery before the stent zone (right), at the coordinate point shown in the left frame. The right frame shows how the flow rate is modified due to the stent in the artery.

The flow rate at the outlet of the artery should have similar characteristics to the inlet flow rate due to the artery being assumed to be a rigid body. However, the flow rate profile at the outlet, as measured at the center of the artery at the coordinate point (−0.03, −0.03, 0.125), is also modified by the stent configuration as shown in the right frame of Figure 10. In the simulation of the artery without stent (Case 1), the flow rate is slightly reduced by the presence of the aneurysm and the change of the transversal area of the artery, and in spite of this the flow looks very similar to that at the inlet. In all cases with stent, the flow rate at the outlet of the artery is seen to increase due to the reduction of the transversal area, while the reduction of the highest flow rates is also smoother in each cardiac cycle as shown in Figure 10.

One purpose of introducing the stent into the artery is to reduce the flow and pressure in the aneurysm. Therefore, it is important to compare the fluid velocity in the aneurysm for the cases when the stent is present to the case when the artery is free of stent. The numerical results indicate that the velocity into the aneurysm decreases by about 88.5% in the cases 2, 4 and 5 compared to case 1 with no stent. The fluid velocity into the aneurysm increases just after the first pulsatile wave. However, it is not seen to decrease in a similar way as occurs close to the inlet as can be observed by comparing Figure 9 and Figure 10. These results suggest that there exists a constant flow into the aneurysm caused by turbulence (see Figure 17) that keeps a constant move in the fluid when it is flowing in the artery. On the other hand, when a stent is introduced in the artery, strong changes in the velocity are seen to occur within the aneurysm. The pulsatile flow in the aneurysm appears in those cases where the stent is not completely extended and disappears when the stent is completely extended, implying a decrease of the velocity field by about 92.8% for case 2. In case 3, the velocity is reduced by about 71.4% due to the design of the stent and the amplitude of the octagonal mesh. In all cases, the fluid velocity and therefore the flow is reduced in the aneurysm when a stent is introduced into the artery. However, the features of the flow are different when the stent is completely extended or not.

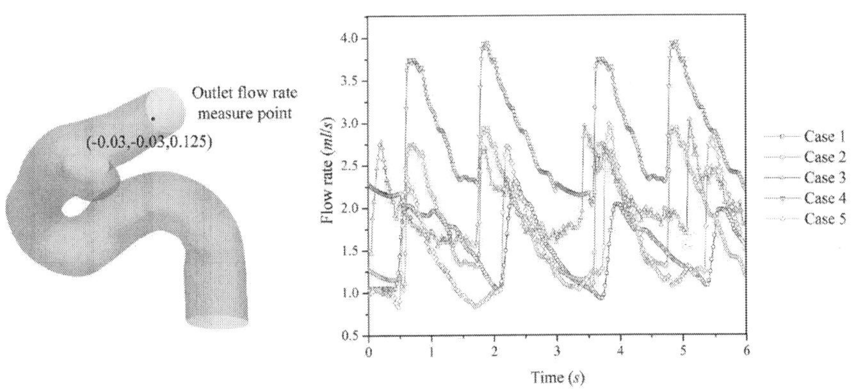

Figure 10: Flow rate profile at the outlet in the center of the artery. The flow is measured at the coordinate point shown in the left frame. The right frame shows how the flow rate is modified due to the stent in the artery.

Since the main purpose of introducing the stent is to reduce the flow in the aneurysm, the results indicate that the stronger reduction is obtained using a completely extended Endeavor stent into the artery. The details of the fluid velocity in the aneurysm are displayed in Figure 11 for all cases of study.

The velocity field within the artery at different times (from 1.0 to 2.5 s as in Figure 8) are shown in Figures 12-16 for cases 1, 2, 3, 4, 5, respectively. In all runs, the highest velocity always occurs near the center of the artery in accordance with the theory of fluid flow in a channel or pipe. When the stent is present, the highest velocity is achieved in the stent zone due to the reduction of the transversal area. The greater increase of the fluid velocity in the stent zone for case 3 is clearly shown in Figure 14. The average velocity of the fluid in the artery is about 35 cm/s. However, in the case where the Endeavor stent is not completely extended the fluid velocity at the center of the stent reaches a velocity as high as 1 m/s. In all cases, the stent is seen to canalize the blood flow, avoiding that it passes through the aneurysm. This is shown in Figures 14-16, where the velocity into the aneurysm is clearly reduced compared to the velocity in the artery. Compared to the cases with stents, the velocity is always greater and even it is seen to increase with time in the aneurysm when the artery is free of stent, as shown in Figure 12.

Figure 17 depicts the velocity vectors in the aneurysm for the case when the artery is free of stent. The formation of a vortical flow is evident, which then increases with time in each cardiac cycle as shown by the size of the vectors in the middle and right frames. Each frame corresponds to one cardiac cycle according to Figure 8. In contrast, when a stent is introduced in the artery, the vortical flow shown in Figure 17 disappears as it is evident from Figures 18-21 corresponding to cases 2 to 5, respectively.

Pressure

The pressure at the inlet of the artery just before the stent zone provides a measure of how the presence of the stent affects the artery internal pressure.

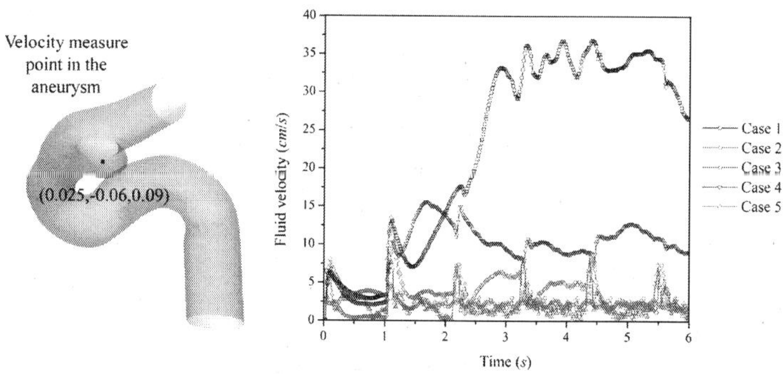

Figure 11: Flow rate profiles in the aneurysm as measured at the coordinate point depicted in the left frame. The right frame shows how the flow rate is modified due to the presence of the stent.

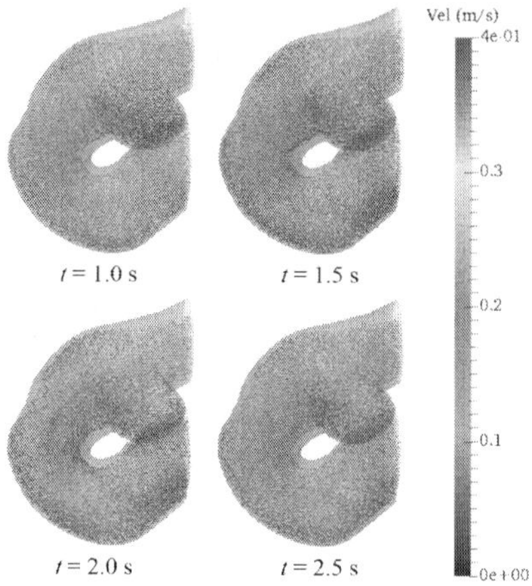

Figure 12: Velocity field in the artery without stent for one cardiac cycle from 1.0 to 2.5 s, for the inlet flow rate shown in Figure 8.

In all cases, the fluid internal pressure increases when the stent is inserted. If it is desired that the internal pressure keeps almost constant before and after the stent zone, then the Endeavor stent is the best option provided that it is completely extended. However, if it is not completely extended this stent represents the worst option because it produces the highest increase in pressure compared with all other cases. The pressure profiles at the inlet of the artery are shown in Figure 22. The pressure at the outlet of the artery provides information on the loss of pressure due to the stent configuration. For case 1, where the artery is free of stent no loss of pressure is seen because the flow is not hindered by an obstacle.

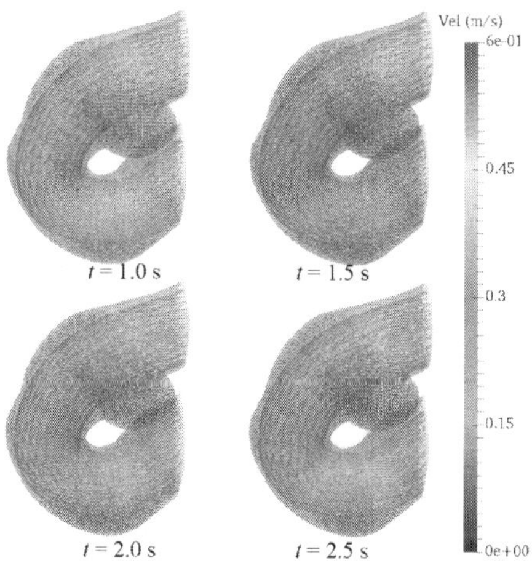

Figure 13: Velocity field into the artery with an Endeavor stent completely extended.

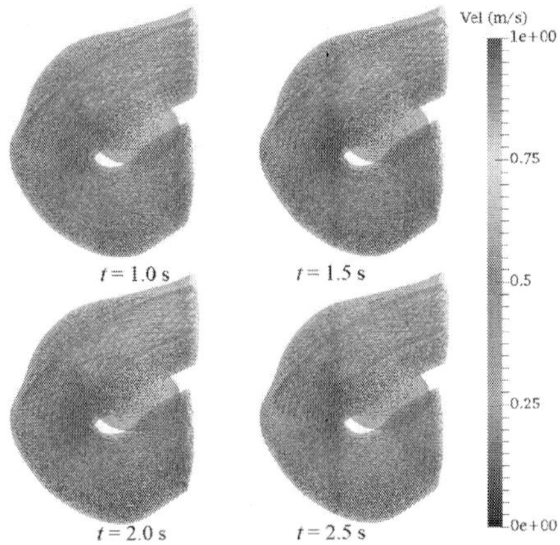

Figure 14: Velocity field into the artery with an Endeavor stent not completely extended.

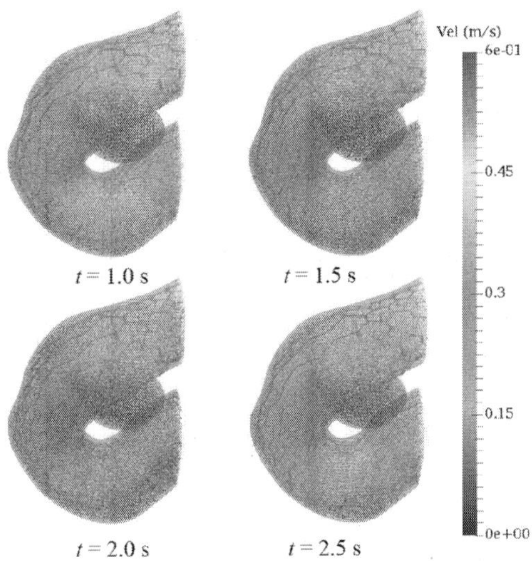

Figure 15: Velocity field into the artery with an octogonal stent completely extended.

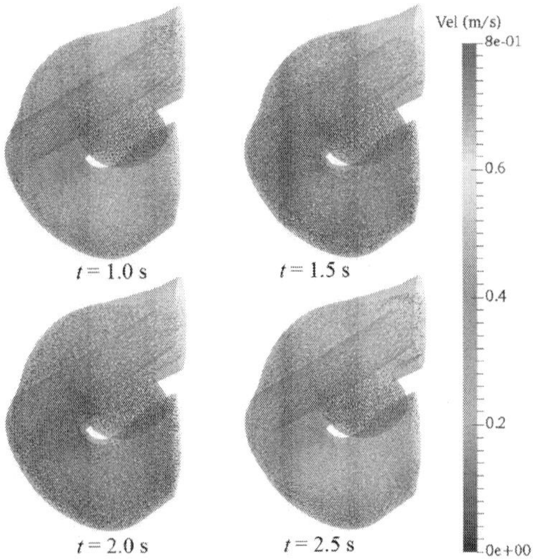

Figure 16: Velocity fields into the artery with an octogonal stent not completely extended.

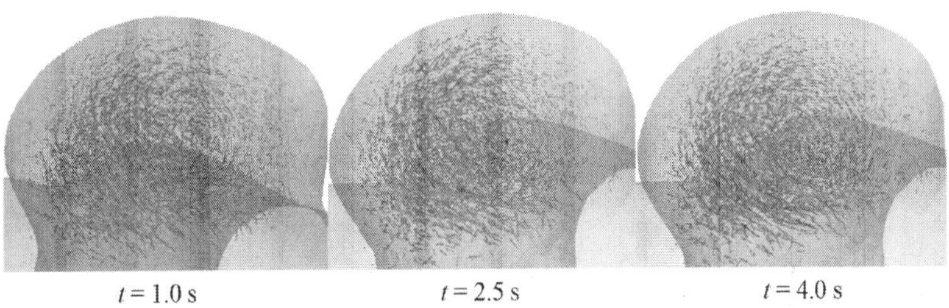

Figure 17: Vorticity field in the aneurysm when the artery is free of stent. The internal velocity increases with time as shown by the size of the velocity vectors.

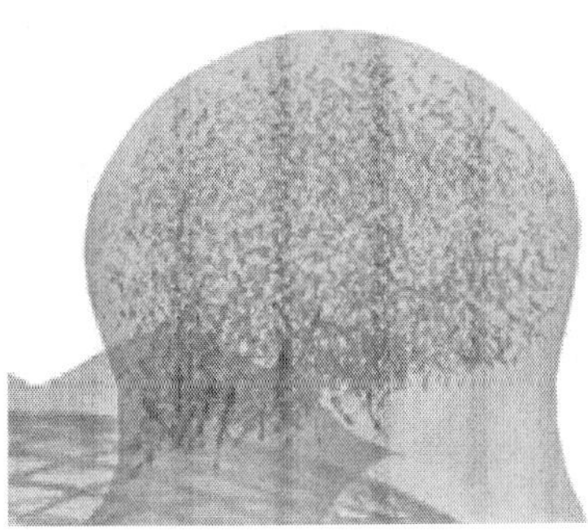

Figure 18: Velocity field in the aneurysm for case 2 at t = 2.5 s.

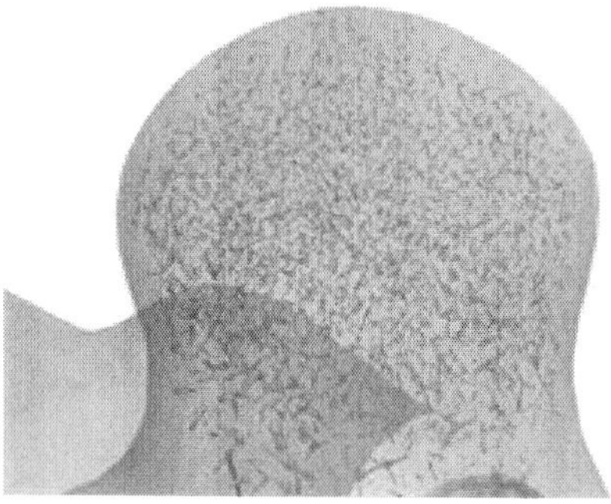

Figure 19: Velocity field in the aneurysm for case 3 at t = 2.5 s.

In case 2 with the Endeavor stent completely extended, the pressure loss is about 300 Pa due to the presence of the stent, while in case 3 with the Endeavor stent not completely extended, the pressure loss increases to about 489 Pa.

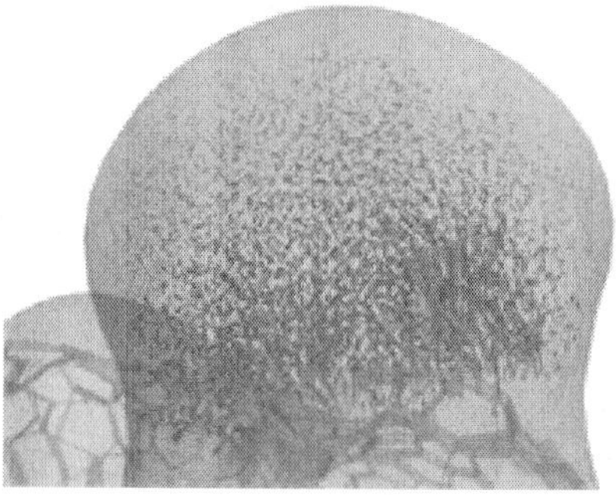

Figure 20: Velocity field in the aneurysm for case 4 at t = 2.5 s.

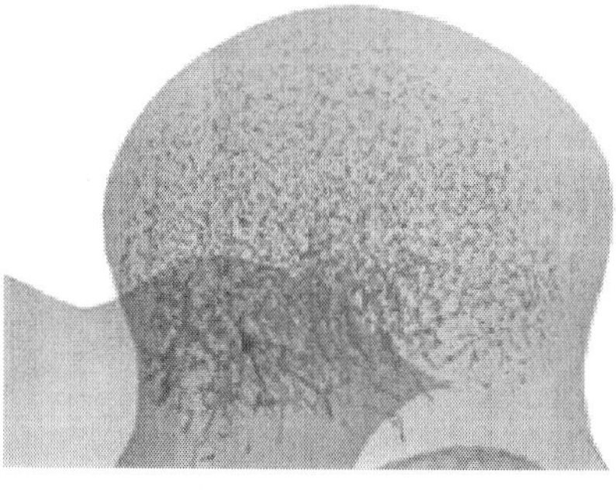

Figure 21: Velocity field in the aneurysm for case 5 at t = 2.5 s.

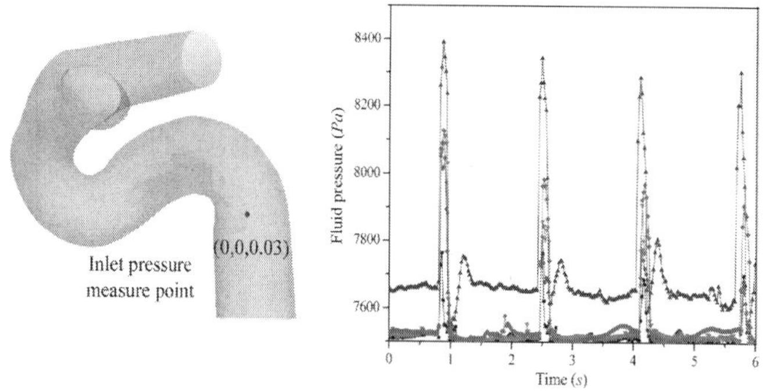

Figure 22: Internal artery fluid pressure just before the stent zone. In all cases, an increase of the pressure occurs when the stent is introduced in the artery.

For comparison, cases 4 and 5 have a pressure loss of about 523 and 572 Pa, respectively. The details of the fluid pressure at the outlet as a function of time are shown in Figure 23. The pressure in the aneurysm has a similar behavior to the pressure at the outlet even in the cases where the stent is present within the artery, which is in good accordance with Pascal's law [25] [26] . The fluid pressure profile into the aneurysm is shown in Figure 24. The resulting pressure field through the artery at three different times into a cardiac cycle is shown in Figures 25-29. The pressure fields shown in these figures are in good agreement with the results described in Section 3.1.

Wall Shear Stress Tensor

For the analysis of the wall shear stress tensor, we follow Tanemura et al. [27] for the calculation of the hemodynamic parameters from the simulated flow fields, namely the wall shear stress (WSS), the oscillatory shear index (OSI), the WSS gradient (WSSG), the relative residence time (RRT) and the flow velocity (FV).

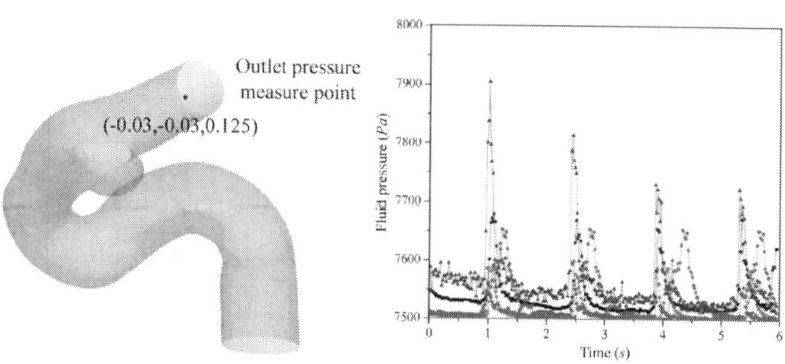

Figure 23: Internal artery fluid pressure at the outlet. In all cases, a pressure loss occurs in the presence of a stent, while no loss is observed without a stent.

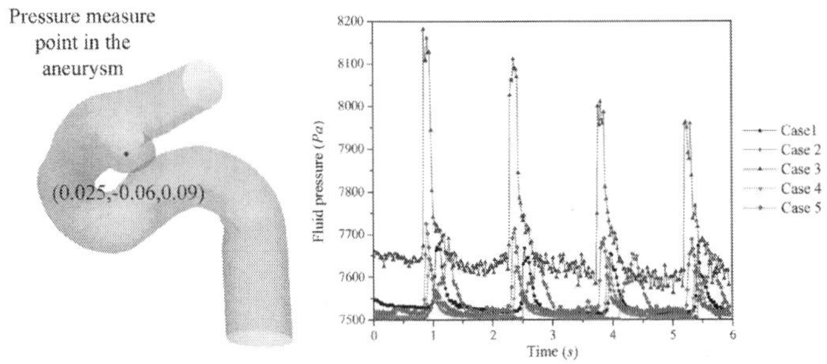

Figure 24: Arterial fluid pressure in the aneurysm. The pressure in the aneurysm has a similar behavior to the outlet pressure.

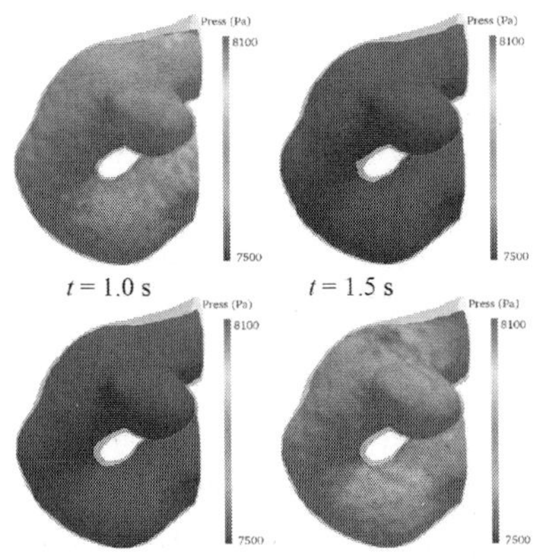

Figure 25: Pressure field in the artery for case 1.

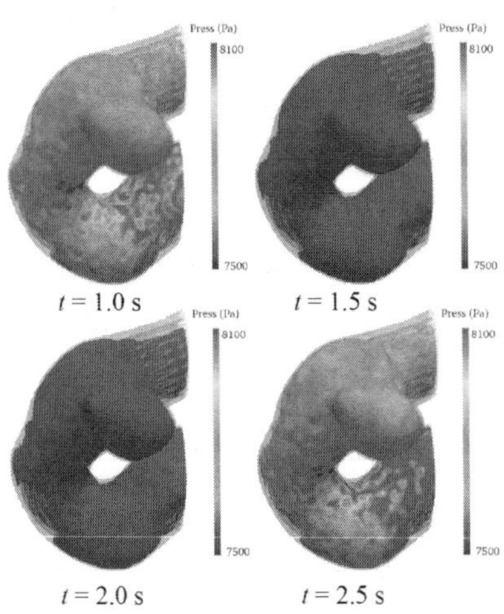

Figure 26: Pressure field in the artery for case 2.

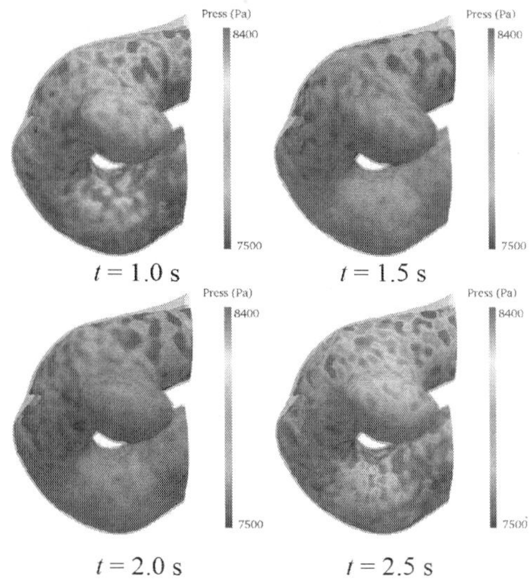

Figure 27: Pressure field in the artery for case 3.

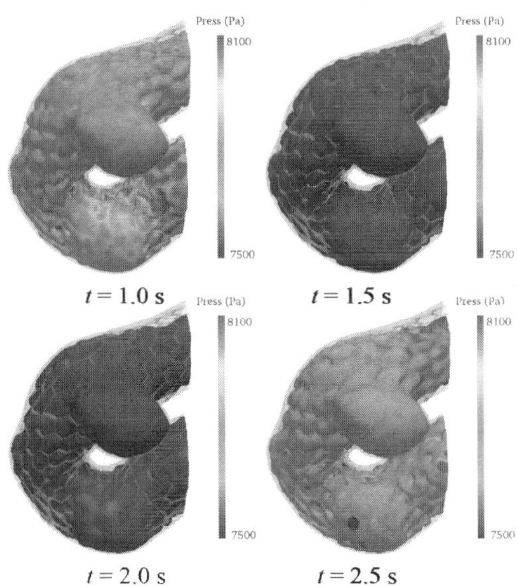

Figure 28. Pressure field in the artery for case 4.

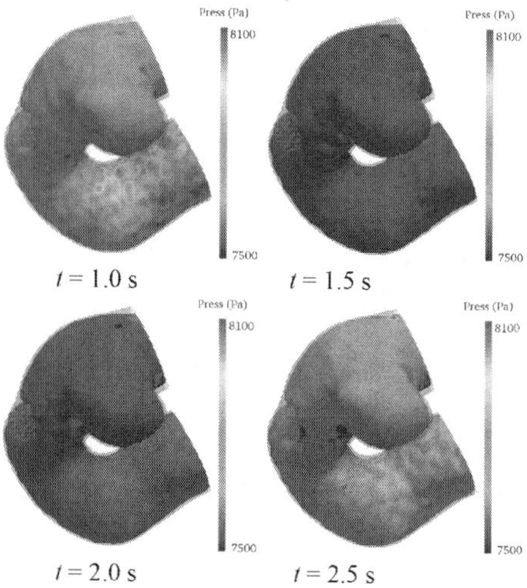

Figure 29: Pressure fields in the artery for case 5.

To evaluate the hemodynamic effect of the stent placement, these hemodynamic parameters were compared for identical models with and without stent. For the details of the calculation of the hemodynamic parameters we refer to Tanemura et al. [27] . The results listed in Table 2 were calculated from the numerical results in the aneurysm region. However, the WSS in the artery is shown in Figure 30 at t = 2.5 s. In all cases, the WSS is reduced when the stent is placed within the artery compared to the case when the artery is free of stent. The WSS value of 7.24 Pa predicted by our calculations for Case 1 with no stent compares qualitatively well with those reported by recent CFD simulations of blood flow in cerebral aneurysms, which range between 10.09 and 5.23 Pa [28] .

In order to analyze the flow of blood around the stent, a singular section was chosen at the inlet of the aneurysm. The singular section is shown in Figure 31 for a completely extended octagonal stent. The numerical results clearly show that a turbulent flow is formed when the blood is blocked by the stent in the streamwise direction. However, this regime becomes laminar just before to be blocked again by the next wall of the stent as is shown in Figure 32. This special case was run with 2 million particles and the same parameters listed in the Table 1.

Table 2: Hemodynamics parameters for each case of study

Parameter	Case 1	Case 2	Case 3	Case 4	Case 5
WSS (Pa)	7.24	3.83	11.1	2.60	0.866
OSI	0.16	0.208	0.269	0.066	0.187
WSSG (Pa/mm)	12.92	7.09	19.4	3.95	1.37
RRT (m^2/N)	0.306	0.671	2.93	0.666	2.77
FV (m/s)	0.589	0.100	0.100	0.263	0.082

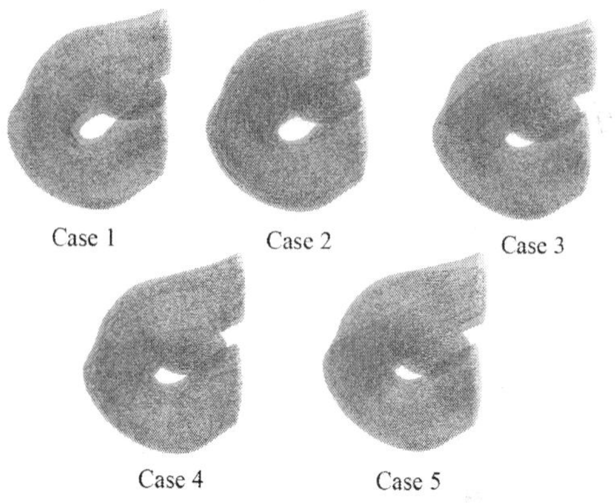

Figure 30: The WSS through the artery. The WSS is reduced into the aneurysm in all cases with stent. However, the value of the WSS increases around the artery in the stent zone.

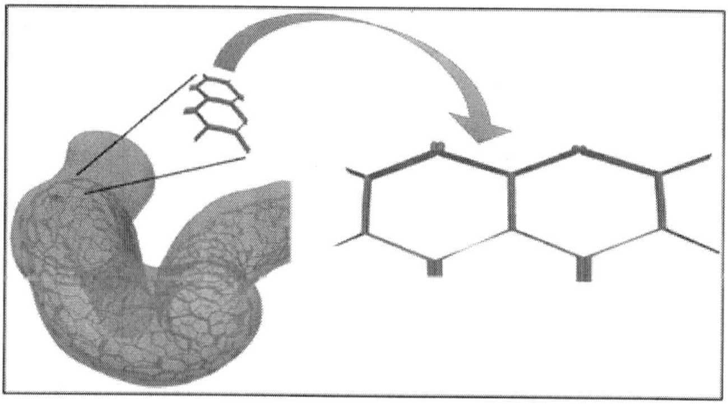

Figure 31: Isolated singular section employed to analyze the fluid flow through the mesh of the stent.

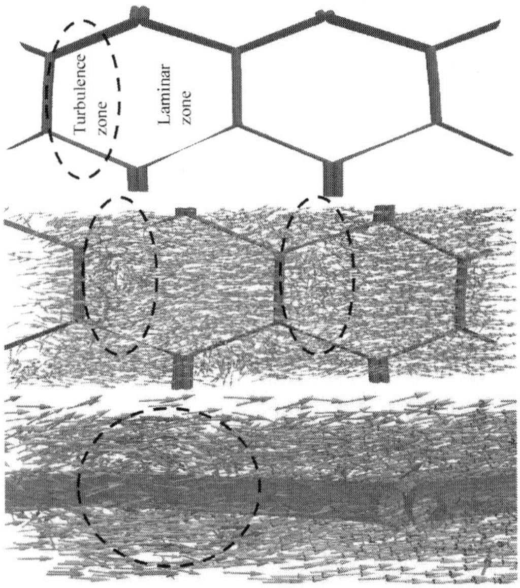

Figure 32: Velocity vectors in the isolated singular section shown in Figure 31.

CONCLUDING REMARKS

We have presented exploratory SPH calculations of blood flow through a brain vascular aneurysm using Smoothed Particle Hydrodynamics (SPH) methods. The initial conditions were constructed from angiographic images of a real patient with an aneurysm. The flow structure within the aneurysm was investigated in terms of the flow velocity, the internal pressure and the wall shear stresses for five model cases of flow with and without stent. The first case corresponds to flow without stent, while the other four cases correspond to flow in the presence of an Endeavor and an octagonal stent, when they are completely extended and partially extended within the artery.

The results indicate that the flow downstream the aneurysm is highly modified by the stent configuration. The thickness of the stent canalizes the flow within the artery due to a reduction of the transversal area, resulting in an increasing flow velocity and a pressure loss. Moreover, when a stent is introduced in the artery, significant changes of the velocity occur in the aneurysm. When the Endeavor stent is completely extended within the artery, the flow in the aneurysm is seen to decrease by more than 90% compared to about 70% for the case of an Endeavor stent not completely extended. Similar drops of the flow velocity were also observed for an octagonal stent. More importantly, the run with no stent predicted the development of turbulent flow within the aneurysm, which was also seen to increase with time. Conversely, this vortical flow was seen to strongly decrease and even disappear when a stent is inserted into the artery. On the other hand, the pressure losses across the aneurysm were seen to be slightly higher for the cases with an octagonal stent. Considering that the main purpose of introducing a stent is that of reducing the flow in the aneurysm, the results indicate that the best choice is to use an Endeavor stent completely extended into the artery. Moreover, if the internal pressure has to remain almost constant before and after the stent, the best option is again the Endeavor stent completely extended. However, if this stent is not completely extended then it becomes the worst option since it produces the highest increase in pressure compared to all other cases.

Future studies in this direction will include models of brain vascular flow with different parameters and different patient aneurysm images. Further testing of the code will also include direct comparison with true medical data.

ACKNOWLEDGEMENTS

We acknowledge the Universidad Autónoma Metropolitana-Azcapotzalco (UAM-A) for partial support.

REFERENCES

1. de Rooij, N.K., Linn, F.H.H., van der Plas, J.A., Algra, A. and Rinkel, G.J.E. (2007) Incidence of Subarachnoid Haemorrhage: A Systematic Review with Emphasis on Region, Age, Gender and Time Trends. Journal of Neurology and Neurosurgery Psychiatry, 78, 1365-1372. https://doi.org/10.1136/jnnp.2007.117655
2. Wiebers, D.O. (2003) Unruptured Intracranial Aneurysms: Natural History, Clinical Outcome, and Risks of Surgical and Endovascular Treatment. The Lancet, 362, 103-110. https://doi.org/10.1016/S0140-6736(03)13860-3
3. Dhar, S., Tremmel, M., Mocco, J., Kim, M., Yamamoto, J., Siddiqui, A.H., et al. (2008) Morphology Parameters for Intracranial Aneurysm Rupture Risk Assessment. Neurosurgery, 63, 185-197. https://doi.org/10.1227/01.NEU.0000316847.64140.81
4. Mocco, J., Brown, R.D., Torner, J.C., Capuano, A.W., Fargen, K.M., Raghavan, M.L., et al. (2018) Aneurysm Morphology and Prediction of Rupture: An International Study of Unruptured Intracranial Aneurysms Analysis. Neurosurgery, 82, 491-496. https://doi.org/10.1093/neuros/nyx226
5. Almefty, K. and Spetzler, R.F. (2014) Management of Giant Internal Carotid Artery Aneurysms. World Neurosurgery, 82, 40-42. https://doi.org/10.1016/j.wneu.2013.03.064
6. Hanel, R.A. and Spetzler, R.F. (2008) Surgical Treatment of Complex Intracranial Aneurysms. Neurosurgery, 62, 1289-1299. https://doi.org/10.1227/01.neu.0000333794.13844.d9
7. Kalani, Y.S.M., Zabramski, J.M., Hu, Y.C. and Spetzler, R.F. (2013) Extracranial-Intracranial Bypass and Vessel Occlusion for the Treatment of Unclippable Giant Middle Cerebral Artery Aneurysms. Neurosurgery, 72, 428-435. https://doi.org/10.1227/NEU.0b013e3182804381
8. Chalouhi, N., Jabbour, P., Tjoumakaris, S., Starke, R.M., Dumont, A.S., Liu, H., et al. (2013) Treatment of Large and Giant Intracranial Aneurysms: Cost Comparison of Flow Diversion and Traditional Embolization Strategies. World Neurosurgery, 82, 1-6.
9. Taussky, P., Tawk, R.G., Miller, D.A., Freeman, W.D. and Hanel, R.A. (2013) New Therapies for Unruptured Intracranial Aneurysms. Neurologic Clinics, 31, 737-747. https://doi.org/10.1016/j.ncl.2013.03.011

10. Auricchio, F., Conti, M., Morganti, S. and Reali, A. (2014) Simulation of Transcatheter Aortic Valve Implantation: A Patient-Specific Finite Element Approach. Computer Methods in Biomechanicsaand Biomedical Engineering, 17, 1347-1357. https://doi.org/10.1080/10255842.2012.746676
11. Mine, B., Bonnet, T., Vazquez-Suarez, J.C., Iosif, C. and Lubicz, B. (2018) Comparison of Stents Used for Endovascular Treatment of Intracranial Aneurysms. Expert Review of Medical Devices, 15, 793-805. https://doi.org/10.1080/17434440.2018.1538779
12. Song, Y., Choe, J., Liu, H., Park, K.J., Yu, H., Lim, O.K., et al. (2016) Virtual Stenting of Intracranial Aneurysms: Application of Hemodynamic Modification Analysis. Acta Radiologica, 57, 992-997. https://doi.org/10.1177/0284185115613653
13. Takizawa, K., Schjodt, K., Puntel, A., Kostov, N. and Tezduyar, T.E. (2012) Patient-Specific Computer Modeling of Blood Flow in Cerebral Arteries with Aneurysm and Stent. Computional Mechanics, 50, 675-686. https://doi.org/10.1007/s00466-012-0760-4
14. Monaghan, J.J. (1992) Smoothed Particle Hydrodynamics. Annual Review of Astronomy and Astrophysics, 30, 543-574. https://doi.org/10.1146/annurev.aa.30.090192.002551
15. Müller, M., Schirm, S. and Teschner, M. (2018) Interactive Blood Simulation for Virtual Surgery Based on Smoothed Particle Hydrodynamics. Technology and Health Care, 12, 25-31.
16. Caballero, A., Mao, W., Liang, L., Oshinski, J., Primiano, C., McKay, R., et al. (2017) Modeling Left Ventricular Blood Flow Using Smoothed Particle Hydrodynamics. Cardiovascular Engineering and Technology, 8, 465-479. https://doi.org/10.1007/s13239-017-0324-z
17. Babiker, M.H., Chong, B., Gonzalez, L.F., Cheema, S. and Frakes, D.H. (2013) Finite Element Modeling of Embolic Coil Deployment: Multifactor Characterization of Treatment Effects on Cerebral Aneurysm Hemodynamics. Journal of Biomechanics, 46, 2809-2816. https://doi.org/10.1016/j.jbiomech.2013.08.021
18. Ishizawa, K., Komori, T., Shimada, T., Arai, E., Imanaka, K., Kyo, S., et al. (2005) Hemodynamic Infarction of the Spinal Cord: Involvement of the Gray Matter plus the Border-Zone between the Central and Peripheral Arteries. Spinal Cord, 43, 306-310. https://doi.org/10.1038/sj.sc.3101671

19. Li, Z. and Kleinstreuer, C. (2005) Fluid-Structure Interaction Effects on Sac-Blood Pressure and Wall Stress in a Stented Aneurysm. Journal of Biomechanical Engineering, 127, 662-671. https://doi.org/10.1115/1.1934040
20. Long, Q., Xu, X.Y., Collins, M.W., Griffith, T.M. and Bourne, M. (1998) The Combination of Magnetic Resonance Angiography and Computational Fluid Dynamics: A Critical Review. Critical Reviews in Biomedical Engineering, 26, 227-274. https://doi.org/10.1615/CritRevBiomedEng.v26.i4.10
21. Qin, Y., Wu, J., Hu, Q., Ghista, D.N. and Wong, K.K.L. (2017) Computational Evaluation of Smoothed Particle Hydrodynamics for Implementing Blood Flow Modelling through CT Reconstructed Arteries. Journal of X-Ray Science and Technology, 25, 213-232. https://doi.org/10.3233/XST-17255
22. Liu, G.R. and Liu, M.B. (2003) Smoothed Particle Hydrodynamics: A Meshfree Particle Method. World Scientific, Singapore. https://doi.org/10.1142/5340
23. Crespo, A.J.C., Gómez-Gesteira, M. and Dalrymple, R.A. (2007) Boundary Conditions Generated by Dynamic Particles in SPH Methods. Computers, Materials & Continua, 5, 173-184.
24. Alvarado-Rodríguez, C.E., Klapp J., Sigalotti, L.Di G., Domínguez, J.M. and de la Cruz Sánchez, E. (2017) Nonreflecting Outlet Boundary Conditions for Incompressible Flows Using SPH. Computers & Fluids, 159, 177-188. https://doi.org/10.1016/j.compfluid.2017.09.020
25. Gotoh, H., Shibahara, T. and Sakai, T. (2001) Sub-Particle-Scale Turbulence Model for the MPS Method—Lagrangian Flow Model for Hydraulic Engineering. Computational Fluid Dynamics Journal, 9, 339-347.
26. Lo, E.Y.M. and Shao, S. (2002) Simulation of Near-Shore Solitary Wave Mechanics by an Incompressible SPH Method. Applied Ocean Research, 24, 275-286.
27. Tanemura, H., Ishida, F., Miura, Y., Umeda, Y., Fukazawa, K., Susuki, H., Sakaida, H., Matsushima, S., Shimosaka, S. and Taky, W. (2013) Change in Hemodynamics after Placing Intercranial Stents. Neurologia Medico-Chirurgica, 53, 171-178.

28. Shishir, S.S., Miah, Md. A.K., Islam, A.K.M.S. and Hasan, A.B.M.T. (2015) Blood Flow Dynamics in Cerebral Aneurysm—A CFD Simulation. Procedia Engineering, 105, 919-927. https://doi.org/10.1016/j.proeng.2015.05.116

Chapter 14

NUMERICAL MODELING OF SOIL WATER FLOW AND NITROGEN DYNAMICS IN A TOMATO FIELD IRRIGATED WITH MUNICIPAL WASTEWATER

Ali Erfani Agah

Department of Water Engineering, Faculty of Agriculture, Shahrood University of Technology, Shahrood, Iran

ABSTRACT

Because of water scarcity, reduction of annual rainfall and the use of wastewater in agriculture, there is a need for research to evaluate the potential impacts of using such sources on hydraulic soil properties and groundwater quality. Nitrate loss from the area under cultivation and regular use of fertilizer and wastewater is a major reason for non-point source

Citation: Ali Erfani Agah, "Numerical Modeling of Soil Water Flow and Nitrogen Dynamics in a Tomato Field Irrigated with Municipal Wastewater", IntechOpen Recent Advances in Numerical Simulations, 2021, DOI: 10.5772/intechopen.98487.

contamination on agricultural lands. Numerical model, Hydrus-1D used to simulate soil nitrate in soil cultivated with tomato-crop during the growing period, in North-East Iran. A randomized completely blocked design with five irrigation treatments with different sources of nitrogen was applied. Comparison between simulated and measured soil moisture content shows that the model can follow the temporal variation of soil water content. However, some over estimation of the measured data was observed during the simulation period. To evaluate the Hydrus model performance with respect to nitrogen transport and transformations, the simulated nitrogen concentrations (NH4-N and NO3-N) are compared for different treatments at different depths of soil profile, (7.5, 22.5, 37.5, 52.5 and 120 cm from soil surface). It takes about 4 days to convert 90% of urea into ammonium and it takes about 70 days to convert 90% of ammonium into nitrate. However, urea concentrations decreased with time between irrigations as a result of hydrolysis. As expected, at 3.73 days, the urea was concentrated near the surface, immediately after fertigation. Ammonium remained concentrated in the immediate in the top soil at all times for all treatments. There was only slight movement, because of soil adsorption and subsequent fast nitrification and/or root uptake. In contrast to ammonium, nitrate moved continuously downwards during the 28-day simulation period, as nitrate is not adsorbed, whereas denitrification was assumed negligible. Leaching percentages were smaller for nitrate wastewater compared to nitrate- fertilizer, and manure. Base on simulation results treated municipal wastewater by an aerated lagoon can be used as a valuable source of irrigation without causing contamination of groundwater.

Keywords: Irrigation, Wastewater, Tomato, Nitrate leaching, Hydrus-1D

INTRODUCTION

Irrigation with wastewater is one of the best options to reduce the stress on limited fresh water available today and to meet the nutrient requirement of crops. There is potential for these nutrients present in recycled water to be used as a fertilizer source when the water is recycled as an irrigation source for agriculture [1]. Nitrogen is a valuable nutrient contained in wastewater [2]. Various studies confirm that municipal wastewater can be useful as an additional water resource for irrigation [3, 4, 5, 6, 7]. Some researchers have shown that the best way to use wastewater after treatment is in agriculture [8].

Understanding the behavior of nitrogen in the soil system helps to maximize crop production while reducing the impacts of N fertilization on the environment. Nitrogen applied as fertilizer or wastewater may be: utilized and stored in the plant; stored as organic nitrogen in the soil; volatilized as ammonia, nitrogen gas or nitrous oxide; lost in runoff; or leached to the groundwater as nitrate [9, 10]. The main processes response for nitrogen transport and transformations in the soil are mass transport of inorganic nitrogen forms, commonly described by the general convection–dispersion equations and both chemical and biological reactions [11].

Nitrate is one of the nitrogen compounds most susceptible to leaching. Three kinds of soil transformation of the N contained in wastewater are important. The first of theses in mineralization:

$$\text{Organic} - \text{N} \rightarrow NH_4{}^{+} \quad (1)$$

Mineralization occurs in soil as microorganisms, both aerobic and anaerobic convert organic nitrogen to inorganic forms. After wastewater application to soil, organic N quickly converts to ammonium nitrogen and then to nitrate nitrogen [12]. The sequel to mineralization is nitrification:

$$NH_4{}^{+} + 3/2\text{O}_2 \rightarrow NO_2{}^{-} + \text{H}_2\text{O} + 2\text{H}^{+} \quad (2)$$

$$No_2{}^{-} + 1/2\text{O}_2 \rightarrow NO_3{}^{-} \quad (3)$$

Microbial activity is also responsible for the two steps of nitrification. Nitrosomonas convert ammonium to nitrite. The second step of nitrification occurs through Nitrobacter species, which convert nitrite to nitrate. This step rapidly follows ammonium conversion to nitrite, and consequently, nitrite concentrations are normally low in soils.

Another important nitrogen transformation in soils is denitrification.

Nitrate, which is the end product of the nitrification process in aerobic soils, it can undergo reduction to NO_2 and finally to N_2 when the soil oxygen content is low and decomposable organic materials are present to furnish energy for the process. The sequence of products is:

$$NO_3{}^{-} \text{---}> NO_2 \text{---}> NO \text{---}> \text{N}_2\text{O} \text{---}> \text{N}_2 \quad (4)$$

This process, which is done by a group of bacteria, is called desalination or denitrification. Denitrification occurs under oxygen-limiting conditions when anaerobic bacteria use nitrate in respiration in the presence of carbon sources such as organic matter. NO_2 and N_2 are both gaseous and emitted from the soil. Factors influencing denitrification control include oxygen restriction and the presence of organic matter. In the case of wastewater irrigation, only wastewater with a high BOD can be a source of organic matter for denitrification [12]. The amount of soil nitrogen losses through denitrification depends on the type of soil and irrigation management applied in the field and may vary between zero and 90% of applied nitrogen [12].

Mineralization and nitrification processes convert the organic N and NH_4^+ into NH_4^+ and NO_3^{2-} respectively, which are absorbed and utilized by crops and termed as available nitrogen [13, 14, 15]. Nitrate is highly mobile and leachable. It has been established that excessive application of nitrogen leads to nitrate pollution of groundwater and surface water [16, 17]. Leaching of NO_3^{2-} below the root zone can be affected by a range of factors, including fertilizer application rates and the timing of applications [18].

Computer models are tools used in science to approximate natural phenomena. Therefore, models that predict flow and transport processes in soils are increasingly being applied to address practical problems. The use of simulation models allows extrapolation, in time and space, of data from leaching experiments and monitoring studies. More recently, computer simulation tools have been applied to predict the fate and transport of contaminants for risk evaluation [19].

In this study we present results of field experimentation and numerical simulations on a loamy soil cultivated with tomato plants, which were used to evaluate the performance of the different component of water and nitrogen dynamic in the soil. Model parameters were either solely derived from laboratory measurements or optimized by the inverse simulation method. The objective of this study was to determine the difference in concentration of nitrate in soil water below the root zone (about 1.5(m) for plots treated with (1) municipal wastewater (2) manure and (3) commercial chemical fertilizers, using HYDRUS-1D model, [31] at the research station of Mashhad in north-east of Iran. Field data, collected on a loamy soil cultivated with tomato plants, were used to evaluate the performance of the different component of water and nitrogen dynamic in the soil.

MATERIAL AND METHODS

Site Description and Measurements

The weather data (daily maximum and minimum temperature, wind speed, humidity, sunshine hour and rainfall data) was collected from metrological station installed 2006 and 2007 at the Mashhad research station site, (36°

13′ latitude, 59°38′ longitude) in Northern east Iran. A soil profile pit was excavated to 120 cm depth and soil samples at different soil texture layers were sampled on 20 March 2009 before tomato sowing and basic properties, including soil water retention and saturated hydraulic conductivity were measured. The soil consists of heterogeneous layers with a deep groundwater ground water table (far below 80 m) and is characterized as sandy loam top soil (0–40 cm) over sandy clay loam (40–65) over sandy clay (65-120 cm). From the rooting depth (120 cm) of tomato crop 100 soil samples were collected and analyzed for various physical and chemical parameters before starting the experiment (Table 1).

Table 1: Physical and chemical properties of soil at initial condition

Physical properties of soil			
Parameters	**Soil layers, cm**		
	0–40	**40–65**	**65–120**
Clay %	11.1	19.4	34.9
Silt %	39.4	35.9	23.8
Sand %	50.5	44.7	42.3
Textural class	Sandy loam	sandy clay loam	sandy clay
Bulk density, g/cm^3	1.55	1.43	1.35
FC, (vol. %)	21.34	27.61	28.22
PWP, (vol. %)	7.18	9.87	11.18
Chemical properties of soil			
pH	7.54	7.54	7.54
EC (ds/m)	1.53	1.53	1.50
Organic carbon (%)	1.82	1.79	1.71
Organic matter, g/100 g soil	1.19	0.67	0.63
Total nitrogen, g/100 g soil	0.059	0.060	0.054
Available P (mg/kg)	54.2	54.2	48

TDR probes and ceramic cup tensiometers were installed at 0–20, 20–40, 40–60 and 60–100 cm soil layers in the investigated area. Water content measurements were taken daily starting in January 2009 and concluded in October 2010. TDR data will be used to assess estimate of shallow soil water content at soil profile. The irrigation scheduling was based on the soil moisture deficit in the root zone at each irrigation event (difference between root zone soil water at field capacity and at irrigation time) with intervals of 10 days. The characteristics of water and wastewater are summarized in Table 2. Total irrigation depth during this period was 23.9 cm. The rainfall at the same period was 11.48 cm and reached 42.84 cm for the whole simulation period of one year.

Table 2: Physico-chemical characteristics of water and treated wastewater

Parameter	Unit	Well Water	Wastewater	Standard value
PH	—	7.9	7.9	6–8.5
EC	dS/m	0.58	1.1	—
Na^+	mg/l	63.7	85.8	—
K^+	mg/l	2.3	24.6	—
Ca^{2+}	mg/l	24	36.5	—
Mg^{2+}	mg/l	15	18.2	100
NH_3-N	mg/l	—	32	
NO_3-N	mg/l	—	0	
Org – N	mg/l	—	14	
P	mg/l	0.18	2.8	—
Cl	mg/l	12.4	118	600
SAR	mg/l	2.45	2.87	—
B	mg/l	0.86	0.9	1
Alkanity	—	—	670	
Hardness	—	—	189	
Total N of Coliform	—	—	1000	—
BOD_5	mg/l		115	100
COD	mg/l		145	200

Experimental Design

Plots were irrigated with either well water or wastewater in a random complete block design (RCBD), with four replications according to the following treatments:

- T1 - Irrigation by treated wastewater during all growing season, (%100 wastewater).
- T2 - Alternate irrigation by treated wastewater and well water.
  - Alternate irrigation of tomato with wastewater and well water during the growing season, (%75 wastewater +%25 well water).
  - Alternate irrigation of tomato with wastewater and well water during the growing season, (%50 wastewater +%50 well water).

– Alternate irrigation of tomato with wastewater and well water during the growing season, (%25 wastewater +%75 well water).

- T3 -Irrigation with well water plus application animal manure.
- T4 - Irrigation with well water plus application of fertilizer.
- T5 - Irrigation with well water only.

To obtain these ratios were used in the operation of the irrigation turn. (Table 3).

Table 3: The proportions of water and wastewater

Irrigation turns	**Mixing ratios of wastewater**				
	1%	**2%**	**3%**	**4%**	**5%**
100s	**75**	**50**	**25**	**0**	
First	Wastewater	Well water	Well water	Well water	Well water
Second	Wastewater	Wastewater	Wastewater	Well water	Well water
Third	Wastewater	Wastewater	Well water	Well water	Well water
Fourth	Wastewater	Wastewater	Wastewater	Wastewater	Well water
Fifth	Wastewater	Well water	Well water	Well water	Well water
Sixth	Wastewater	Wastewater	Wastewater	Well water	Well water
Seventh	Wastewater	Wastewater	Well water	Well water	Well water
Eighth	Wastewater	Wastewater	Wastewater	Wastewater	Well water
Ninth	Wastewater	Wastewater	Wastewater	Well water	Well water
Tenth	Wastewater	Well water	Well water	Wastewater	Well water

The experiments were carried out on 20 plots, and each experiment included five irrigated furrows 4 m in width and 4.2 m in length (along the crop rows). Each plot consisted of 5 crop rows with a plant row spacing of 75 cm. The plots (T3) were grazing prior sowing with 3000 kg ha^{-1} or (3 kg m^{-2}) animal manure. The chemical analysis of animal manure has been showed in Table 4.

Table 4: Chemical analyses of animal manure

pH	**EC**	**N**	**P**	**K**	**OC**	**C:N**	**Fe**	**Mn**	**Cu**	**Zn**
—	ds/m^{-1}	%				—	Mgkg^{-1}			
7.73	13.6	2.4	1.02	0.81	61	25.4	1611	72	4	54

The plots (T4) were fertilized based on soil sample tests with 300 kg ha^{-1} or (30 gr m^{-2})of triple super phosphate, broadcast at seedbed preparation, and 110 kg ha^{-1} of net nitrogen or (200 kg ha^{-1}) of urea at tillage time (at two equal section) and 6 weeks after panting. In this study, the total manure was applied prior sowing and for chemical fertilizer, 50% N and total P fertilizers were applied to the sowing seeds. Tomato was seeded on first week of May of each year in the plots at a plant spacing of 75 cm; Weed, diseases and insect control were uniformly managed during the growing season. After planting, irrigation was applied as required with well water until green stage and then treatments and irrigation applied as required during the growing season.

Model Selection: HYDRUS 1D

The HYDRUS-1D software package uses numerical methods to solve the Richards' equation for saturated–unsaturated water flow and the convection–dispersion equation for solute transport [20]. In this study, we used HYDRUS-1D to analyze water flow and nitrogen transport through tomato field irrigated with wastewater and soil surface management strategies. The measured data used are taken from completed research projects in field study. The data measurements were realized by [3, 21, 22] and were combined with additional measurements. Before simulation, the model was calibrated with field data.

Boundary Conditions

As the all the plots were at field capacity during the transplantation, therefore, the initial condition for volumetric soil water content was between 0.2–0.3 $cm^3\ cm^{-3}$ for all simulations. The upper boundary soil condition was the atmospheric boundary with a surface layer at which rainfall and evaporation occurred. The upper and lower soil boundary conditions (BC) for solute transport were considered as flux BC and zero concentration gradient.

Soil Hydraulic Properties

In this model, soil hydraulic properties, concerning soil moisture retention characteristics, $\theta(h)$, and saturated hydraulic conductivity, $K_{sat,}$ were measured in the field. The parameters of the [23] model were evaluated by fitting on $\theta(h)$ data using the Curve RETC code. The average values of Van Genuchten parameters for study at different soil depths are given in Table 5.

Table 5: Parameters of Van Genuchten equation for the soil moisture retention characteristics and the hydraulic conductivity function

Depth	θ_r	θ_s	a	n	m	K_{sat}
(cm)	($cm^3\ cm^{-3}$)	($cm^3\ cm^{-3}$)	cm^{-1}	—	—	cm $hr.^{-1}$
25	0.067	0.412	0.0073	1.86	0.414	3.896
60	0.095	0.375	0.0075	1.317	0.5310	2.160
85	0.185	0.421	0.0068	1.657	0.3916	0.131
110	0.048	0.473	0.0298	1.751	0.4259	18.922

Note: Van Genuchten model:θ(h)=θr+(θs−θr)/[1+(a|h|)n]mθ(h)=θr+(θs−θr)/[1+(a|h|)n]m, m = 1−1/n.

$K(h) = KsatSe^{1/2}[1\text{-}(1\text{-}Se^{1/m})^m]^2$ with $Se = (\theta - \theta_r)/(\theta_s - \theta_r)$, θ_s and θ_r are saturated and residual water content, respectively, K_{sat} is saturated hydraulic conductivity, a and n empirical parameters.

Parameter values

Initial soil water contents for tomato in different soil depths were 0.20–0.30 $cm^3\ cm^{-3}$ (giving a mean value of 0.27 $cm^3\ cm^{-3}$). Transport parameters were the model inputs. They were modified to calibrate the model. The modified longitudinal dispersivity and molecular diffusion coefficients of NO_3–N in free water (Do) were used as/set at 1.0 cm and 1.65 $cm^2\ d^{-1}$), respectively. Urea and $NO3^-$ were assumed to be present only in the dissolved phase (i.e., Kd = 0 $cm^3\ g^{-1}$) soil. The first-order decay coefficient, μ, for urea, representing hydrolysis, was set at 0.38 day^{-1}. Again, similar values were used in the literature, for example by [24] and by [25] who all considered hydrolysis to be in the range of 0.36 or 0.38 to 0.56 day^{-1}. Nitrification from NO_4^+ to NO_3^- was modeled using the rate coefficient of 0.2 day^{-1}, which represents the center of the range of values reported in the literature, e.g., 0.2 day^{-1} ([24, 26], 0.02–0.5 day^{-1} [27], 0.226–0.432 day^{-1} [28], 0.15–0.25 day^{-1} [25], and 0.24–0.72 day^{-1} [29]. It is further assumed that the maximum rooting depth increases logistically with time (increases from 2 cm at germination at 60 cm at harvest), and that there is an exponential root distribution with depth. Uptake of nitrate and ammonium is by passive uptake only. That means that, e.g. NO_3-N uptake at a given time and depth is equal to the water uptake multiply nitrate concentration in neglected. But assume that the maximum allowed concentration for solute is (50 ppm N 550 mgN/L). Assume the soil profile is initially solute free.

Model Testing

The model was evaluated by comparing measured and simulated values over time and depth using both qualitative and quantitative procedures. The qualitative procedures consisted of visually comparison between measured and simulated values over time and depth. For quantitative procedures, statistical analysis were used to calculate the average error (AE), the root mean square error (RMSE), the coefficient of residual mass (CRM), and the modeling efficiency (EF) between the measured and simulated values of water content in the soil during the study period [30, 31, 32].

4The (AE) is the average difference between the simulated and the measured values. The AE with a positive or negative sign indicates whether the model tends to overestimate or underestimate the measured values. The RMSE statistical index shows the mean difference between simulated and observational data. The RMSE coefficient is equal to the variance of the remaining error and the lower the value, the higher the accuracy of the model. In the Nash-Sutcliffe criterion (EF), the numerical value of one indicates the complete conformity of the simulated and observational data. The CRM also shows the difference between experimental and estimated values. Positive CRM values indicate that the proposed model estimates the values less than its actual value, and vice versa. In the most optimal case, the RMSE and CRM values are equal to zero, in which case the proposed model estimates the values with the highest possible accuracy. Wilmott agreement statistical index (d) with a value it is between zero and one that the value of one indicates the best fit. The value of Willmott's index (d) reflects the degree of agreement, and $d = 1$ indicates perfect agreement between the measured and simulated values.

The closer the calculated values are to zero, the better the approximation of the simulated data to the field data [33]. The optimum values of AE, RMSE, EF and CRM criteria are 0, 0, 1, and 0, respectively. Positive values of CRM indicate that the model underestimates the measurements and negative values for CRM indicate a tendency to overestimate them. If EF is less than zero, the models' predicted values are worse than simply using the observed mean. The average error and root mean square error are calculated as outlined in [33]:

The average error is defined as:

$$Averageerror\,(AE) = \frac{\left(\sum_{i=1}^{n} S_i\text{-}Q_i\right)}{n} \tag{5}$$

The Root Mean Square Error is defined as:

$$Rootmeansquareerror\,(RMSE) = \sqrt{\frac{\left(\sum_{i=1}^{n} S_i - Q_i\right)^2}{n}} \tag{6}$$

The Nash-Sutcliffe Efficiency is defined as:

$$EF = \left[1 - \frac{\sum_{i=1}^{n} (S_i - O_i)^2}{\sum_{i=1}^{n} \left(O_i - \overline{O_i}\right)^2}\right] \tag{7}$$

The coefficient of residual mass (CRM) is defined as:

$$CRM = \left[\frac{\sum_{i=1}^{n} (Q_i - S_i)}{\sum_{i=1}^{n} (O_i)}\right] \tag{8}$$

Where n is the number of observations, $\overline{O_i}$ is the average of the observed values and S_i and O_i are the simulated and measured values, respectively [34].

RESULTS AND DISCUSSION

Water Flow Simulation

Simulated soil water content in the soil profile are shown in Figure 1. However, the results in this section are presented to have an idea about the water regime in the soil profile with respect to the day of planting and harvesting the tomato. There were several rainfall events during the simulation period; however, more rainfall events were registered in the first part of the simulation period. As a result, the soil water content showed at all depths fluctuated more frequently in the first part of the simulation compared to the second and third part.

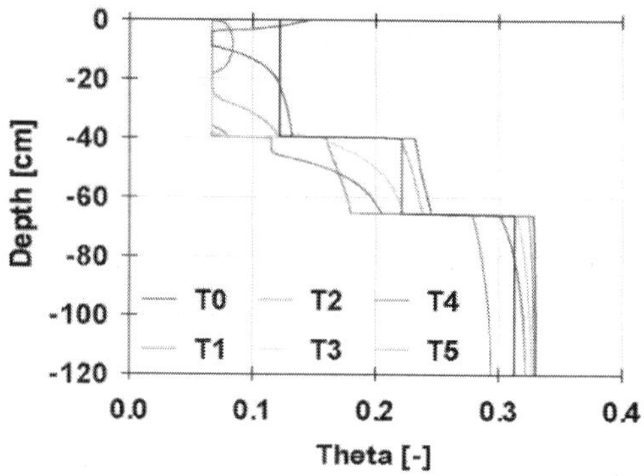

Figure 1: Simulated soil water content at different times of the experiment in the soil profile. With (T0=50 days, T1=150 days, T2=200 days, T3= 240 days, T4=300 days and T5=339 days).

HYDRUS 1-D was also compared to the water content from field data collected from each treatment (by TDR) and simulated data for the soil profile over the growing season. The following observations are based on visual assessment of model fit compared with observed values of moisture contents of soil. The simulated and measured water contents at 20, 40, 60 and 100 cm are shown in Figure 2A and B, Figure 3C and Figure 4D, respectively. The predicted water contents at 20 cm depth agree well with the measured values during growing season. The simulation closely match the measured moisture dynamics, except in the (wet) spring and winter of 2010 when the model at times underestimates the soil water content in the top soil. The simulated water contents did not agree well with the measured data at 40, 60 and 100 cm depths, the response of the model was lower than measured, especially deeper in the soil profile. At all depths, a close agreement between the measured and simulated data was registered during (wet) winter period. The difference between simulated and measured water contents varied with depth from −0.045 to 0.152 $cm^3\ cm^{-3}$. For the deeper positions (40, 60 and 100 cm), the model systematically underestimates the measured water content by 0.04 to 0.110 cm^3/cm^3 over the entire growing season at deeper depths, potentially due to under-estimation in the amount of free drainage and an over-estimation of the soil porosity, although the dynamics (water depletion in summer, replenishment in winter) is well simulated. Given that the underestimation is not just limited to the growing season, but is also evident in winter periods when there is little evapotranspiration and the entire soil profile is draining suggests that the problem is not with the crop parameters or evapotranspiration, but rather with the soil hydraulic properties of the deeper soil horizons: the parameters of the van Genuchten-Mualem K-h-θ relationship control the equilibrium water contents in winter ('field capacity').

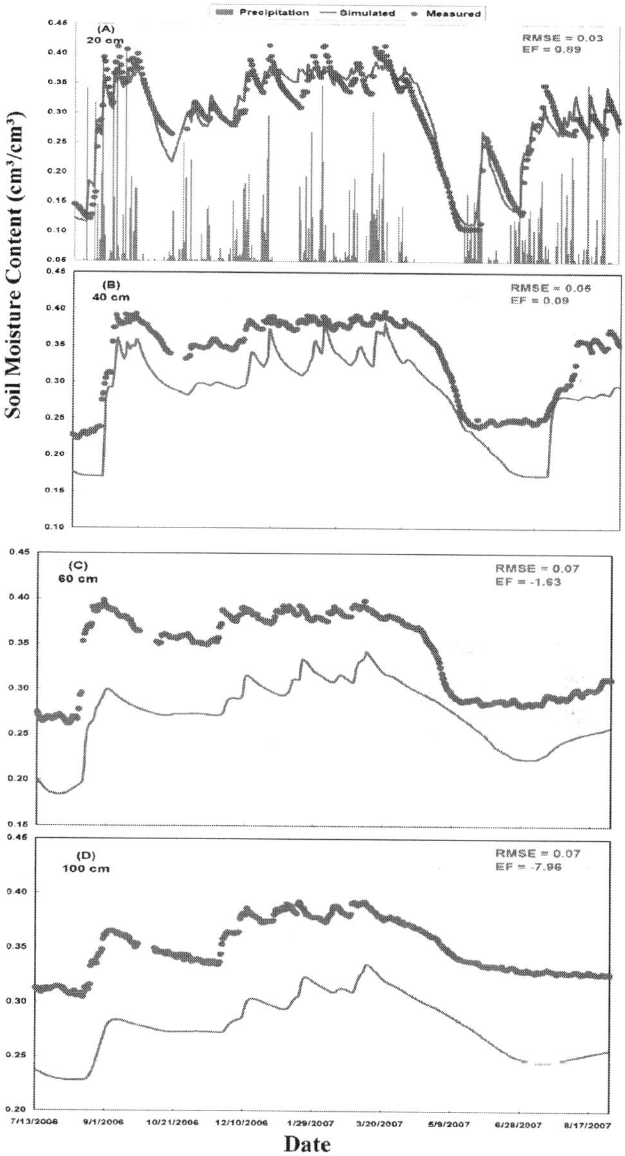

Figure 2: Daily rainfall (solid bars, A) [cm], and measured (circles) and simulated (lines) soil water contents at 20 cm (A), 40 cm (B), 60 cm (C) and 100 cm (D) depths.

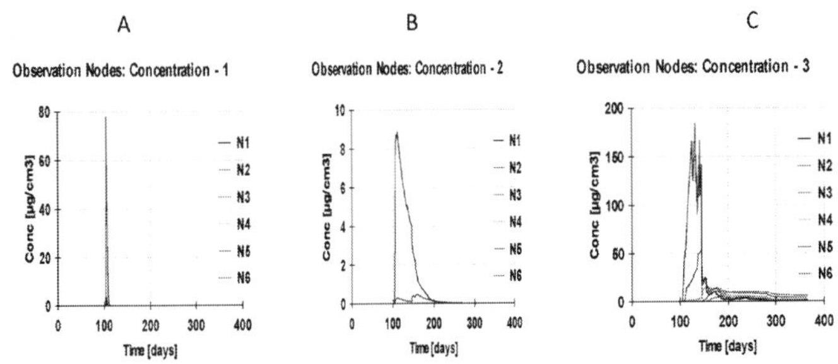

Figure 3: Variation of concentration Urea, Ammonium and Nitate in different observational nodes of experiment in the soil profile. With (N1=20cm , N2=40cm, N3=60cm, N4=100cm and N5=120cm).

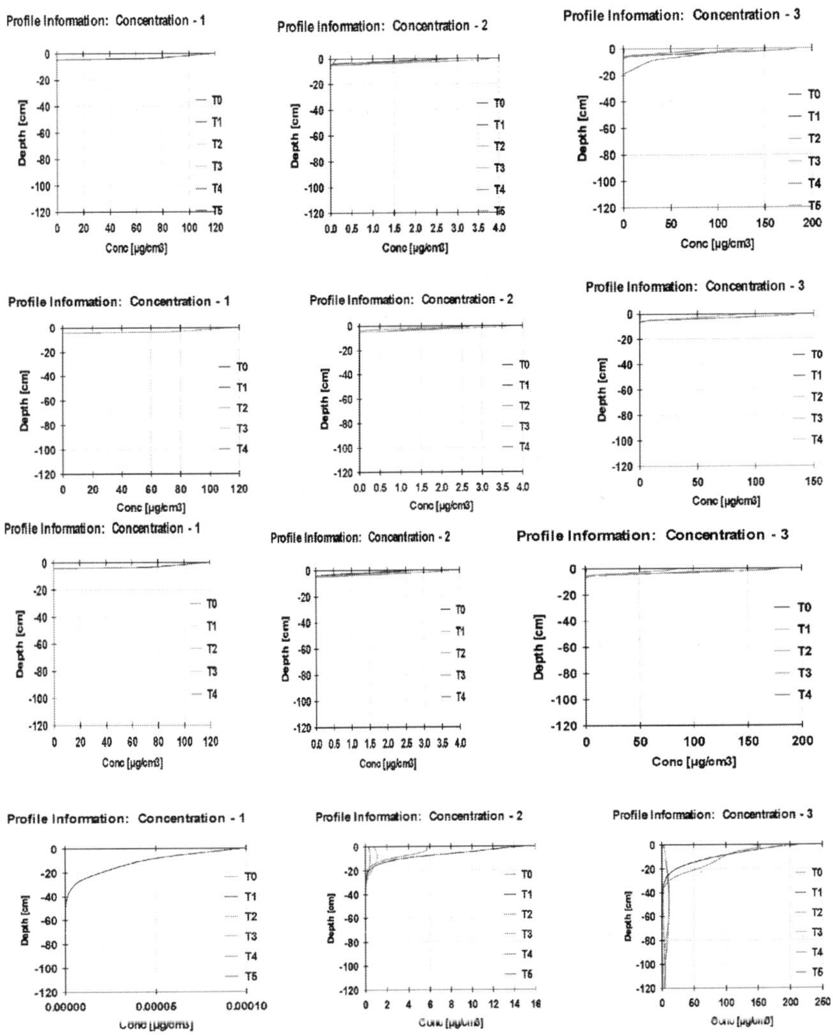

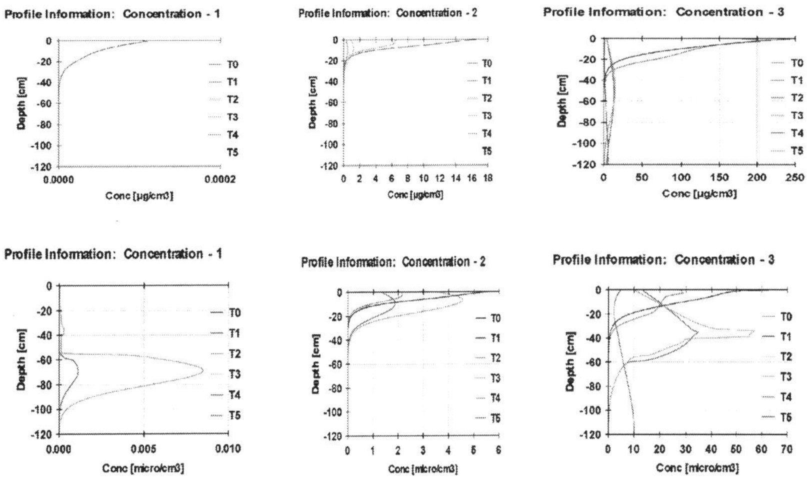

Figure 4: Variation of concentration urea, ammonium and Nitate with depths for different times and depths of experiment for different treatment. With (T0=50 days, T1=150 days, T2=200 days, T3= 240 days , T4=300 days and T5=339 days).

The statistical criteria of quantitative model evaluation between simulated and measured soil water content are summarized in Table 6. Overall, the values calculated demonstrate a good correlation of the model to field data. The results of the simulations may be affected by the value of the saturated hydraulic conductivity (Ks). Therefore, optimizing this parameter for all the three layers using inverse modeling of the Hydrus-1D, would slightly improve the simulation results. So the predicted water contents at −40, −60 and -100 cm are indeed much closer to the measured value, and this parameter change does not affect the (good) match observed for -20 cm. For further improvement, other hydraulic parameters (e.g. θs and α) also should be optimized. In addition, changing the matric pressure head may lead to good results.

Table 6: Statistical criteria for the simulated and measured soil water content

Depth	**Statistical criteria**				
(cm)	**n**	**AE (cm^3 cm^{-3})**	**RMSE (%)**	**EF**	**CRM**
20	365	−0.078	0.115	0.953	0.277
40	365	−0.094	0.115	0.958	0.303
60	365	−0.085	0.106	0.964	0.274
100	365	−0.032	0.058	0.989	0.098

Note: n = number of measurements, AE = average error, RMSE = relative root mean square error, EF = modeling efficient and CRM= coefficient of residual mass.

Fate of Nitrogen Sources

We determined the effects of different sources of nitrogen on the soil distribution of urea, ammonium, and nitrate during of growing season tomato field. To evaluate the Hydrus model performance with respect to nitrogen transport and transformations, the simulated nitrogen concentrations (NH4-N and NO3-N) are compared for different treatments at different depths of soil profile, (7.5, 22.5, 37.5, 52.5 and 120 cm from soil surface). Figure 3(A–C) gives the daily variations in the simulated Urea-N, NH4-N and NO3-N concentrations respectively. It takes about 4 days to convert 90% of urea into ammonium and it takes about 70 days to convert 90% of ammonium into nitrate. Urea fertilizer is easily dissolved in water and transferred to the soil. After fertilization, urea is hydrolysed in the soil a urea concentration decreased over time between irrigations and ammonium is formed and then, during the nitration process by bacteria in the soil, convert ammonium to nitrite and then to nitrate. Immediately after fertigation, at 3.73 days, the urea was concentrated near the soil surface.

For all treatments ammonium accumulated in the topsoil immediately (Figure 4). Because of soil adsorption and subsequent fast nitrification and/or root uptake, there was only a slight movement of ammonium in the soil profile. The results obtained in this study indicated that nitrate moved continuously downwards during the 28-day of growing season simulation. Also, nitrate is easily exposed to leaching due to its high mobility and is not adsorbed to the soil, therefore denitrification was assumed negligible.

Nitrogen is applied to the soil solution by fertilizer application, treated wastewater irrigation and animal manure.

Ammonium is usually ionically exchanged and stabilized in the surface of clay minerals. It is found in small amounts in soil solution and can be retained by the negative charges of clay mineral particles and organic particles. Therefore, the mobility of ammonium ions is lower than that of nitrate ions. The NH4-N then transformed to nitrate by the nitrification process, which is the most soluble form of nitrogen in the soil for uptake by crops.

Table 7 shows the different components of nitrogen balance during the simulation period. Slightly smaller leaching percentages were computed for the urea–ammonium–nitrate wastewater compared to the nitrate- fertilizer and manure. Fertilizer use efficiency ranged from 54% (treatmentT4) to 84.9% (treatment T1). The results reported from nitrogen balance components show that nitrate leaching losses (0%, 23% and 25%) in treatments T1, T3, T4 respectively, and mainly occurred during the winter period. The reduced level of leaching is explained by low amount of drainage water, low nitrogen concentration of irrigation wastewater and excessive nitrogen uptake by the crop. Since the nitrate transport through the soil profile and out into field drains or deep groundwater, is usually controlled by water movement. The effect of irrigating different on the grain yield of tomato was also significant ($P < 5\%$) (Table 7). The results showed an increase in the mean of fresh and dry forage yield (8.25% fresh forage and 23.14% of dry forage) (Table 7). Because treated wastewater is an important source of plant nutrients and can be reused for irrigation to increase forage crop production.

Table 7: Components of nitrogen balance at the end of simulation period in kg N ha^{-1} for a soil depth of 150 cm

N balance components Applied – Inflow kg N ha^{-1}	**Losses – Outflow kg N ha^{-1}**			**Corn grain yield kg ha^{-1}**
	Crop uptake		**Leaching**	
N applied by wastewater	**Ammonium-N**	**Nitrate-N**	**Nitrate-N**	
T1 = 100	5.3	47	-	82510
T2 = 75	5.2	47		81830
T2 = 50	4.1	39		7812
T2 = 25	1.7	23		63253
T3 = N-manure =100	4.2	43	21	78956
T4 = N-fertilizer =90	5.8	44	25	77582
T4 = split application	4.9	43	23	78962
T5 = well water = 0	—	—	—	51254

CONCLUSIONS

The HYDRUS-1D software was performed to simulated water flow and nitrogen transport in tomato crop soil for wastewater irrigation and fertilization. Based on the study carried out in the field, the ability of the model to predict the moisture in the soil at various depths is accurate. This can be due to an acceptable method in the simulation model.

The results reported from nitrogen balance components show that nitrate leaching losses (0%, 23% and 25%) in treatments T1, T3, T4 respectively and mainly occurred during the winter period. The reduced level of leaching is explained by low amount of drainage water, and excessive nitrogen uptake by the crop. Since the nitrate transport through the soil profile and out into field drains or deep groundwater, is usually controlled by water movement. It was fund that the slightly smaller leaching percentages for the urea–ammonium–nitrate wastewater compared to the nitrate- fertilizer and manure. Fertilizer use efficiency ranged from 54% (treatment T4) to 84.9% (treatment T1). Based on these results we conclude that nitrogen from wastewater has smaller nitrate leaching compared to nitrogen from animal manure and commonly fertilizer. Nevertheless, our simulation results provide guidance on the appropriate fertigation strategy for use of waste water in irrigation.

REFERENCES

1. Erfani Agah A Environmental aspects use of treated municipal wastewater in irrigation. Published and presented in Iranian National Committee on Irrigation and Drainage (IRNCID) In, 1996.
2. Erfani Agah A, Haghnia G, Alizade A. Effect of irrigation by treated wastewater on the yield and quality of tomato. Journal of Agricultural Sciences and Technology. 2001;15:65-70
3. Erfani Agah A, Haghnia G, Alizade A. Effect of irrigation with wastewater on yield and lettuce qualitative and soil properties Esfahan Agric Tec And Sci Natural Resourses J. 2002;5:72-77
4. Kaboosi K. The assessment of treated wastewater quality and the effects of mid-term irrigation on soil physical and chemical properties (case study: Bandargaz-treated wastewater) applied water science 7:2385-2396. 2017. DOI: 10.1007/s13201-016-0420-5
5. Kurian M, Reddy VR, Dietz T, Brdjanovic D. Wastewater re-use for peri-urban agriculture: A viable option for adaptive water management? Sustainability Science. 2013;8:47-59. DOI: 10.1007/s11625-012-0178-0
6. Lyu SD, Chen WP, Zhang WL, Fan YP, Jiao WT. Wastewater reclamation and reuse in China: Opportunities and challenges Journal of Environmental Sciences. 2016;39:86-96. DOI: 10.1016/j.jes.2015.11.012
7. Lavrnic S, Zapater-Pereyra M, Mancini ML. Water Scarcity and Wastewater Reuse Standards in Southern Europe: Focus on Agriculture Water Air and Soil Pollution. 2017:228. DOI: 10.1007/s11270-017-3425-2
8. Pescod MB (1992) Wastewater treatment and use in agriculture. In: Drainage FIa (ed). Rome, p 47
9. Mtshali JS, Tiruneh AT, Fadiran AO. Characterization of sewage sludge generated from wastewater treatment plants in Swaziland in relation to agricultural uses. Res Environ. 2014;4:190-199
10. Kominko H, Gorazda K, Wzorek Z. The possibility of Organo-mineral fertilizer production from sewage sludge. Waste and Biomass Valorization. 2017;8:1781-1791
11. Hazen and Sawyer (2010) Literature review of nitrogen fate and transport modeling. Florida Department of Health .Division of

Environmental Health. Bureau of Onsite Sewage Programs 4042 Bald Cypress Way Bin #A-08. Tallahassee, FL 32399-1713.

12. Pettygrove, Asano (1990) California State Water Resources Control Board., "Irrigation with reclaimed municipal wastewater – guidance manual". Edited by: G. Stuart Pettygrove and Takashi Asano. Prepared by: Department of Land, Air and Water Resources. University of California. Pub: Lewis Publishers.
13. Robertson GP, Groffman PM. Nitrogen transformations. In: Paul EA, editor. Soil Microbiology, Ecology and Biochemistry. Fourth ed. Burlington, Massachusetts, USA: Academic Press; 2015. pp. 421-446
14. Kim DY, Burger JA. Nitrogen transformations and soil processes in a wastewater-irrigated, mature Appalachian hardwood forest Forest ecology and management 90:1-11. 1997. DOI: https://doi.org/10.1016/S0378-1127(96)03889-3
15. Wlodarczyk T, Kotowska U. Nitrogen Transformations and Redox Potential Changes in Irrigated Organic Soil. Lublin: Institute of Agrophysics PAS; 2005
16. Almasri MN. Kaluarachchi JJ. Modeling nitrate contamination of groundwater in agricultural watersheds Journal of Hydrology. 2007;343:211-229. DOI: 10.1016/j.jhydrol.2007.06.016
17. Waskom R. Best management practices for irrigation management. Colorado State Univ. Coop. Ext. Bul. 1994;XCM-173
18. Mekala C, Nambi IM. Experimental and simulation studies on nitrogen dynamics in unsaturated and saturated soil using HYDRUS-2D Procedia technology 25:122-129. 2016. DOI: https://doi.org/10.1016/j.protcy.2016.08.089
19. Šimůnek J, van Genuchten MT, Šejna M. Recent Developments and Applications of the HYDRUS Computer Software Packages Vadose Zone Journal. 2016:15. DOI: 10.2136/vzj2016.04.0033
20. Šimůnek J, Šejna M, Saito H, M VG (2013) The Hydrus-1D Software Package for Simulating the Movement of Water, Heat, and Multiple Solutes in Variably Saturated Media, Version 4.17, HYDRUS Software Series 3,
21. Erfani Agah A, Wyseure G. Experimental investigation of water flow and solute transport in unsaturated columns International Journal of AgriScience. 2013;3:228-239

22. Erfani Agah A, Meire P, Deckere Ed (2016) Simulation of Phosphorus Transport in Soil Under Municipal Wastewater Application Using Hydrus-1D. In: Larramendy ML, Soloneski S (eds) Soil Contamination - Current Consequences and Further Solutions. InTech, Rijeka, p Ch. 09. doi:10.5772/66214
23. Van Genuchten MT (1980) A closed-form equation for predicting the hydraulic conductivity of unsaturated soils soil science Society of America Journal 44:892-898
24. Hanson BR, Simunek J, Hopmans JW. Evaluation of urea-ammonium-nitrate fertigation with drip irrigation using numerical modeling Agricultural Water Management. 2006;86:102-113. DOI: 10.1016/j.agwat.2006.06.013
25. Ling G, El-Kadi A. A lumped parameter model for nitrogen transformation in the unsaturated zone. Water Resources Research. 1998;34:203-212. DOI: 10.1029/97WR02683
26. Johnsson H, Bergstrom L, Jansson P-E, Paustian K. Simulated nitrogen dynamics and losses in a layered agricultural soil agriculture. Ecosystems & Environment. 1987;18:333-356. DOI: https://doi.org/10.1016/0167-8809(87)90099-5
27. Lotse EG, Jabro JD, Simmons KE, Baker DE. Simulation of nitrogen dynamics and leaching from arable soils Journal of Contaminant Hydrology. 1992;10:183-196. DOI: https://doi.org/10.1016/0169-7722(92)90060-R
28. Selim HM, Iskandar IK. Modeling nitrogen transport and transformations in soils: 1. Theoretical considerations. Soil Science. 1981;131:233-241
29. Misra C, Nielsen DR, Biggar JW. Nitrogen transformations in soil during leaching; I. theoretical Considerations1 soil science Society of America Journal 38:289-293. 1974. DOI: 10.2136/sssaj1974.03615995003800020024x
30. Vanclooster M, Ducheyne S, Dust M, Vereecken H. Evaluation of pesticide dynamics of the WAVE model. Agric Water Manage. 2000;44:371-388
31. Loague K, Green RE. Statistical and graphical methods for evaluating solute transport models: Overview and application. Validation of flow and transport models for the unsaturated zone. J Contam Hydrol. 1991;7:51-73

32. Antonopoulos V. Simulation of soil moisture dynamics on irrigated cotton in semi-arid climates. Agric Water Manage. 1997;34:233-246
33. Olinski AJ. Verification of Drainmod ver. 5.1 for estimating water balance and nitrogen transport through soils in southern Ontario. M.Sc. Thesis. University of Guelph, Guelph. Ontario. 2005;3:24-27
34. Moriasi DN, Arnold JG, Van Liew MW, Bingner RL, Harmel RD. Veith TL. Model evaluation guidelines for systematic quantification of accuracy in watershed simulations Transactions of the Asabe. 2007;50:885-900

SECTION 4: SYMBOLIC COMPUTING APPLICATIONS

Chapter 15

SYMBOLIC TIME SERIES ANALYSIS AND ITS APPLICATION IN SOCIAL SCIENCES

Wiston Adrián Risso

Institute of Economics (IECON), University of the Republic, Montevideo, Uruguay

ABSTRACT

The present chapter intents to present the symbolic time series analysis (STSA) reviewing the recent developments in sciences. Even if there are very few works applied to social sciences, STSA has a potential to be developed. In particular, due to the limitations about historical data, fields such as Economics and Finance need to develop statistical tests to prove their hypotheses. An independence test and a causality test based on STSA are reviewed. They seem to be more powerful, detecting different kinds of

Citation: Wiston Adrián Risso, "Symbolic Time Series Analysis and Its Application in Social Sciences", IntechOpen Time Series Analysis and Applications, 2017, DOI: 10.5772/intechopen.70826.

nonlinearities compared with the classical tests, usually applied in social sciences. However, there is much work to do with STSA, and social sciences are a fertile field for the development of new powerful tools.

Keywords: STSA, information theory, entropy, independence, causality

INTRODUCTION

Time series has a long history in social sciences, especially in economics and finance. As it is well known, much of economics and finances are concerned with modeling dynamics, and systematization of data over time was a subject that appeared early. In particular, two empirical topics become important when working with time series in social sciences: inferences and forecasting. The cumulated historical data permitted to applied statistical methods in order to find evidence of causation between social variables, finding some support to social theories. Considering the nonexperimental nature of the social sciences, this also encourages the development of statistical techniques. In fact, while in physics, it is relatively easy to get hundreds of thousands of data for a given time series, in economics there are often only 50 or 100 data for a time series, and maybe we can obtain thousands of data in financial series. For this reason, much of the statistical effort, in particular econometric effort was focused on developing powerful statistical tests, considering the availability of small samples. This is an important different approach between econometrics and for example, statistical mechanics in theoretical physics.

We can identify two main groups in time series econometrics: univariate time series analysis concerning with techniques for the analysis of dependence in adjacent observations. It has increased importance since 1970 based on the main ideas underlying in [1]; multivariate time series analysis based on the vector autoregressive (VAR) models, made popular by [2]. In the first group, we find all the autoregressive integrated moving average (ARIMA) models and the related generalized autoregressive conditional heteroscedasticity (GARCH) models developed by [3]. The second group is a generalization of the AR models and we can find two important developments based on this: cointegration proposed by [4] focusing on finding a statistical relationship between variables; and noncausality test developed by [5], which takes the concept of predetermination try to test if a variable causes another. Much of the development in time series econometrics is found in books such as [6–18].

In summary, dependence and causation are two important topics in time series econometrics and time series analysis. These topics are related with the importance of inference and forecasting in social sciences. Econometrics has been focused in developing powerful test considering the available small samples. Most of these developments are based on linear models even if there are some developments considering nonlinearities; see for instance [19, 20].

Time series analysis in econometrics is mostly based on observations belonging to the set of the real numbers. Some variables can be categorical such as dummy variables. However, in this chapter, we will talk about a different approach that is known as symbolic time series analysis (STSA). It has been originally applied to physics and engineering as a statistical methodology to detect the very dynamic of highly noise time series. The application to social sciences such as economics or finance is very recent and there are some novel developments.

As mentioned before, the application of STSA in social sciences requires a different approach due to data limitation. In this sense, the design of powerful test considering the availability of data is crucial. As abovementioned, dependence and causation are two important topics. In this sense, we review an independence test and a first approach on testing noncausality, both based on STSA. The information theory was adopted as an approach to analyze the symbolic time series and the approximation of Shannon Entropy as an important measure, applied to test design.

The chapter is organized as follows. Section 2 presents the symbolic time series approach and its relation with the symbolic dynamics. In Section 3, we review some of the literature of STSA applied to the sciences. In Section 4, the information theory approach and Shannon Entropy measure is explained. Section 5 presents a review of the independence symbolic test. Section 6 focuses on causality test based on STSA. Section 7 discusses the difference between the proposed symbolic noncausality test and the traditional and well-known Granger noncausality test. Finally, in Section 8, we draw some conclusions and present some future lines of research.

SYMBOLIC TIME SERIES ANALYSIS

The concept of symbolization has its roots in dynamical systems theory, particularly in the study of nonlinear systems, which can exhibit bifurcation and chaos. In [21], it is asserted that symbolic dynamics is a method for studying nonlinear discrete-time systems by taking a previously codified trajectory using strings of symbols from a finite set, also called an alphabet. According to [22], symbolic dynamics and symbolic analysis are connected but are different concepts. In fact, the former is the practice of modeling a dynamical system by a discrete space. However, the latter is an empirical approach to characterize highly noisy data by considering a partition, discretizing the data, and obtaining a string representing the very dynamic of the process.

As asserted by [23], symbolization involves transformation of raw time series measurements into a series of discretized symbols that are processed to extract information about the generating process. In this way, we can search for nonrandom patterns and dependence by transforming a given time series $\{x_1, x_2, \ldots, x_T\}$ into a symbolic string $\{s_1, s_2, \ldots, s_T\}$.

The STSA approach is easy to apply but the definition of the right partition is the most difficult thing to do. Generally, it applied an equiprobable partition implying to take the empirical distribution of a given time series $\{x_1, x_2, \ldots, x_T\}$ and establishing two or more equally probable regions. For instance, for a Gaussian time series, we can define two equally probable regions considering as partition the mean equal to zero. After that, we can assign the symbol $s_i = 0$ for negative values and $s_i = 1$ for positive ones. In this way, we transform a continuously random series into a discrete string similar to the outcomes from flipping a coin.

STSA IN APPLIED SCIENCES

In [23], the applications of STSA techniques to the different fields of science are reviewed. According to the authors, the different applications suggest that symbolization can increase the efficiency of finding and quantifying information from the systems. Mechanical systems were one of the first applications where symbolic analysis was successfully used to characterize complex dynamics. In [24–26], symbolic methods to the analysis of experimental combustion data from internal combustion engines are applied. Their objective was to study the onset of combustion instabilities as the fueling mixture was leaned. STSA has also been applied in Astrophysics and Geophysics. For instance, [27] analyzes weak-reflected radar signals from the planet Venus to measure the rotational period. In [28], a binary symbolization to analyze solar flare events is utilized. Biology and Medicine is another field where STSA has been applied. There have been many recent applications of symbolic analysis for biological systems, most notably for laboratory measurements of neural systems and clinical diagnosis of neural pathologies. STSA has been applied in neurosciences. In [29, 30], symbolization data is applied to equal-sized interval to partition EEG signals to identify seizure precursors in electroencephalograms. [31] proposed a new damage localization method based on STSA to detect and localize a gradually evolving deterioration in the system. They assert that this method could be demanded for implementation in real-time observation application such as structural health monitoring. In [32], the STSA is used in human gait dynamics. The results of this study can have implication modeling physiological control mechanism and for quantifying human gait dynamics in physiological and stressed conditions. In [33], the heart-rate dynamics is studied by using partitions aligned on the data mean and ±1 and ±2 sample standard deviations, for a symbol-set size of 6. In [34], the prevalence of irreversibility in human heartbeat is analyzed applying STSA.

Application of symbolization to fluid flow measurements has spanned a wide range of data types from global measurements of flow and pressure drop, to formation and coalescence of bubbles and drops, to spatiotemporal measurements of turbulence. In [35], an approach for transforming images of complex flow fields (as well as other textured fields) into a symbolic representation is developed. In [36], STSA is applied to the networks of genes, which is important underlying the normal development and function of organisms. Information about the structure of the genome of humans and other organisms is increasing exponentially. In [37], equiprobable symbols are used for analyzing measurements from free liquid jets in order to readily discriminate between random and nonrandom behavior. In [38], STSA is applied to the detection of incipient fault in commercial aircraft gas turbine engines. In [39], combustion instability in a swirl-stabilized combustor is investigated using STSA. Chemistry-related applications of symbolic techniques have been developed for chemical systems involving spontaneous oscillations or propagating reaction fronts. In [40], a type of symbolization for improving the performance of Fourier-transform ion-cyclotron mass spectrometry is applied. Artificial Intelligence, Control, and Communication are fields where symbolization has been incorporated. In [41], a phase-space partitioning to model communication is used. An example application of symbolization to communication is found in [42], utilizing small perturbations to encode messages in oscillations of the Belousov-Zhabotinsky (BZ) reaction. In robotics, a symbolic time series–based statistical learning method to construct the generative models of the gaits (i.e., the modes of walking) for a robot, see [43], has been developed. Efficacy of the proposed algorithm is demonstrated by laboratory experimentation to model and then infer the hidden dynamics of different gaits for the T-hex walking robot. In [44], an algorithm to intuitively cluster groups of agent trails from networks based on STSA is proposed. The authors assert that temporal trails generated by agents traveling to various locations at different time epochs are becoming more prevalent in large social networks. The algorithm was applied to real world network trails obtained from merchant marine ships GPS locations. It is able to intuitively detect and extract the underlying patterns in the trails and form clusters of similar trails.

The methods of data symbolization have also been applied for data mining, classification, and rule discovery. In [45], rule discovery techniques to real-valued time series via a process of symbolization are applied. Finally, we find some applications of STSA in Social Science. In [46–48], STSA and minimal spanning tree (MST) are applied to construct cluster of financial asset with application to portfolio theory. Utilizing a similar methodology, in [49], the dynamics of exchange market is studied, and in [50], the international hotel industry in Spain is analyzed. In [51, 52], STSA and entropy are applied to measure informational efficiency in financial markets.

INFORMATION THEORY AND SHANNON ENTROPY

The term entropy was first used by Rudolf Clausius in [53] related to the second law of thermodynamics. Subsequently, the communication theory [54] used the Shannon entropy as a measure of uncertainty where the maximum entropy corresponds to the maximum degree of uncertainty. In this sense, a random process will take the maximum entropy value. In fact, English language is not a random process; some patterns such as "THE" are more probable than sequences such as "DXC". Note, that in a random process, the two sequences should have the same probability. This principle is very relevant because if a symbolic string is random, the entropy should be the maximum.

The entropy measure (H) must meet the following conditions:

- $H(P)$ should be a function of the probability distribution of the n events expressed as the vector $P = (p_1, p_2, \ldots, p_n)$.
- (*Continuity*), $H(P)$ should be a continuous function of vector P.

- (*Symmetry*), the measure should be unchanged if the outcomes p_i are re-ordered.
- (*Expansible*), Event of probability zero should not contribute to the entropy, $H(p_1, p_2, \ldots, p_n, 0) = H(p_1, p_2, \ldots, p_n)$.
- (*Minimum*), the measure should take value 0 when there is not uncertainty.
- (*Maximum*), the measure should be maximal if all the outcomes are equally likely. It means $p_1 = p_2 = \ldots = p_n = 1/n$.
- For equiprobable events, the entropy increases with the number of outcomes. $H(p_1 = 1/(n + 1), \ldots, p_{n+1} = 1/(n + 1)) > H(p_1 = 1/n, \ldots, p_n = 1/n)$.

In [54], the Shannon entropy function is proposed:

$$H_n(P) = -\sum p_i \, log_2(p_i) \qquad (1)$$

The entropy is frequently measured in bits by using log base 2 satisfying all the properties already mentioned. Note that the maximum property is confirmed solving the following Lagrangian expression (2).

$$-\sum p_i \, log_2(p_i) - \lambda\left(\sum p_i = 1\right) \qquad (2)$$

The Shannon entropy is concaved with a global maximum when all the probabilities are equal. In addition, when $p_i = 0$, the convention that $0.log0 = 0$ is used. Thus, adding zero, probability terms do not change the entropy value.

In order to clarify the concept of Shannon, consider two possible events and their respective probabilities p and $q = 1-p$. The Shannon entropy will be defined by Eq. (3).

$$H = -(p.\,log(p) + q.\,log(q)) \qquad (3)$$

Figure 1 shows graphically the function shape, note that the maximum is obtained when the probability is 0.5 for each event. This case corresponds to a random event; on the other hand, note that a certain event (when probability of one event is 1) will produce entropy equal to 0.

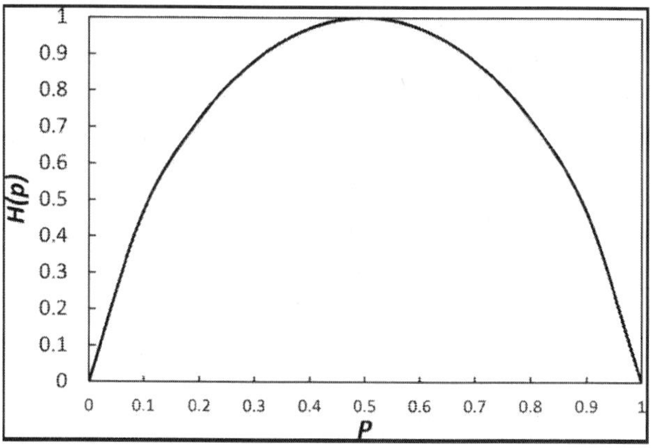

Figure 1: Shape of the Shannon entropy function. Note that maximum happens when the process is random (p = 0.5).

In general, [55] showed that any measure satisfying all the properties must take the following form:

$$-c \sum p_i \, log_2 \, (p_i) \qquad (4)$$

In order to normalize the Shannon entropy, c usually takes the value $1/log_2(n)$ allowing to compare events of different sizes.

SYMBOLIC INDEPENDENCE TEST

STSA seems to present a good performance when detecting independence in time series. A variety of dynamical processes are present in economics. Linearity, nonlinearity, deterministic chaos, and stochastic models have been applied when modeling a complex reality. In [56], a runs test is designed, asserting that the problem of testing randomness arises frequently in quality control of manufactured products. It is remarked that detecting dependence in time series is an essential task for econometricians and applied economist. In [57], the well-known BDS test is introduced, considered as a powerful test to detect nonlinearity. In [58], a simple and powerful test based on STSA is proposed and the results are compared with the BDS and runs test. On one hand, it is found that BDS is not able to detect processes such as the chaotic Anosov and the stochastic processes nonlinear sign model (NLSIGN), nonlinear autoregressive model (NLAR), and nonlinear moving average model (NLMA). On the other hand, runs test cannot detect the chaotic Anosov, the logistic process, the bilinear, the NLAR, and the NLMA stochastic processes. The experiments show that the test based on STSA has no problem detecting all these dynamics. It is concluded that proposed test is simple, easy to compute, and is powerful with respect to the other two tests. In particular, for small samples, it is the only one able to detect models such as chaotic Anosov and nonlinear moving average (NLMA). Besides, the test is applied to financial time series to detect nonlinearity on the residuals after applying a GARCH model. In this case, the BDS rejected the independence few times whereas the SRS test still detects nonlinearity in the residuals. It seems that BDS considers that the GARCH(1,1) model is a good model most of the time. However, the symbolic test suggests that GARCH(1,1) would not be a good model considering all the nonlinear components.

Here, we review briefly the test and repeat some experiments comparing the results with the well-known BDS and runs tests. At first, let us consider a finite time series generated by an independent or random process-sized T^* $\{x_t\}_{t=1,2,\ldots,T^*}$. Define a partition in the series in "a" equiprobable regions obtaining the symbolized time series $\{s_t\}_{t=1,2,\ldots,T^*,}$ where each symbol s_t takes a symbolic value from the alphabet $A = \{A_1,A_2,\ldots,A_a\}$. Since, we want to derive a general statistic for different alphabet sizes a and different subsequences lengths w, we have to make two considerations: (1) from now, we will call n to the quantity of possible events. That is $n = a^w$, where for the simplest case ($w = 1$) implies $n = a,$ then the quantity of events is equal to the symbol-set size; (2) in practice, we have a finite sample size T^*, there is no problem for $w = 1$, but when we compute subsequences or time-windows w of consecutive symbols we loss observations. For example, when we compute the frequency for two consecutive symbols, we have a total sample size T^*-1. In general, we can define the sample size $T = T^* + w-1$, again for the trivial case $w = 1$, $T^* = T$

Note that defining S_i for $i = 1,2,\ldots,n$ as the sum of the total i events in the time series, we can derive the multidimensional variable $S = \{S_i/T\}$ being distributed as a multinomial with $E(S_i/T) = (1/n)$, $Var(S_i/T) = (1/n)(n\text{-}1)/nT$ and $Cov(S_i/T,S_j/T) = -(1/n)(1/nT)\ \forall i \neq j$. As we will see, frequencies of the events should be important in the statistic and the vector of the n frequencies S_i/T could be approximated by a multivariate normal distribution $N(1/n,\sigma^2\Sigma)$ where σ^2 is $(1/nT)$ and Σ is a idempotent matrix as in (5)

$$\Sigma_{nxn} \equiv \begin{bmatrix} (n-1)/n & -1/n & \dots & -1/n \\ -1/n & (n-1)/n & \dots & -1/n \\ \vdots & \vdots & \ddots & \vdots \\ -1/n & -1/n & \dots & (n-1)/n \end{bmatrix} \tag{5}$$

For convenience, we can define the normalized vector variable $\{\varepsilon_i\} = \{(S_i/T)\text{-}(1/n)\}_{i=1,2,\dots,n}$ having a multivariate normal distribution $N(\o,\sigma^2\Sigma)$, being ø, the null vector. Then, the statistic can be defined as a quadratic form in random normal variables (6).

$$\left\{ \frac{\sum_{i=1}^{i=n} \varepsilon_i^2}{\sigma^2} \right\} \tag{6}$$

In [58] is applied the distribution of quadratic forms in normal variables presented in [59]. $X = (\varepsilon_1/\sigma, \varepsilon_2/\sigma, \dots, \varepsilon_n/\sigma)$ is distributed multivariate normal $N(\o,\Sigma)$. The theorem indicates that $tr(A\Sigma) = n-1$, and thus $X'AX$ distributes Chi-square with $(n-1)$ degrees of freedom. In this case, A is the identity matrix I, and Σ is symmetric, singular, and idempotent. Remembering that $\sigma^2 = (1/nT)$, then we obtain that the distribution of the symbolic randomness statistic (SRS) as in (7).

$$SRS \equiv Tn \left\{ \sum_{i=1}^{i=n} \left(\frac{S_i}{T} - \frac{1}{n} \right)^2 \right\} \textit{ assymptotically distributes } \chi^2_{n-1} \tag{7}$$

Note that in practice computing the statistic is very simple. We just have to consider the symbols (a) and subsequences or length (w) and compute the frequencies for each event ($n = a^w$) in the time series.

The algorithm to compute the test is as follows:

Step 1: Considering time series $\{x_t\}_{t=1,2,\dots,T^*}$, compute the empirical distribution, and define equiprobable regions according to the quantity of symbols or the alphabet size.

Step 2: According to the partition, translate $\{x_t\}_{t=1,2,\dots,T^*}$ into $\{s_t\}_{t=1,2,\dots,T^*}$, the symbolic time series when $w = 1$.

Step 3: Compute different symbolic time series for different lengths w, remember that the obtained series in step 2 corresponds to $w = 1$.

Step 4: For each w, compute the frequency of the n different events S_i/T for $i = 1,2,\dots,n$.

Step 5: For each *w*, compute the $SRS(a,w) = Tn\{\Sigma(S_i/T - 1/n)^2\}$ as shown in Eq. (7).

Step 6: Compare the *SRS*(*a*,*w*) with the Chi-2 with *n*-1 degree of freedom at 0.05 of significance, under the independence null hypothesis. When *SRS*(*a*,*w*) is larger than the critical value we reject the null hypothesis.

In [58], it is found that the statistic introduced in (7) is related to the Shannon entropy (H). We can derive the approximation expressed in Eq. (8).

$$SRS \approx (1 - H).T.\ln(n) \quad (8)$$

Note the generalization implied in STSA permits to study different dynamical process. For instance, consider a string of the first 3000 letters from the book "*A Christmas Carol*", s_1 = {*marleywasdeadtobeginwith... scroogecar*} and a random string of 3000 letters from an alphabet of 26, s_2 = {*iskynbmhjp...vbbihjfkk*}. Imagine testing this kind of process with BDS or runs test. However, note that would be easy to test this dynamics with the symbolic test. In this case, we can define an alphabet of 26 letters and the string. On the one hand, applying the *SRS*(26,1) and *SRS*(26,2) for the s_1, we obtain the following values 2102.40 and 12331.26, respectively. On the other hand, SRS(26,1) and SRS(26,2) for the string s_2 are 25.79 and 690.26, respectively. Considering that a Chi-2 with 25 degree of freedom at 95% is 37.65 and a Chi-2 with 675 degree of freedom (26^2–1) at 95% is 736.55. Since, the statistics for s_1 are large than the critical value, we can conclude that the process is not random. However, since the statistics for s_2 are less than critical values, we cannot reject the hypothesis of independence.

In [58] is shown that the test is conservative, rejecting the null hypothesis less time than expected. However, it is powerful in detecting nonrandom and nonlinear processes. Considering the four sample sizes, selecting two symbols and length 4 presents decent results in most of the cases. Selecting three symbols seems to be a relative good option for size of 200 or larger and three symbols for a sample size of 500 or larger. The best result is given for a sample of 2000 applying three symbols and length 4. Table 1 presents the experiments using 1000 Monte Carlo simulations on Normal, Logistic, NLMA, Anosov, and NLSIGN processes reproducing the experiments in [58].

Table 1: Simulated size of the SRS, Runs and BDS statistics

Sample size	Test	Normal (%)	Logistic (%)	NLMA (%)	Anosov (%)	NLSIGN (%)
$T = 50$	SRS(2,3)	1.20	41.00	1.30	2.90	0.20
	SRS(3,2)	0.70	100.00	0.40	0.80	0.50
	BDS	9.70	68.10	7.60	18.10	12.00
	RUN test	2.90	23.90	1.30	3.90	2.20
$T = 500$	SRS(2,3)	2.10	19.30	13.50	2.30	9.90
	SRS(3,2)	0.40	100.00	1.70	0.80	1.10
	SRS(4,3)	3.20	100.00	97.50	93.80	17.40
	BDS	3.60	66.40	7.20	5.30	4.40
	RUN test	3.80	14.70	16.30	5.50	24.10
$T = 2000$	SRS(2,3)	1.90	14.40	62.90	2.50	64.80
	SRS(3,2)	0.30	100.00	25.30	1.00	3.80
	SRS(4,3)	2.50	100.00	100.00	100.00	92.30
	SRS(5,3)	1.70	100.00	100.00	100.00	92.20
	BDS	2.80	80.40	14.60	3.60	4.00
	RUN test	4.40	12.20	50.00	5.70	84.30

Note that the symbolic test is more conservative than BDS and Runs test when rejecting independence in a normal random process. However, the symbolic test is powerful in detecting nonlinearities in the studied processes. For a sample of 50, Logistic model is detected 100% by the symbolic test, but BDS detects 68%, and Runs test rejects independence 23.90% of the time. Logistic model is still hard to be detected by the run test when sample increases to 2000. Note that NLMA model is detected by the symbolic test when sample is 500 or larger, but it is not detected by BDS and Runs test. It is interesting to note that the chaotic process of Anosov is detected by the symbolic test for a sample larger than 500 but both BDS and Runs tests reject independence less than 6% of the cases. NLSIGN is hard to be detected, for a sample of 2000 the symbolic test detects more than 90% of the cases and Runs test detects 84% of the cases. However, BDS cannot detect the NLSIGN process. In [58] similar results are obtained, the proposed SRS is the only one that is able to detect chaotic Anosov and nonlinear process NLMA when T = 2000.

SYMBOLIC NONCAUSALITY TEST

The present section reviews the symbolic noncausality test (SNC) and discusses the differences with the classical Granger noncausality test. As in the case of independence test, the main idea here is to derive the asymptotic distribution for the statistic when there is no causality between the series. A full explanation of the test is shown in [60].

Let us consider that *X* and *Y* are two independent random time series sized T + 1 and the symbolized time series can be expressed as $Sx = \{sx_1, sx_2, .., sx_{T+1}\}$ and $Sy = \{sy_1, sy_2, ..., sy_{T+1}\}$. To test causality, we have to define two new series, grouping *Sx* and *Sy* in the following way:

(1) $Sxy = \{(sx_1, sy_2), (sx_2, sy_3), ..., (sx_{t-1}, sy_t), ..., (sx_T, sx_{T+1})\}$

(2) $Syx = \{(sx_1, sy_2), (sx_2, sy_3), ..., (sx_{t-1}, sy_t), ..., (sx_T, sx_{T+1})\}$

If the alphabet is composed by three symbols, the combination (sx_{t-1}, sy_t) takes a value from the set of nine possible events {(1,1), (1,2), (1,3), (2,1), (2,2),(2,3),(3,1),(3,2),(3,3)}. Note that each event should be independent with probability 1/9 (*Sx* and *Sy* are random). Only if at least one event were deviated from 1/9, would there be evidence of noncausality.

An alphabet of $a = 3$ symbols determines $n = 3^2 = 9$ possible events in the set of pairs $\{(x_{t-1}, y_t)\}$ or $\{(y_{t-1}, x_t)\}$. Considering "*a*" symbols and the events $n = a^2$, the vector of the *n* frequencies Exy_i/T and Eyx_i/T could be approximated by a multivariate normal distribution $N(1/n, \sigma^2\Omega)$ where σ^2 is $(1/nT)$ and Ω is a idempotent matrix as in (9).

$$\Omega_{nxn} \equiv \begin{bmatrix} (n-1)/n & -1/n & \dots & -1/n \\ -1/n & (n-1)/n & \dots & -1/n \\ \vdots & \vdots & \ddots & \vdots \\ -1/n & -1/n & \dots & (n-1)/n \end{bmatrix} \tag{9}$$

Following a similar approach as in Section 5, the statistics for the both hypothesis can be defined as in (10) and (11).

$$\left\{ \frac{\sum_{i=1}^{i=n} \varepsilon xy_i^2}{\sigma^2} \right\} \tag{10}$$

$$\left\{ \frac{\sum_{i=1}^{i=n} \varepsilon yx_i^2}{\sigma^2} \right\} \tag{11}$$

The term in brackets in (10), (11) are quadratic forms in random normal variables. Applying the theorem presented in [59], in the present case where vector $X = (\varepsilon_1/\sigma, \varepsilon_2/\sigma, \ldots, \varepsilon_n/\sigma)$ is distributed multivariate normal $N(\emptyset, \Omega)$. As mentioned in Section 5, $tr(A\Omega) = n$-1, thus $X'AX$ distributes Chi-square with $(n-1)$ degrees of freedom. In this case, A is the identity matrix I and Ω is symmetric, singular, and idempotent.

Note that we derive the test assuming that X and Y are random processes. However, we can apply the test for stationary time series and optionally apply an autoregressive process if we want to remove linear dependence and testing the noncausality between the residuals of the two series.

$$x_t = \alpha_0 + \alpha_1 x_{t-1} + ux_t \quad (12)$$

$$y_t = \beta_0 + \beta_1 y_{t-1} + uy_t \quad (13)$$

Finally, the statistics of noncausality SNC(X → Y) and SNC(Y → X) are defined as in (14) and (15).

$$SNC(X \to Y) \equiv nT\left\{\sum_{i=1}^{i=n}\left(\frac{Exy_i}{T} - \frac{1}{n}\right)^2\right\} \textit{assymptotically distrib} \quad (14)$$

$$SNC(Y \to X) \equiv nT\left\{\sum_{i=1}^{i=n}\left(\frac{Eyx_i}{T} - \frac{1}{n}\right)^2\right\} \textit{assymptotically distributes } \chi^2_{n-1} \quad (15)$$

Note that in practice, computing the statistic is very simple. In summary, the test works as follows:

Step 1: Consider time series $\{x_t\}_{t=1,2,\ldots,T+2}$ and $\{y_t\}_{t=1,2,\ldots,T+2}$ we can optionally apply an AR(1) to both series as in (12) and (13) in order to eliminate autocorrelation and define the new residuals time series $\{ux_t\}_{t=1,2,\ldots,T+1}$ and $\{uy_t\}_{t=1,2,\ldots,T+1}$. Note that 1 observation is lost after applying AR(1).

Step 2: In $\{ux_t\}_{t=1,2,\ldots,T+1}$ and $\{uy_t\}_{t=1,2,\ldots,T+1}$ apply a partition in "a" equiprobable regions and translate the series into $\{sx_t\}_{t=1,2,\ldots,T+1}$ and $\{sy_t\}_{t=1,2,\ldots,T+1}$.

Step 3: According to the two hypothesis, $X \to Y$ and $Y \to X$ define the two sets $Sxy = \{(sx_1, sy_2), (sx_2, sy_3), \ldots, (sx_{t-1}, sy_t), \ldots, (sx_T, sx_{T+1})\}$ and $Syx = \{(sx_1, sy_2), (sx_2, sy_3), \ldots, (sx_{t-1}, sy_t), \ldots, (sx_T, sx_{T+1})\}$.

Step 4: For Sxy and Syx, compute the frequency of the $n = a^2$ different events Exy_i/T and Eyx_i/T considering $i = 1,2,\ldots, a^2$.

Step 5: Taking into account Eqs. (14) and (15) compute the $SNC(X \to Y) = nT\{\Sigma[(Exy_i/T){-}(1/n)]^2\}$ and $SNC(Y \to X) = nT\{\Sigma[(Eyx_i/T) - (1/n)]^2\}$.

Step 6: Finally, two null hypotheses must be contrasted: X does not cause Y, and Y does not cause X. In the first case $SNC(X \rightarrow Y)$ should be compared with a Chi-2 with n-1 degree of freedom at 0.05 of significance, if $SNC(X \rightarrow Y)$ is larger than the critical value the null hypothesis is rejected. The same should be done with $SNC(Y \rightarrow X)$.

SYMBOLIC NONCAUSALITY AND GRANGER NONCAUSALITY

The concept of causality into the experimental practice is due to Clive Granger. The classical approach of Granger causality is based on temporal properties. Although the principle was formulated for wide classes of systems, the autoregressive modeling framework proposed by Granger was basically a linear model, and as mentioned in [61] the choice was made due to practical reasons. Granger noncausality test is among the most applied tool testing causality. Three limitations should be noted: (1) the classical test has a good performance when the process is linear. This is because it is based on the vector autoregressive model (VAR); (2) there are extension of the classical test to consider nonlinear causality but they are related with a particular nonlinear model; (3) some authors assert that empirical time series are generally contaminated with noise producing what is known as spurious causality or not allowing to detect the causality.

SCN test presented in [60] is a nonparametric noncausality test based on the symbolic time series analysis. The idea is to develop a complementary test to the Granger noncausality, showing strengths in the points where the Granger test is weak. In this sense, the proposed SNC test performs well detecting nonlinear processes, in particular the chaotic processes. In addition, the mentioned problem related with spurious causality should be alleviated. In fact, according to some experiments nonlinear models such as NLAR model, Lorenz map, and models with exponential terms are not detected by Granger test but the SNC identifies these processes. The test is based on information theory considering an approximation of the entropy as the measure of uncertainty of a random variable. Information theory is considered to be a subset of communication theory. However, in [62] is consider that it is much more. It has fundamental contributions to make in statistical physics, computer science, and statistical inference, and in probability and statistics. It is important to highlight and is an important idea relating symbolic analysis, information theory, and the concept of noise. Information theory considers that communication between A and B is a physical process in an imperfect ambient contaminated by noise. Another important concept is the discrete channel, defined as a system consisting of an input alphabet X and output alphabet Y and a probability transition matrix $p(y|x)$ that expresses the probability of observing the output symbol y given that we send the symbol x.

To compare the performance between the classical Granger noncausality and the proposed SNC test, the following stochastic and deterministic models were simulated:

- AR(1). We consider two independent series generated by autoregressive (AR) processes: $X_t = 0.2 + 0.45X_{t-1} + \varepsilon_{1t}$ and $Y_t = 0.8 + 0.5Y_{t-1} + \varepsilon_{2t}$. Where ε_{1t} and ε_{2t} are i.i.d. and normally distributed (0,1).
- Nonlinear with exponential component. $X_t = 1.4–0.5X_{t-1}e^{Y_{t-1}} + \varepsilon_{1t}$ and $Y_t = 0.4 + 0.23Y_{t-1} + \varepsilon_{2t}$; where ε_{1t} and ε_{2t} are i.i.d. normal(0,1).
- NLAR (Autoregressive Nonlinear). $X_t = 0.2|X_{t-1}|/(2 + |X_{t-1}|) + \varepsilon_{1t}$ and $Y_t = 0.7|Y_{t-1}|/(1 + |X_{t-1}|) + \varepsilon_{2t}$; where ε_{1t} and ε_{2t} are i.i.d. normal(0,1).
- Lorenz: $X_t = 1.96X_{t-1}−0.8X_{t-1}Y_{t-1}$; $Y_t = 0.2Y_{t-1} + 0.8X^2_{t-1}$; with initial conditions X_1, Y_1 generated randomly. This is a discrete version of the Lorenz process as in [63].

Table 2 shows the results of the power experiments applying the SNC and the Granger noncausality test to 10,000 Monte Carlo simulations for the four models and for different sample sizes ($T = 50$, 100, 500, 1000, and 5000).

Table 2: Simulated power of the SNC and the Granger non causality statistic

Sample size	Model	Symbolic noncausality		Granger noncausality	
		$X \to Y$	$Y \to X$	$X \to Y$	$Y \to X$
$T = 50$	**AR(1)** (None)	0.40	0.45	5.66	5.25
$T = 100$		0.42	0.37	5.28	5.09
$T = 500$		0.41	0.27	5.47	5.14
$T = 1000$		0.39	0.42	5.18	5.14
$T = 5000$		0.34	0.46	5.30	5.09

$T = 50$	**NLAR** (X → Y)	0.01	0.01	0.05	0.05
$T = 100$		0.73	0.34	4.82	4.74
$T = 500$		5.96	0.31	4.94	5.16
$T = 1000$		17.87	0.29	5.01	5.05
$T = 5000$		98.02	0.39	6.51	5.00
$T = 50$	**Nonlinear exponential** (Y → X)	0.51	3.76	2.89	16.89
$T = 100$		0.28	11.85	2.78	13.36
$T = 500$		0.43	89.50	2.53	11.48
$T = 1000$		0.40	99.22	2.73	11.29
$T = 5000$		1.42	100.00	2.67	11.19
$T = 50$	**Lorenz** (X → Y, Y → X)	96.61	31.90	30.77	13.86
$T = 100$		99.99	90.49	28.52	12.64
$T = 500$		100.00	100.00	23.60	11.74
$T = 1000$		100.00	100.00	24.42	11.69
$T = 5000$		100.00	100.00	23.87	11.52

Following [60], a 60% acceptance or rejection of the null hypothesis is considered as a threshold. SNC and Granger noncausality correctly identifies noncausality in AR(1) process. Table 2 suggests that SNC is more conservative in the rejection of causality with percentages less than 5%. The nonlinear model with an exponential component implies causality from Y to X. Note that SNC detects the causality when the sample size is 500 or larger. However, Granger test does not detect causality in any case. As asserted by [58] the NLAR process is very difficult to detect. Note that SCN is the only one detecting the causality when $T = 5000$. The Lorenz discrete map is also chaotic, and it is detected by SNC starting from $T = 100$. However, note that Granger test never detects the causality. In particular, is highlighted that Granger test is not able to detect the model with an exponential component, the NLAR model and the chaotic Lorenz map.

Finally, we compare both tests with real data from US. In particular, we consider two well-known relationships in economics: the Phillips curve [64] about the relation between unemployment and inflation rates, the Okun's law [65] establishing a relation between unemployment and economic rate. We take annual data for the US unemployment rate, inflation rate, and economic growth for the period 1948–2016 representing a total of 69 observations. Table 3 shows the results of the Granger noncausality test and the symbolic test considering a partition of two symbols.

Table 3: SNC and the Granger non causality for the Phillips Curve and Okun's Law in US

Null hypothesis	**Granger**	**SNC(2 symbols)**
Phillips curve		
Unemployment does not cause inflation	0.04	1.53
Inflation does not cause unemployment	16.90*	9.41*
Okun's law		
Unemployment does not cause economic growth	3.37	2.94
Economic growth does not cause unemployment	61.01*	9.65*

*Indicates rejection of the null hypothesis at the 5% level significance.

The results are similar for both tests. On one hand, Granger and symbolic tests detect causality from inflation to unemployment in the Phillips curve. On the other hand, the two tests detect causality running from economic growth to unemployment in the Okun's law. The economic theory suggests that inflation increases unemployment while economic growth reduces it. Note that STSA allows thinking about causality in a more general way, whereas Granger noncausality needs to think of continuous measured variables, this should not be a problem for STSA. Let us consider the following example; we now can test the hypothesis of causality from economic growth (G) and inflation (P) to unemployment (U). The main problem is that we have to test causality from a two-dimensional variable to a one dimensional. Symbolization permits to transform the two-dimensional problem in one dimensional and then to apply the symbolic test as explained. We can follow a similar approach as in [66] where STSA is applied to dynamic regimes. Figure 2 shows the transformation of the variable (G, P) in a symbolic variable with an alphabet of four symbols (I: low economic growth and low inflation, II: low economic growth and high inflation, III: high economic growth and high inflation, IV: high economic growth and low inflation) considering as partition the mean of each variable. Note that now the application of symbolic causality is easy, the hypothesis that the economic regime (G, P) does not cause unemployment is rejected since the SNC is 31.76 and Chi-2 with 15 degree of freedom (4^2–1) at 95% is 25.00. The opposite hypothesis is not rejected because the SNC is 24.71. It is not possible to test this type of causality with the traditional Granger noncausality test.

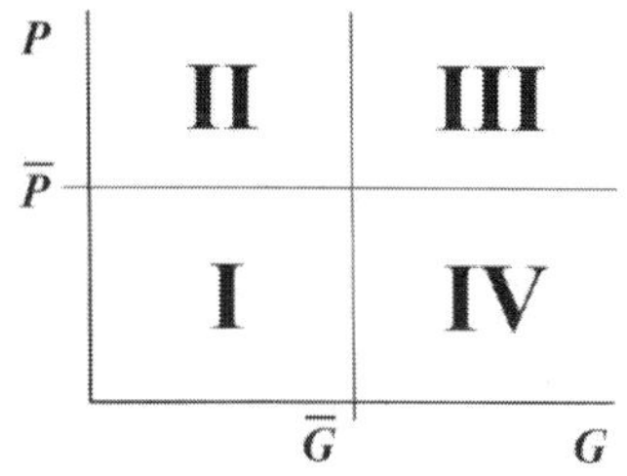

Figure 2: Two-dimensional variable (economic growth and inflation) is transformed into a four symbol variable.

CONCLUSION

STSA is a powerful tool being applied to many scientific fields. There are recent applications in robotic, biology, medicine, communication, and engineering. However, applications in social sciences are very recent. The main difficult is the few historical data produced by the social processes. Social sciences are used to applied statistical tests for proving their hypothesis. However, there is much work to do in developing statistical tests based on STSA to be applied in social sciences. There are some very recent efforts applied to economics and finance using STSA. In particular, we present a symbolic independence test, which seems to be powerful in detecting nonlinearities compared with well-known BDS and runs test. The symbolic test is better detecting models such as the chaotic Anosov and Logistic or some stochastic models such as NLMA or NLSIGN. A second symbolic test about causality detects complex processes such as NLAR, nonlinear exponential, or the Lorenz chaotic process when the traditional Granger noncausality cannot. The symbolic causality also enables causality to be tested in a more general perspective. The application of test from a two-dimensional economic variable to a one-dimensional economic variable is a clear example of the potential of STSA in economics and social sciences in general. One future research line could be to develop a powerful nonlinear test for multidimensional variables. As it was explained, STSA permits to transform a multidimensional time series in a one-dimensional time series simplifying the analysis. This could have important applications in relationships involving vector functions. A more general line of research is to find methodologies to define the optimal partition. As mentioned before, equiprobable partition is generally applied but to find the right partition is still a theoretical and practical weakness in STSA.

REFERENCES

1. Box G, Jenkins G. Time Series Analysis: Forecasting and Control. San Francisco: Holden-Day ed; 1970
2. Sims C. Macroeconomics and reality. Econometrica: Journal of the Econometric Society. 1980;48(1):1-48
3. Engle R. Autoregressive conditional heteroscedasticity with estimates of the variance of United Kingdom inflation. Econometrica: Journal of the Econometric Society. 1982;50(4):987-1007
4. Engle R, Granger C. Co-integration and error correction: Representation, estimation, and testing. Econometrica: Journal of the Econometric Society. 1987;55(2):251-276
5. Granger C. Investigating causal relations by econometrics models and cross-spectral methods. Econometrica. 1969;37(3):424-438
6. 6.Box G, Jenkins G, Reinsel G, Ljung G. Time Series Analysis: Forecasting and Control. 5th ed. John Wiley & Sons: New Jersey, US; 2015
7. Granger C, Newbold P. Forecasting Economic Time Series. Academic Press: New York, US; 2014
8. Hamilton J. Time Series Analysis. Princeton: Princeton University; 1994
9. Harris R, Sollis R. Applied Time Series Modelling and Forecasting. Wiley: Chichester, UK; 2003
10. Franses P. Time Series Models for Business and Economic Forecasting. Cambridge University Press: Cambridge, UK; 1998
11. Chatfield C. Time-Series Forecasting. CRC Press: Florida, US; 2000
12. Lütkepohl H. New Introduction to Multiple Time Series Analysis. Springer Science & Business Media: Berlin, Germany; 2005
13. Clements M, Hendry D. Forecasting Economic Time Series. Cambridge University Press: Cambridge, UK; 1998
14. Harris R. Using Cointegration Analysis in Econometric Modelling (Vol. 82). London: Prentice Hall; 1995
15. Banerjee A, Dolado JJ, Galbraith JW, Hendry D. Co-integration, Error Correction, and The Econometric Analysis of Non-Stationary Data. OUP Catalogue: Oxford, UK; 1993

16. Juselius K. The Cointegrated VAR Model: Methodology and Applications. Oxford University Press: Oxford, UK; 2006
17. Johansen S. Likelihood-Based Inference in Cointegrated Vector Autoregressive Models. Oxford University Press: Oxford, UK; 1995
18. Tsay R. Analysis of Financial Time Series (Vol. 543). John Wiley & Sons: New York, US; 2005
19. Terasvirta T, Tjostheim D, Granger CW. Modelling Nonlinear Economic Time Series. OUP Catalogue: Oxford, UK; 2010
20. Kantz H, Schreiber T. Nonlinear Time Series Analysis (Vol. 7). Cambridge University Press: Cambridge, UK; 2004
21. Williams S. Symbolic dynamics and its applications. Proceeding of Symposia in Applied Mathematics. 2004;60:150
22. Piccardi C. On the control of chaotic systems via symbolic time series analysis. Chaos. 2004;14(4):1026-1034
23. Daw C, Finney C, Tracy E. A review of symbolic analysis of experimental data. Review of Scientific Instruments. 2003;74(2):915-930
24. Daw CS, Kennel MB, Finney CEA, Connolly F. Observing and modeling nonlinear dynamics in an internal combustion engine. Physical Review E. 1998;57(3):2811
25. Daw C, Green J, Wagner R, Finney C, Connolly F. Synchronization of combustion variations in a multicylinder spark ignition engine. Proceedings of the Combustion Institute. 2000;28(1):1249-1255
26. Daw C, Finney C, Kennel M. Symbolic approach for measuring temporal "irreversibility". Physical Review E. 2000;62(2):1912
27. Goldstein R. A technique for the measurement of the power spectra of very weak signals. IRE Transactions on Space Electronics and Telemetry. 1962;2:170-173
28. Schwarz U, Benz A, Kurths J, Witt A. Analysis of solar spike events by means of symbolic dynamics methods. Astronomy and Astrophysics. 1993;277:215
29. Hively L, Gailey P, Protopopescu V. Detecting dynamical change in nonlinear time series. Physics Letters A. 1999;258(2):103-114
30. Hively L, Protopopescu V, Gailey P. Timely detection of dynamical change in scalp EEG signals. Chaos: An Interdisciplinary Journal of Nonlinear Science. 2000;10(4):864-875

31. Makki M, Samali B, Li J. Damage localization based on symbolic time series analysis. Structural Control and Health Monitoring. 2015;22(2):374-393
32. Qumar A, Aziz W, Saeed S, Ahmed I, Hussain L. Comparative study of multiscale entropy analysis and symbolic time series analysis when applied to human gait dynamics. In: 2013 International Conference on Open Source Systems and Technologies (ICOSST). IEEE: Lahore, Pakistan; December, 2013; 2013.
33. Kim J, Park J, Seo J, Lee W, Kim H, Noh J, Yum M. Decreased entropy of symbolic heart rate dynamics during daily activity as a predictor of positive head-up tilt test in patients with alleged neurocardiogenic syncope. Physics in Medicine and Biology. 2003;45(11):3403
34. Cammarota C, Rogora E. Time reversal, symbolic series and irreversibility of human heartbeat. Chaos, Solitons & Fractals. 2007;32(5):1649-1654
35. Rao A, Jain R. Computerized flow field analysis: Oriented texture fields. IEEE Transactions on Pattern Analysis and Machine Intelligence. 1992;14(7):693-709
36. Edwards R, Siegelmann H, Aziza K, Glass L. Symbolic dynamics and computation in model gene networks. Chaos: An Interdisciplinary Journal of Nonlinear Science. 2001;11(1):160-169
37. Godelle J, Letellier C. Symbolic sequence statistical analysis for free liquid jets. Physical Review E. 2000;62(6):7973
38. Sarkar S, Mukherjee K, Sarkar S, Ray A. Symbolic transient time-series analysis for fault detection in aircraft gas turbine engines. In: American Control Conference (ACC); June 2002; 2002. pp. 5132–5137.
39. Ramanan V, Chakravarthy SR, Sarkar S, Ray A Investigation of combustion instability in a swirl-stabilized combustor using symbolic time series analysis. In: American Society of Mechanical Engineers, editor. ASME 2014 Gas Turbine India Conference; December 2014; 2014. p. V001T03A012-V001T03A012.
40. Hsu A, Marshall A, Ricca T. Clipped representations of fourier-transform ion-cyclotron resonance mass spectra. Analytica Chimica Acta. 1985;178:27-41
41. Baptista M, Rosa E, Grebogi C. Communication through chaotic modeling of languages. Physical Review E. 2000;61(4):3590

42. Dolnik M, Bollt E. Communication with chemical chaos in the presence of noise. Chaos: An Interdisciplinary Journal of Nonlinear Science. 1998;8(3):702-710
43. Seto Y, Takahashi N, Jha D, Virani N, Ray A Data-driven robot gait modeling via symbolic time series analysis. In: American Control Conference (ACC); July 2016; 2016. p. 3904–3909.
44. Gullapalli A, Carley K. Extracting ordinal temporal trail clusters in networks using symbolic time-series analysis. Social Network Analysis and Mining. 2013;3(4):1179-1194
45. Das G, Lin KI, Mannila H, Renganathan G, Smyth P. Rule discovery from time series. KKD. 1998;98(1):16-22
46. Brida J, Risso W. Dynamics and structure of the main italian companies. International Journal of Modern Physics C. 2007;18(11):1783-1793
47. Brida J, Risso W. Multidimensional minimal spanning tree: The dow jones case. Physica A: Statistical Mechanics and its Applications. 2008;387(21):5205-5210
48. Brida J, Risso W. Dynamic and structure of the Italian Stock Market based on returns and volume trading. Economics Bulletin. 2009;29(3):2417-2423
49. Brida J, Gomez D, Risso W. Symbolic hierarchical analysis in currency markets. An application to contagion in currency crises. Expert Systems with Applications. 2009;36(4):7721-7728
50. Brida J, Esteban L, Risso W, Such S. The international hotel industry in Spain: Its hierarchical structure. Tourism Management. 2010;31(1):57-73
51. Risso W. The informational efficiency and the financial crashes. Research in International Business and Finance. 2008;22(3):396-408
52. Risso W. The informational efficiency: The emerging markets versus the developed markets. Applied Economics Letters. 2009;16(5):485-487
53. Clausius R. The nature of the motion we call heat. In: Brush SG, editor. Kinetic Theory. (Selected Readings in Physics). Vol. Vol. 2. Pergamon Press: Oxford, UK; 1965. p. 111-134
54. Shannon C. A mathematical theory of communication. Bell System Technical Journal. 1948;27:379-423
55. Khinchin A. Mathematical Foundations of Information Theory. Courier Dover Publications: New York, US; 1957

56. Wald A, Wolfowitz J. An exact test for randomness in the non-parametric case based on serial correlation. The Annals of Mathematical Statistics. 1943;14(4):378-388
57. Brock W, Dechert W, LeBaron B, Scheinkman J. A test for independence based on the correlation dimension. Econometric Reviews. 1996;15:197-235
58. Risso W. An independence test based on symbolic time series. International Journal of Statistical Mechanics. 2014;2014:809383
59. Mathai A, Provost S. Quadratic Forms in Random Variables: Theory and Applications. Marcel Dekker, Inc.; 1992
60. Risso W. A first approach on testing non-causality with symbolic time series. Economic Computation and Economic Cybernetics Studies and Research. 2015;49(3):123-142
61. Hlaváčková-Schindler K, Paluš M, Vejmelka M, Bhattacharya J. Causality detection based on information-theoretic approaches in time series analysis. Physics Reports. 2007;441(1):1-46
62. Cover T, Thomas J. Elements of Information Theory, Wiley. 2nd ed2006
63. Stork, M., Hrusak, J., Mayer, D.. Discrete-time chaotic systems impulsive synchronization and data transmission. In: Proc. 13th WSEAS Int. Conf. System; 2009.
64. Phillips A. The relationship between unemployment and the rate of change of money wages in the United Kingdom 1861–1957. Economica. 1958;25(100):283-299. DOI: 10.1111/j.1468-0335.1958.tb00003.x
65. Okun A. Potential GNP, its measurement and significance. Proceedings of the Business and Economics Statistics Section of the American Statistical Association. 1962:98-104
66. Brida J, Punzo L. Symbolic time series analysis and dynamic regimes. Structural Change and Economic Dynamics. 2003;14(2):159-183

Chapter

APPLICATION OF SYMBOLIC COMPUTATION IN NONLINEAR DIFFERENTIAL-DIFFERENCE EQUATIONS

16

Fuding Xie[1,2], Zhen Wang[3], and Min Ji[1]

[1]Department of Computer Science, Liaoning Normal University, Liaoning, Dalian 116081, China
[2]Key Laboratory of Mathematics and Mechanization (KLMM), Academy of Mathematics and Systems Science, Chinese Academy of Science, Beijing 100080, China
[3]School of Physics and Electronic Technology, Liaoning Normal University, Liaoning, Dalian 116029, China

ABSTRACT

A method is proposed to construct closed-form solutions of nonlinear differential-difference equations. For the variety of nonlinearities, this method only deals with such equations which are written in polynomials in function and its derivative. Some closed-form solutions of Hybrid lattice,

Citation: Fuding Xie, Zhen Wang, Min Ji, "Application of Symbolic Computation in Nonlinear Differential-Difference Equations", Discrete Dynamics in Nature and Society, vol. 2009, Article ID 158142, 8 pages, 2009. https://doi.org/10.1155/2009/158142.

Discrete mKdV lattice, and modified Volterra lattice are obtained by using the proposed method. The travelling wave solutions of nonlinear differential-difference equations in polynomial in function tanh are included in these solutions. This implies that the proposed method is more powerful than the one introduced by Baldwin et al. The results obtained in this paper show the validity of the proposal.

INTRODUCTION

Wadati [1] introduced the following nonlinear differential-difference equation (NDDE):

$$\frac{du_n(t)}{dt} = \left(\alpha + \beta u_n + \gamma u_n^2\right)(u_{n-1} - u_{n+1}), \tag{1.1}$$

where α, β, and γ ≠ 0 are constants.

However, (1.1) can be thought as a discrete version of a nonlinear partial differential equation:

$$u_t + 6\alpha u u_x + 6\beta u^2 u_x + u_{xxx} = 0, \tag{1.2}$$

which can be solved by the inverse scattering method [2].

However, (1.1) obviously includes the following famous NDDEs, namely,

(a) hybrid lattice [3]:

$$\frac{du_n(t)}{dt} = \left(1 + \beta u_n + \gamma u_n^2\right)(u_{n-1} - u_{n+1}), \tag{1.3}$$

(b) discrete mKdV lattice [1, 4]:

$$\frac{du_n(t)}{dt} = \left(1 + u_n^2\right)(u_{n+1} - u_{n-1}), \tag{1.4}$$

(c) modified Volterra lattice [5]:

$$\frac{du_n(t)}{dt} = u_n^2(u_{n+1} - u_{n-1}). \tag{1.5}$$

The travelling wave solutions of (1.3) in polynomial in function tanh are reported in [3]. Ü. Göktas and W. Hereman [4] investigated the conservations laws of (1.4). However, (1.5) is a very well studied integrable model. It is a bi-Hamiltonian, possesses a Lax pair, recursion operator, local master-symmetry, infinite hierarchy of higher symmetries, and conservation laws

[5]. In this study, searching for the closed-form solutions, especially solitary wave solutions and periodic solutions of (1.4) and (1.5), is considered.

In the theory of lattice-soliton, there existed several classical methods to seek for solutions of NDDEs, such as the inverse scattering method [6, 7], bilinear form [8, 9], symmetries [10], and numerical methods [11]. As far as we know, little work is being done to find closed-form solutions of NDDEs by using of symbolic computation. Baldwin et al. [3] recently presented an adaptation of the tanh-method to solve NDDEs. Some analytical (closed-form) solutions of several lattices in polynomial in function tanh have been obtained [3]. Their work may be thought as a breakthrough in solving NDDEs symbolically.

In this study, the method proposed in [3] where tanh-solutions are only considered is firstly generalized, and then is applied to solve (1.1). As a result, a variety of closed-form solutions of (1.1) have been found in terms of trigonometric and Jacobi elliptic functions. The proposed approach allows us to exactly solve NDDEs with the aid of symbolic computation. The solutions presented here not only cover the known one presented by Baldwin et al.[3], but also introduce new solutions for some NDDEs.

The rest of the paper is organized as follows: inthe following section, an improved method is proposed and how to obtain the solitary wave solutions and periodic solutions to NDDEs is depicted. Section 3 is devoted to illustrating the application of the proposal in exactly solving (1.1). As a result, some new solitary wave solutions and periodic solutions of Hybrid lattice, discrete mKdV lattice, and modified Volterra lattice have been obtained. The final is conclusions.

THE IMPROVED METHOD

To solve NDDEs directly, in this section, one would like to describe the improved method and its algorithm. Suppose the NDDE we study in this work is in the following form

$$P\Big(u_{n+p_1}(t), u_{n+p_2}(t), \ldots, u_{n+p_s}(t), u'_{n+p_1}(t), u'_{n+p_2}(t), \ldots, u'_{n+p_s}(t), \ldots,$$
$$u^{(k)}_{n+p_1}(t), u^{(k)}_{n+p_2}(t), \ldots, u^{(k)}_{n+p_s}(t)\Big) = 0, \tag{2.1}$$

where P is a polynomial; $U_n(t)$ is a dependent variable; t is a continuous variable; the superscript denotes the order of derivative; n, $p_i \in Z$.

To compute the travelling wave solutions of (2.1), we first set $u_n(t) = u(\xi_n)$ and

$$\xi_n = dn + ct + \xi_0, \tag{2.2}$$

where d and c are constants to be determined later and ξ_0 is a constant.

Step 1. We assume that the travelling wave solutions of (2.1) we are looking for are in the following frame:

$$u_n(t) = a_0 + \sum_{i=1}^{m} g^{i-1}(\xi_n)\left[a_i f(\xi_n) + b_j g(\xi_n)\right], \tag{2.3}$$

with

$$f(\xi_n) = \frac{\sinh(\xi_n)}{\cosh(\xi_n) + r},$$

$$g(\xi_n) = \frac{1}{\cosh(\xi_n) + r}, \tag{2.4}$$

where a_i, b_j are all constants to be determined later and r a constant.

The degree m in (2.3) can be determined by balancing the highest nonlinear term and the highest-order derivative term in (2.1) as in the continuous case.

The functions $f(\xi_n)$ and $g(\xi_n)$ in (2.4) satisfy the following equations:

$$f'(\xi_n) = \left(1 + r^2\right)g^2(\xi_n) - rg(\xi_n), \quad g'(\xi_n) = -f(\xi_n)g(\xi_n),$$

$$f^2(\xi_n) = (1 - rg(\xi_n))^2 - g^2(\xi_n). \tag{2.5}$$

However, (2.5) reveals that $f^j(\xi_n)\ (j \geq 2)$ can be expressed by $f^l(\xi_n)g^h(\xi_n)\ (l = 0, 1)$.

Step 2 (to derive and solve the algebraic system). The following identity is easily obtained in terms of (2.2):

$$\xi_{n+p_i} = d(n + p_i) + ct + \xi_0 = \xi_n + dp_i. \tag{2.6}$$

It says the relationship between ξ_{n+p_i} and ξ_n. To find the solutions of NDDEs, one also utilizes the following formulae:

$$\sinh(x \pm y) = \sinh(x)\cosh(y) \pm \cosh(x)\sinh(y),$$
$$\cosh(x \pm y) = \cosh(x)\cosh(y) \pm \sinh(x)\sinh(y). \tag{2.7}$$

If the ξ_n in (2.3) is replaced by ξ_{n+p_i}, then expression of u_{n+p_i} in $\sinh(\xi_n)$ and $\cosh(\xi_n)$ will be obtained in terms of (2.7).

With the aid of symbolic computation software Maple 8, we substitute u_{n+p_i} into (2.1) and apply the rule

$$\sinh^2(x) = \cosh^2(x) - 1, \tag{2.8}$$

to simplify the expression. Clearing the denominator and collecting the coefficients of $\cosh^k(\xi_n)\sinh^l(\xi_n)$ $(k = 0,1,\ldots,h;\ l = 0,1)$ and setting them to zero, we can obtain the nonlinear algebraic equations. The values of unknowns will be found by using the Wu›s method [12] to solve algebraic system.

Step 3 (construct and test the exact solutions). Substitute the values obtained in Step 2 along with (2.2) and (2.4) into (2.3), one can find the solutions of (2.1). To assure the correctness of the solutions, it is necessary to substitute them into the original equation.

Remark 2.1. If we set r=0 in (2.4), the travelling wave solutions of (2.1) in polynomial in tanh and sech will be found. If one further requires that b_j=0, then the polynomial travelling wave solutions in tanh will be obtained. In this sense, we can say that this method covers the one given in [3].

Remark 2.2. More important, if the hyperbolic functions in (2.4) are replaced by the trigonometric function, that is,

$$f(\xi_n) = \frac{\sin(\xi_n)}{\cos(\xi_n) + r},$$
$$g(\xi_n) = \frac{1}{\cos(\xi_n) + r}, \tag{2.9}$$

the periodic solutions to (2.1) will be obtained. While doing so, it is necessary to modify the formulae (2.7) and (2.8) properly.

Remark 2.3. We can obviously write (2.4) into the following solitary wave form which have physical relevance:

$$f(\xi_n) = \frac{\tanh(\xi_n)}{1 + r\operatorname{sech}(\xi_n)} = \frac{1}{\coth(\xi_n) + r\operatorname{csch}(\xi_n)},$$
$$g(\xi_n) = \frac{\operatorname{sech}(\xi_n)}{1 + r\operatorname{sech}(\xi_n)} = \frac{\operatorname{csch}(\xi_n)}{\coth(\xi_n) + r\operatorname{csch}(\xi_n)}. \tag{2.10}$$

In what follows, we will apply this method to solve (1.1). As a result, their abundant exact solutions have been derived.

SOLITON WAVE SOLUTIONS AND PERIODIC SOLUTIONS OF (1.1)

By balancing the highest nonlinear term u_n^2 and the highest-order derivative term $du_n(t)/dt$ in (1.1), we have m + 1 = 2m, that is, m = 1.

Therefore, the following formal solutions for (1.1) can be assumed:

$$u_n(t) = u(\xi_n) = a_0 + \frac{a_1 \sinh(\xi_n)}{\cosh(\xi_n) + r} + \frac{b_1}{\cosh(\xi_n) + r}, \tag{3.1}$$

with ξ_n as (2.2).

Starting from (2.6), we have

$$u_{n\pm1}(t) = u(\xi_{n\pm1}) = u(\xi_n \pm d). \tag{3.2}$$

The expressions of $u_{n+1}(t)$ and $u_{n-1}(t)$ in $\sinh(\xi_n)$ and $\cosh(\xi_n)$ are obtained by (2.7), (3.1), and (3.2). Substituting them and (3.1) into (1.1), clearing the denominator, and setting coefficients of the terms $\sinh^l(\xi_n)\cosh^k(\xi_n)$ $(k = 0, 1, \ldots, 3;\ l = 0, 1)$ to zero give

$$a_1\left[2\sinh(d)\left(\gamma r a_1^2 + \beta a_0 r + \alpha r - \beta b_1 - 2\gamma a_0 b_1 + \gamma a_0^2 r\right) + cr\right] = 0,$$

$$2\sinh(d)\left(2\gamma a_0 a_1^2 r - \gamma a_1^2 b_1 - \alpha b_1 - \beta a_0 b_1 - \gamma a_0^2 b_1 + \beta a_1^2 r\right) - b_1 c = 0,$$

$$a_1\Big[c + 2\sinh(d)\cosh(d)\left(\gamma a_1^2 + \gamma a_0^2 + \beta a_0 + \alpha\right) + 4\sinh(d)\left(\gamma a_0^2 r^2 - \gamma b_1^2 + \beta a_0 r^2 + \alpha r^2\right)$$
$$+2cr^2\cosh(d)\Big] = 0,$$

$$\sinh(d)\left(\beta a_1^2 r^2 - 2\gamma a_0 b_1^2 - \beta b_1^2 - 2\gamma a_0^2 r b_1 - 2\alpha r b_1 + 2\gamma a_0 r^2 a_1^2 - 2\beta a_0 r b_1 + \gamma a_1^2 b_1 r\right)$$
$$+ 2a_1^2 \sinh(d)\cosh(d)(2\gamma\, a_0 + \beta) - r b_1 c\cosh(d) = 0,$$

$$a_1\Big[2\sinh(d)\cosh(d)\left(2\alpha r + \beta b_1 + 2\beta\, a_0 r + \gamma a_0 b_1 + \gamma a_0^2 r\right)$$
$$+ 2\sinh(d)\left(-\gamma a_1^2 r + 2\gamma a_0 b_1 + \gamma b_1^2 r + \gamma a_0^2 r^3 + \beta b_1 r^2 + \alpha r^3 + \beta a_0 r^3 + \beta b_1 + 2\gamma a_0 r^2 b_1\right)$$
$$+2cr\cosh(d) + cr\cosh^2(d) + cr^3 - cr\Big] = 0,$$

$$2b_1\sinh(d)\left(-\gamma b_1{}^2-\gamma a_0{}^2r^2+\gamma a_1{}^2-\alpha r^2-\beta a_0r^2-2\gamma a_0rb_1-\beta b_1r\right)$$

$$+2a_1{}^2\sinh(d)\cosh(d)(2\gamma a_0r+2\gamma b_1+\beta r)-b_1c\cosh^2(d)+b_1c-b_1cr^2=0,$$

$$a_1\Big[2\sinh(d)\cosh(d)\left(\beta a_0r^2-\gamma a_1{}^2+\beta b_1r+\gamma a_0{}^2r^2+2\gamma a_0rb_1+\alpha r^2+\gamma b_1{}^2\right)$$

$$+2b_1\sinh(d)(2\gamma b_1+\beta r)+4\gamma a_0ra_1b_1a_1c\cosh^2(d)+cr^2-c\Big]=0. \tag{3.3}$$

It is difficult to solve this system by hand. Thus, we fall back on symbolic computation software Maple 8 and Wu's method which is a powerful tool to deal with nonlinear algebraic equations. With the aid of them, we can find the solutions to the above system as follows:

Case 1.

$$a_1=0,\qquad a_0=-\frac{\beta}{2\gamma}\mp\frac{r\sqrt{(\beta^2-4\alpha\gamma)(r^2-1)}\tanh(d/2)}{2\gamma(r^2-1)}$$

$$b_1=\pm\frac{[2r^2-1-\cosh(d)]\sqrt{(\beta^2-4\alpha\gamma)(r^2-1)}\tanh(d/2)}{2\gamma(r^2-1)},$$

$$c=\frac{[2r^2-1-\cosh(d)](\beta^2-4\alpha\gamma)\tanh(d/2)}{2\gamma(r^2-1)}; \tag{3.4}$$

Case 2.

$$a_0=-\frac{\beta}{2\gamma},\qquad a_1=\pm\frac{\sqrt{\beta^2-4\alpha\gamma}\tanh(d/2)}{2\gamma},$$

$$b_1=\pm\frac{\sqrt{(\beta^2-4\alpha\gamma)(r^2-1)}\tanh(d/2)}{2\gamma},\qquad c=\frac{(\beta^2-4\alpha\gamma)\tanh(d/2)}{\gamma}; \tag{3.5}$$

Case 3.

$$b_1=0,\qquad r=0,\qquad a_0=-\frac{\beta}{2\gamma},$$

$$a_1=\pm\frac{\sqrt{\beta^2-4\alpha\gamma}\tanh(d)}{2\gamma},\qquad c=\frac{(\beta^2-4\alpha\gamma)\tanh(d)}{2\gamma}. \tag{3.6}$$

Thus the solitary wave solutions of (1.1) are

$$u_n(t) = \frac{\beta}{2\gamma} \mp \frac{\sqrt{(\beta^2 - 4\alpha\gamma)(r^2-1)}\ \tanh(d/2)}{2\gamma(r^2-1)}\left(r - \frac{2r^2 - 1 - \cosh(d)}{\cosh(\xi_n) + r}\right), \tag{3.7}$$

where $\xi_n = nd + ([2r^2 - 1 - \cosh(d)](\beta^2 - 4\alpha\gamma)\tanh(d/2)/2\gamma(r^2-1))t + \xi_0$;

$$u_n(t) = -\frac{\beta}{2\gamma} \pm \frac{\sqrt{\beta^2 - 4\alpha\gamma}\ \tanh(d/2)}{2\gamma}\frac{\sinh(\xi_n) + \sqrt{r^2-1}}{\cosh(\xi_n) + r}, \tag{3.8}$$

where $\xi_n = nd + ((\beta^2 - 4\alpha\gamma)\tanh(d/2)/\gamma)t + \xi_0$;

$$u_n(t) = -\frac{\beta}{2\gamma} \pm \frac{\sqrt{\beta^2 - 4\alpha\gamma}\ \tanh(d)}{2\gamma}\tanh\left(nd + \frac{(\beta^2 - 4\alpha\gamma)\tanh(d)}{2\gamma}t + \xi_0\right). \tag{3.9}$$

Similarly, we can also assume that (1.1) possesses the following form solution:

$$u_n(t) = s_0 + \frac{s_1 \sin(\xi_n)}{\cos(\xi_n) + r} + \frac{s_2}{\cos(\xi_n) + r}, \tag{3.10}$$

where s_i $(i = 0, 1, 2)$ is a constant to be determined later.

Repeating the above process and properly modifying the formulae (2.7) and (2.8), we can derive the periodic solutions for (1.1) as follows. For brevity, the procedure of seeking for periodic solutions to (1.1) is omitted:

$$u_n(t) = -\frac{\beta}{2\gamma} \pm \frac{\sqrt{(\beta^2 - 4\alpha\gamma)(r^2-1)}\tan(d/2)}{2\gamma(r^2-1)}\left(r + \frac{1 + \cosh(d) - 2r^2}{\cos(\xi_n) + r}\right), \tag{3.11}$$

where $\xi_n = nd + ([2r^2 - 1 - \cos(d)](\beta^2 - 4\alpha\gamma)\tan(d/2)/2\gamma(r^2-1))t + \xi_0$.;

$$u_n(t) = -\frac{\beta}{2\gamma} \pm \frac{\sqrt{\beta^2 - 4\alpha\gamma}\tan(d/2)}{2\gamma}\frac{\sin(\xi_n) + \sqrt{r^2-1}}{\cos(\xi_n) + r}, \tag{3.12}$$

where $\xi_n = nd + ((\beta^2 - 4\alpha\gamma)\tan(d/2)/\gamma)t + \xi_0$;

$$u_n(t) = -\frac{\beta}{2\gamma} \pm \frac{\sqrt{\beta^2 - 4\alpha\gamma}\tan(d)}{2\gamma}\tan\left(nd + \frac{(\beta^2 - 4\alpha\gamma)\tan(d)}{2\gamma}t + \xi_0\right). \tag{3.13}$$

If we properly set the parameters α, β, and γ, then it is easily to obtain closed-form solutions of Hybrid lattice, discrete mKdV lattice, and modified Volterra lattice, respectively.

In fact, the solution given in [3] for (1.3) is nothing but the solution (3.9). As far as we know, no works have been reported on the closed form solutions of (1.4) and (1.5). Thus, in this sense, these results are novel.

CONCLUSIONS

A method is proposed to find the solitary wave solutions and periodic solutions for NDDEs. The new closed-form solutions of Hybrid lattice, discrete mKdV lattice, and modified Volterra lattice have been found by using the proposal and symbolic computation. This study reveals that computer algebra plays an important role in exactly solving NDDEs. Meanwhile, it is worthwhile to point out that the proposed method may be applicable to other NDDEs to seek for their travelling wave solutions.

ACKNOWLEDGMENT

The authors would like to thank the anonymous reviewer for his/her valuable suggestions. This work was supported by the National Natural Science Foundation of China (Grant no. 10771092).

REFERENCES

1. M. Wadati, "Transformation theories for nonlinear discrete systems," *Progress of Theoretical Physics Supplement*, vol. 59, pp. 36–63, 1976.
2. M. Wadati, "Wave propagation in nonlinear lattice—I," *Journal of the Physical Society of Japan*, vol. 38, pp. 673–680, 1975. | MathSciNet
3. D. Baldwin, Ü Göktas, and W. Hereman, "Symbolic computation of hyperbolic tangent solutions for nonlinear differential-difference equations," *Computer Physics Communications*, vol. 162, no. 3, pp. 203–217, 2004.
4. Ü Göktas and W. Hereman, "Computation of conservation laws for nonlinear lattices," *Physica D*, vol. 123, no. 1–4, pp. 425–436, 1998. View at: Google Scholar
5. V. E. Adler, S. I. Svinolupov, and R. I. Yamilov, "Multi-component Volterra and Toda type integrable equations," *Physics Letters A*, vol. 254, no. 1-2, pp. 24–36, 1999.
6. M. J. Ablowitz and J. F. Ladik, "Nonlinear differential-difference equations," *Journal of Mathematical Physics*, vol. 16, no. 3, pp. 598–603, 1975.
7. M. Toda, *Theory of Nonlinear Lattices*, Springer, Berlin, Germany, 1988.
8. X.-B. Hu and W.-X. Ma, "Application of Hirota's bilinear formalism to the Toeplitz lattice—some special soliton-like solutions," *Physics Letters A*, vol. 293, no. 3-4, pp. 161–165, 2002.
9. H.-W. Tam and X.-B. Hu, "Soliton solutions and Bäcklund transformation for the Kupershmidt five-field lattice: a bilinear approach," *Applied Mathematics Letters*, vol. 15, no. 8, pp. 987–993, 2002.
10. S. Y. Lou, "Generalized symmetries and W∞ algebras in three-dimensional Toda field theory," *Physical Review Letters*, vol. 71, no. 25, pp. 4099–4102, 1993.
11. C. E. Elmer and E. S. Van Vleck, "A variant of Newton's method for the computation of traveling waves of bistable differential-difference equations," *Journal of Dynamics and Differential Equations*, vol. 14, no. 3, pp. 493–517, 2002.
12. W. T. Wu, *Polynomial Equation-Solving and Its Application, Algorithms and Computation*, Springer, Berlin, Germany, 1994.

Chapter

SYMBOLIC MODELLING OF DYNAMIC HUMAN MOTIONS

17

David Stirling, Amir Hesami, Christian Ritz, Kevin Adistambha and Fazel Naghdy

The University of Wollongong Australia

INTRODUCTION

Numerous psychological studies have shown that humans develop various stylistic patterns of motion behaviour, or dynamic signatures, which can be in general, or in some cases uniquely, associated with an individual. In a broad sense, such motion features provide a basis for non-verbal communication (NVC), or body language, and in more specific circumstances they combine to form a Dynamic Finger Print (DFP) of an individual, such as their gait, or walking pattern.

Citation: David Stirling, Amir Hesami, Christian Ritz, Kevin Adistambha and Fazel Naghdy, "Symbolic Modelling of Dynamic Human Motions", IntechOpen Biosensors, 2010, DOI: 10.5772/7215.

Human gait has been studied scientifically for over a century. Some researchers such as Marey (1880) attached white tape to the limbs of a walker dressed in a black body stocking.

Humans are able to derive rich and varied information from the different ways in which people walk and move. This study aims at automating this process. Later Braune and Fischer (1904) used a similar approach to study human motion but instead of attaching white tapes to the limbs of an individual, light rods were attached. Johansson (1973) used MLDs (Moving Light Displays; a method of using markers attached to joints or points of interests) in psychophysical experiments to show that humans can recognize gaits representing different activities such as walking, stair climbing, etc. The Identification of an individual from his/her biometric information has always been desirable in various applications and a challenge to be achieved. Various methods have been developed in response to this need including fingerprints and pupil identification. Such methods have proved to be partially reliable. Studies in psychology indicate that it is possible to identify an individual through non-verbal gestures and body movements and the way they walk.

A new modelling and classification approach for spatiotemporal human motions is proposed, and in particular the walking gait. The movements are obtained through a full body inertial motion capture suit, allowing unconstrained freedom of movements in natural environments. This involves a network of 16 miniature inertial sensors distributed around the body via a suit worn by the individual. Each inertial sensor provides (wirelessly) multiple streams of measurements of its spatial orientation, plus energy related: velocity, acceleration, angular velocity and angular acceleration. These are also subsequently transformed and interpreted as features of a dynamic biomechanical model with 23 degrees of freedom (DOF).

This scheme provides an unparalleled array of ground-truth information with which to further model dynamic human motions compared to the traditional optically-based motion capture technologies. Using a subset of the available multidimensional features, several successful classification models were developed through a supervised machine learning approach.

This chapter describes the approach, methods used together with several successful outcomes demonstrating: plausible DFP models amongst several individuals performing the same tasks, models of common motion tasks performed by several individuals, and finally a model to differentiate abnormal from normal motion behaviour.

Future developments are also discussed by extending the range of features to also include the energy related attributes. In doing so, valuable future extensions are also possible in modelling, beyond the objective pose and dynamic motions of a human, to include the intent associated with each motion. This has become a key research area for the perception of motion within video multimedia, for improved Human Computer Interfaces (HCI), as well as its application directions to better animate more realistic behaviours for synthesised avatars.

DYNAMIC HUMAN MOTIONS USED IN BODILY COMMUNICATION

Bodily communication or non–verbal communication (NVC) plays a central part in human social behaviour. Non-verbal communication is also referred to as the communication without words. Face, hands, shrugs, head movements and so on, are considered as the NVC. These sorts of movements are often subconscious and are mostly used for:

- Expressing emotions
- Conveying attitudes
- Demonstrating personality traits
- Supporting verbal communication (McNeil, 205)

Body language is a subset of NVC. Body language is used when one is communicating using body movements or gestures plus, or instead of, vocal or verbal communication. As mentioned previously these movements are subconscious, and so many people are not aware of them although they are sending and receiving these all the time. Researchers have also shown that up to 80% of all communications is body language. Mehrabian (1971) reported that only 7% of communication comes from spoken works, 38% is from tone of the voice, and 55% comes from body language.

A commonly identified range of NVC signals have been identified (Argyle, 1988) such as:

- Facial expression
- Bodily contact
- Gaze and pupil direction
- Gesture and other bodily movements

- Posture
- Spatial behaviour
- Non–verbal vocalizations
- Smell
- Clothes, and other aspects of appearance

In addition to this as Argyle described the meaning of a non–verbal signal can be different from sender or receiver's points of view. To a sender it might be his emotion, or the message he intends to send and to the receiver can be found in his interpretation. Some NVC signals are common among all the different cultures where some others might have different meanings in different cultures. According to Schmidt and Cohn (2002) and Donato et al. (1999) there are 6 universally recognized facial expressions:

- Disgust
- Fear
- Joy
- Surprise
- Sadness
- Anger

But there are other emotions that could be recognized through body movements including anxiety, nervousness, embarrassment, lying, aggression, boredom, interest, tiredness, defensive, curiosity, agreement, disagreement, and even some states such as thinking and judging. Some emotions are expressed as a sequence of movements, so one will need to use prior or posterior information from movements in order to be able to recognize such specific emotions.

Body Parts and Related Emotions

Certain movements of one body part often need to be associated with the movements of various other parts in order to be interpreted as an emotion. Table 1 details a basic list of the parts that one is is able to acquire data from their movements and the emotions related to those movements are described.

Table 1: Noted emotions for associated body movements (Straker, 2008)

member	movement	interpretation
head	lowering	defensive or tiredness.
	raising	interest, visual thinking.
	tilting	interest, curiosity.
	oscillating up & down	agreement.
	oscillating left & right	disagreement.
	touching	thinking.
arms	expanding	aggression
	crossing	anxiety
hands	holding behind	lying, self confidence
	palms up or down	asking
	rubbing together	extreme happiness.
	repetitive movements	anxiety, impatience..
neck	touching	fear.
shoulder	raised	tension, anxiety or fear.
	lowered	relax
chest	rubbing	tension and stress.
belly	Rubbing or holding	tension
legs	standing with feet together	anxiety
	crossing	tension and anxiety
	repetitive movements	anxiety, impatience
thighs	touching	readiness
feet	curling	extreme pleasure
	stamping	anger and aggression
	moving	anxiety, impatience, lying

These interpretations are acquired from different psychological researches through different web sites and dissertations. Interpretation would clearly depend on cultural and other context.

Table 1 infers a highly complex multidimensional space in which a human body can relay emotional expressions as various spatial articulations at any point in time. This together with any associated temporal sequence surrounding an observed postural state, combine to provide an extremely challenging context in which to capture and further model the dynamics of human motions. A rich array of initial, contributory intentions further obfuscate matters. The decidedly successful analysis of facial micro expressions by Ekman and others (Ekman, 1999) has proven insightful for identifying the underlying emotions and intent of a subject. In a related but possibly more prosaic manner, it is the intended to establish three basic goals from the analysis and modeling of dynamic motions of a human body, these are to:

- develop a sufficient model of dynamic finger printing between several individuals
- model distinctive motion tasks between individuals
- formulate a model to identify motion pretence (acting) as well as normal and abnormal motion behaviours

Successfully achieving some or all of these goals would provide invaluable outcomes for human behavioural aspects in surveillance and the detection of possible terrorism events as well as medical applications involving dysfunction of the body's motor control.

MOTION CAPTURE DATA

Given the three distinct task areas it became prudent to utilise, were ever possible, any existing general motion capture data that may be available, as well as record specific motion data that addressed more specific task needs. To this end the Carnegie Mellon University (CMU) Motion Capture Database (2007) has been utilized explore the second goal, that is to investigate plausible models for the identification of distinctive motion tasks between individuals. This database was created with funding from NSF EIA-0196217, and has become a significant resource providing a rich array of motion behaviours that have been recoded over a prolonged period. Alternatively, the first and last goal objectives require more specific, or specialised captured motion data. For these areas, a motion capture system based on a network array of inertial wireless sensors, as opposed to the more traditional, optical multiple camera based system.

Inertial Motion Capture

Data recorded from this technology is being acquired using an inertial movement suit, Moven® from Xsens Technologies, which provides data on 23 different segments of the body kinematics such as position, orientation, velocity, acceleration, angular velocity and angular acceleration as shown in Figure 1.

In capturing human body motion no external emitters or cameras are required. As explained by Roetenberg et al. (2007) mechanical trackers use Goniometers which are worn by the user to provide joint angle data to kinematic algorithms for determining body posture. Full 6DOF tracking of the body segments are determined using connected inertial sensor modules (MTx), where each body segment's orientation, position, velocity, acceleration, angular velocity and angular acceleration can be estimated. The kinematics data is saved in an MVNX file format which is subsequently read and used, using an intermediate program coded in MATLAB.

Using the extracted features, a DFP (Dynamic Finger Print) can be generated for each individual. DFP is used to identify the individual or detect departure from his/her expected pattern of behaviour. Using this comparison, it is possible to find the smoothness or stiffness of the movement and find out if the person is concealing an object. In order to recognize identity of an individual, different measurements will be made to extract the unique

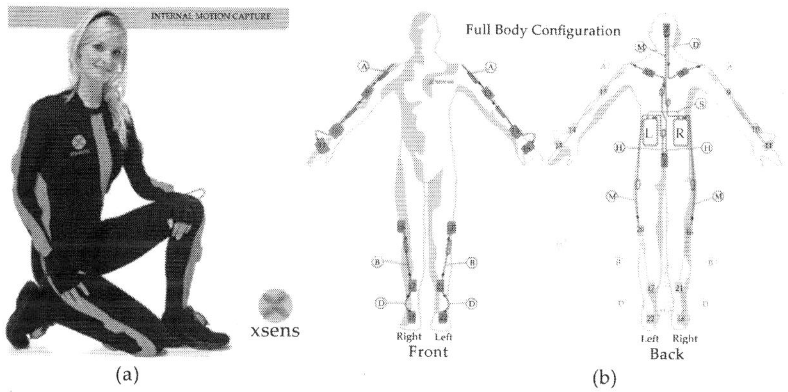

Figure 1: Inertial Motion Capture: (a) Moven®, light weight latex motion suit housing a network of 16 MTx inertial sensors (b) distribution of MTx sensors including the L and R aggregation and wireless transmitter units— adapted from (Xsens Technologies, 2007).

Dynamic Finger Print (DFP) for that individual. The data produced by the suit consists of kinematics information associated with 23 segments of the body. The position, velocity, acceleration data for each segment will be then analyzed and a set of feature of derived will be used in classification system.

Feature Extraction

The determination/selection and extraction of appropriate features is an important aspect of the research. All the classification results would be based on the extracted features. The features should be easy to extract and also must contain enough information about the dynamics of the motion. The selected features should be independent of the location, direction and trajectory of the motion studied. In the case of a sequence of walking motions (or gait) it would be reasonable to deduce that the most decisive/important facets to consider would be the legs, feet and arms. Features are extracted in a gait cycle for each individual. The gait cycle is a complete stride with both legs stepping, starting with the right leg as shown in Figure 2. A typical recording session of a participant wearing the suit is shown in Figure 3.

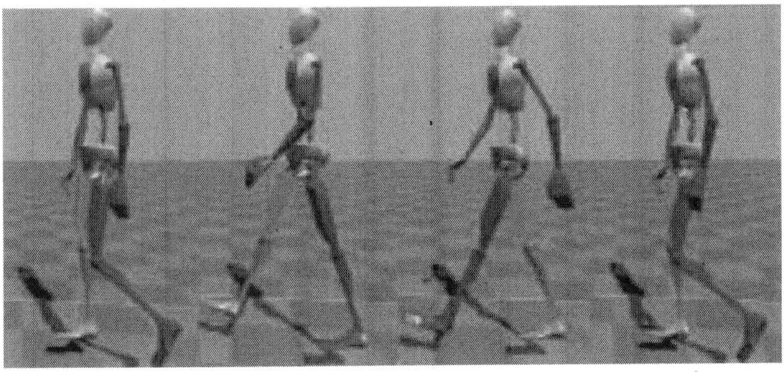

Figure 2: A sample gait cycle: as received from the wireless inertial motion suit and animated on a 23 DOF avatar within the Moven Studio software.

The data produced by the Moven system is stored in rich detail within an MVNX (Moven Open XML format) file which contains 3D position, 3D orientation, 3D acceleration, 3D velocity, 3D angular rate and 3D angular acceleration of each segment in an XML format (ASCII). The orientation output is represented by quaternion formalism.

Figure 3: Recording of the Body Motions; on average, each participant walked between ground markers, white to black, and return in some seven seconds.

The extracted features chosen are the subtended angles of the following body elements:

- Left and Right Foot Orientation,
- Left and Right Foot,
- Left and Right Knee,
- Left and Right Thigh,
- Left and Right Elbow,
- Left and Right Arm.

In total 12 features per individual was extracted, were each angle is given in radians. The location and interpretation of these features is illustrated on the animated motion avatar in Figure 4.

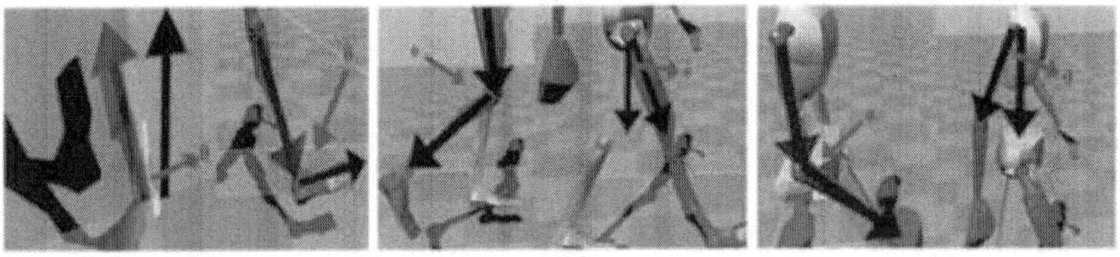

(a) (b) (c)

Figure 4: Selected features annotated of the Moven avatar; (a) Foot Orientation Angle and Foot Angle, (b) Knee Angle and Thigh Angle (c) Elbow Angle and Arm Angle.

An example plot combining all of the 12 selected features, for five participants (p6-p10), can be seen in Figure 5. These have been concatenated together for comparison; the extent of each individual is delineated by grey vertical lines—each individual marking some 3 to 4 gait cycles in-between. This amounted to some 3 to 4 seconds for a subject to walk from one marker to the other, and for a sample rate of 120Hz this equates to some 360 to 480 captured data frames per person.

One can readily appreciate several various differences in gait amongst these participants—such as the marked variations in angular extent of foot orientations (Left Foot O, Right Foot O), and their associated temporal behaviour. Despite this array of other differences the leg period of each remains approximately similar as their variation of height is not significant, nor the distance each travelled between the markers during the recording sessions.

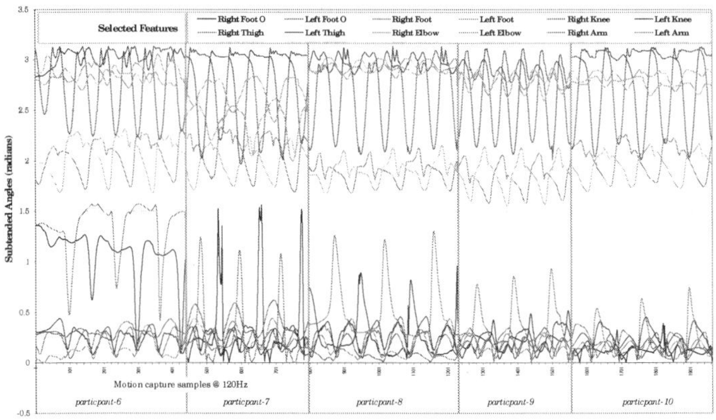

Figure 5: Temporal trends for the 12 selected features across participants p6—p10.

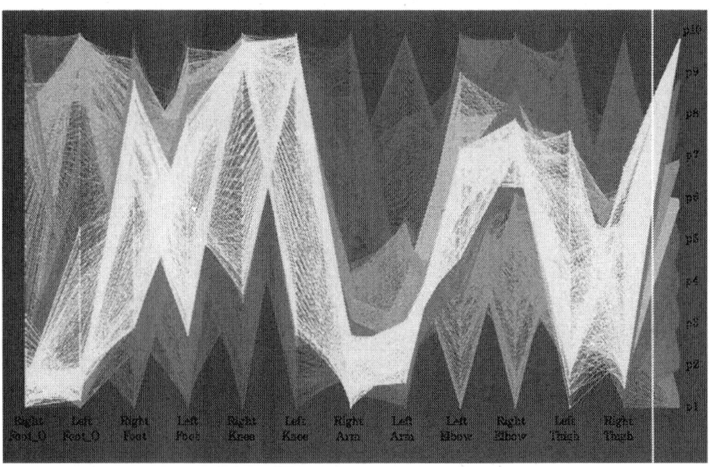

Figure 6: Parallel Coordinate Plot: providing visualisation of all selected features, for all participants (p1-p10) —covering here, 3837 data frames.

Although there degrees of diversities between the trends in Figure 5 of all selected features, one may still remain unconvinced that a set of dynamic finger prints ultimately exists, and if so how could they possibly be reliably extracted? Part of this difficulty arises from observing the distinct feature dissimilarities as a function of time. A more pragmatic approach would be to transform these into alternative domains such as FFT or Wavelets. However, an alternative to either of these might be to visualise the features through a

Parallel Coordinate Plot (PCP), as illustrated in Figure 6, in order to explore the multivariate data without the coupling effect of time.

The PCP of Figure 6 obtained via a visualisation tool Ggobi (Cook and Swayne, 2007), here, arranges a series of parallel coordinates axes, one for each feature, scaled to represent the normalised range of each. The right-most axis of this plot further provides a numerically ordered array of the 10 participants. Every frame of the motion capture data, although constrained to the 12 selected features, is represented by a distinct line that intersects each feature coordinate axis at an appropriate (normalised) value. By colour coding (brushing) the data fames for each participant, one can more readily appreciate potentially unique signatures of profile patterns (or DFP) across the combined feature space. In comparison, both Figure 5 and Figure 6 are derived from the same data; however the participants in the former are essentially contrasted with each other (but only half of these for clarity) in the temporal domain. However, in the latter case of Figure 6 all participants are explicitly compared with each other solely in the feature domain, which also reveals strong visual evidence for the existence of motion signatures amongst the various individuals.

SYMBOLIC MODELLING OF DFP

The principal benefit of symbolic machine-learning (modelling), as opposed to other approaches such as physical modelling (or knowledge-driven modelling), is that it is essentially an empirical, or data driven, modelling process which endeavours to represent only the patterns of relationships or process behaviours (here human movements). Hence, it is readily able to cope with significantly higher dimensionality of data. Non-symbolic machine learning approaches, such as artificial neural networks also address such problems, but lack the major benefits offered by symbolic modelling — these being the transparency of learnt outcomes or patterns, plus an adaptive process of the model structure to scale to accommodate data. These abilities are necessary in order to critique and understand patterns and knowledge that may be discovered.

In order to examine the Dynamic Finger Print hypothesis, the ten individuals wearing the Moven suit, undertook four repetitions of a simple walking task. From these tasks, the selected features, across the individuals were collected and recorded for an identification trial. For this trial, the goal was to clearly identify an individual based purely on a combination of the subtended joint angles. In addressing this recognition challenge, the machine learning, rule induction system known as See5 (RuleQuest, 2007) was used. This system, being a supervised learning algorithm was utilised to induce symbolic classification models, such as decision trees, and or rule sets, based on the range of chosen features (attributes), including a priori known classes. The final decision trees and rule sets were created through adjustment of the various pruning options, but primarily through the (major) pruning control for the minimum number of cases option (M).

Essentially a large tree is first grown to fit the data closely and then pruned by removing parts that are predicated to have relatively high error rate. The pruning option, M, is essentially a stopping criterion to arrest the expansion formation of a decision tree and any associated rule set derived from it. It specifies the minimum number of cases that are required before any leaf classification node is formed and essentially constrains the degree to which the induced model can fit the data. In order to obtain a more reliable estimate of the predictive accuracy of the symbolic model *nfold* cross validation is used as illustrated in Figure 7.

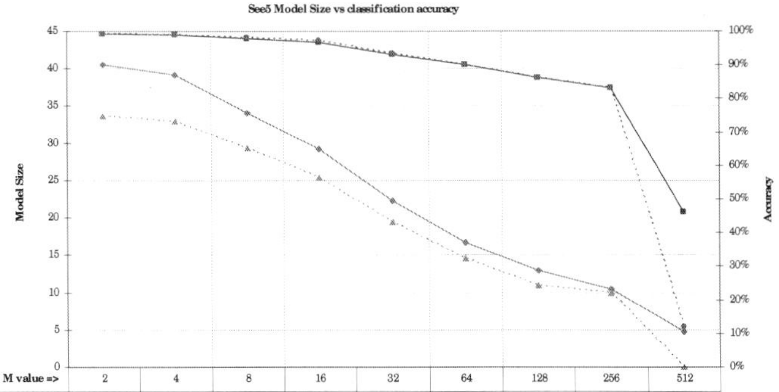

M value =>	2	4	8	16	32	64	128	256	512
Tree size	40.5	39.1	34	29.2	22.3	16.6	12.9	10.4	4.7
Rules size	33.7	32.9	29.3	25.4	19.4	14.5	11	10	0
Tree accuracy	99.3%	98.9%	97.9%	96.8%	93.1%	90.0%	86.1%	83.1%	46.0%
Rules accuracy	99%	99%	98%	97%	93%	90%	86%	83%	12%

Figure 7: Model size and accuracy variations as measured by 10fold cross validation.

The cases in the feature data file are divided into *nblocks* of approximately the same size and class distribution. For each block in turn, a classifier model is induced from the cases in the remaining blocks and tested on the cases in the holdout block. In this manner, every data frame is used just once as a test case. The error rate of a See5 classifier produced from all the cases is then estimated as the ratio of the total number of errors on the holdout cases to the total number of cases (See5, 2002). Here, the number of folds has been set to 10.

As can been seen in Figure 7 there is a nonlinear trade-off between model size and accuracy. Given that the intended use of the model can be guided as to the most dominant factor. Which at the two extremes can be either; a greater generalisation with a reduced model size or, alternatively, a larger, more sensitive model that is less likely to produce miss-classifications. The objective in this task was to model potential motion signatures, and as an example we have chosen a model size that generally reflects a 90~95% accuracy, here M=64.

Once a suitable classifier performance level has been identified using the cross validation trends, the resultant model is generated as illustrated by the rule set model in Figure 8.

For this task we are seeking to establish an individual motion signature for all participants, thus there are ten classes p1p10. Participants undertaking the experiments were 5 males and 5 females between 18 to 40 years of age. According to Figure 8, the average error rate achieved is some 6.8% and number of rules is 18.

Rule 1: (1119/ 728, lift 3.3) Left Foot O > 1.124812 Right Elbow <= 2.901795 => class p1 [0.350]	Rule 2: (296/ 28, lift 9.7) Left Foot O > 1.124812 Right Elbow > 2.901795 Left Elbow > 2.918272 => class p2 [0.903]	Rule 3: (66/ 28, lift 6.2) Right Foot O > 1.260007 Left Foot O <= 1.124812 Right Elbow > 2.640656 => class p2 [0.574]	Rule 4: (225/ 3, lift 10.7) Left Foot O > 1.124812 Right Foot <= 2.100667 Right Knee <= 2.866177 Right Elbow <= 2.901795 Left Elbow > 2.795459 Left Arm > 0.1387282 => class p3 [0.982]
Rule 5: (191/ 21, lift 9.6) Left Foot O > 1.124812 Right Foot <= 2.100667 Right Knee <= 2.866177 Right Elbow <= 2.901795 Right Arm <= 0.2898046 => class p3 [0.886]	Rule 6: (65/ 25, lift 6.7) Right Foot O > 1.053137 Left Foot O <= 1.124812 Right Elbow <= 2.640656 => class p3 [0.612]	Rule 7: (350, lift 10.9) Left Foot O <= 1.124812 Left Arm > 0.4538144 => class p4 [0.997]	Rule 8: (395, lift 9.4) Left Foot O > 1.124812 Right Elbow > 2.901795 Left Elbow <= 2.918272 => class p5 [0.997]
Rule 9: (224/ 15, lift 8.1) Left Foot O > 1.124812 Right Knee > 2.866177 Right Elbow <= 2.901795 Left Elbow > 2.795459 => class p6 [0.929]	Rule 10: (188/ 15, lift 8.0) Left Foot O > 1.124812 Right Elbow <= 2.901795 Left Elbow > 2.795459 Right Arm > 0.2898046 Left Arm <= 0.1387282 => class p6 [0.916]	Rule 11: (80/ 13, lift 7.2) Right Foot O > 1.00804 Right Foot O <= 1.260007 Left Foot O <= 1.124812 Right Elbow > 2.640656 => class p6 [0.829]	Rule 12: (615/ 311, lift 4.3) Left Foot O > 1.124812 Right Elbow <= 2.901795 Right Arm <= 0.3535621 => class p6 [0.494]
Rule 13: (326, lift 10.7) Right Foot O <= 1.053137 Left Foot O <= 1.124812 Right Elbow <= 2.640656 => class p7 [0.997]	Rule 14: (838/ 435, lift 4.3) Right Foot O <= 1.00804 Left Foot O > 0.1827743 Left Foot O <= 1.124812 Right Elbow > 2.640656 Left Arm <= 0.4538144 => class p8 [0.481]	Rule 15: (295/ 16, lift 11.1) Right Foot O <= 1.00804 Left Foot O > 0.1827743 Left Foot O <= 1.124812 Right Elbow > 2.640656 Left Elbow <= 2.852491 Left Arm <= 0.4538144 => class p9 [0.943]	Rule 16: (169/ 28, lift 9.7) Right Foot O <= 1.00804 Left Foot O <= 1.124812 Left Knee <= 3.004622 Right Elbow > 2.838296 Left Elbow <= 2.879424 Left Arm <= 0.4538144 => class p9 [0.830]
Rule 17: (302, lift 9.3) Right Foot O <= 1.00804 Left Foot O <= 0.1827743 Right Elbow > 2.640656 Left Arm <= 0.4538144 => class p10 [0.997]	Rule 18: (228/ 28, lift 8.2) Right Foot O <= 1.00804 Left Foot O <= 1.124812 Right Elbow <= 2.838296 Left Elbow > 2.852491 Left Arm <= 0.4538144 => class p10 [0.874]	Default class: p6 Evaluation on training data (3837 motion frames) Decision Tree: Size 18, Errors 270(7.0%)	Rules: No 18, Errors 261(6.8%)

Figure 8: An example motion signature model for participants, p1p10.

Each rule in Figure 8 consists of an identification number plus some basic statistics such as *(n, lift x)* or *(n/m, lift x)* these, in fact, summarize the performance of each rule. Here, *n,* is the number of training cases covered by the rule and *m*, where it appears, indicates how many of the cases do not belong to the class predicted by the rule. The accuracy of each rule is estimated by the Laplace ratio *(n − m +1)/(n + 2)*. The *lift x,* factor is the result of dividing a rule's estimated accuracy by the relative frequency of the predicted class in the training set. Each rule has one or more antecedent conditions that must all be satisfied if the rule consequence is to be applicable. The class predicted by the rule is show after the conditions, and a value between 0 and 1 that indicates the confidence with which this prediction is made is here shown in square brackets (See5, 2002).

The overall performance of the signature model can be readily observed in the confusion matrix of Figure 9 which details all resultant classifications and miss-classifications within the trial. The sum of values in each row of this matrix represents the total number of true motion frames that are derived from the associated participant (p1p10). Any off-diagonal values in Figure 9 represent miss-classification errors, such as 13 motion frames of participant p5 was very similar to those exhibited by p2. Here an ideal classifier would register only diagonal values in Figure 9.

All extracted features were available to the induction algorithm as it constructed its various classifier models, however not all of these were ultimately utilised in the final rules. For example considering the model of Figure 8, the number of times that each feature has been referred in the rules, which reflects its importance in classifying a person, is shown in Table 2. According to Table 2 the features, Left Foot, plus the, Left Thigh and Right Thigh, angles have not been used in classifier at all, and the two most important features are angle of the Left Foot Orientation and that of the Right Elbow.

(p1)	*(p2)*	*(p3)*	*(p4)*	*(p5)*	*(p6)*	*(p7)*	*(p8)*	*(p9)*	*(p10)*	<= *classified as*
384	15				7					p1
	306	21			29					p2
		311			42					p3
			350							p4
	13			395						p5
		3			436					p6
6		25				326				p7
	28						347	28	28	p8
								327		p9
								16	394	p10

Figure 9: Confusion matrix analysis of the motion signature model for participants p1p10.

Table 2: Usage of features, highlighting three redundant attributes

Feature	Usage	Percentage of usage of all feature s
Left Foot O	18	26.1%
Right Elbow	17	24.6%
Right Foot O	9	13.0%
Left Arm	8	11.6%
Left Elbow	8	11.6%
Right Arm	3	4.3%
Right Knee	3	4.3%
Right Foot	2	2.9%
Left Knee	1	1.4%
Left Foot	0	0%
Right Thigh	0	0%
Left Thigh	0	0%

Although we had originally included all of the apparently, seemingly important bodily attributes, the induced model has found these, Left to be redundant. These leads to an obvious suggestion of not manually selecting or limiting the range of available attributes, but rather allow the algorithm to choose an appropriate sub-set of these. This in fact is one of the specific approaches employed in Section 5.

Ultimately the various rules in such classifiers all define specific hyper-cubes within the multidimensional feature space. As an example, four rules from an initial version of the signature model are overlaid on a 2-dimensional projection of the 12-dimensional feature space. This was observed in some preliminary data visualisation work carried out on the motion data using Ggobi (Cook & Swayne, 2007). Using projection pursuit visualization, the rotating projection was paused whenever a significant 2D segmentation could be observed. Here, in Figure 10 one can clearly identify participants 9 and 10, and also conceptualize four hyper-cubes encompassing the array of these points (motion data frames) with rules 13, 14, 23 and 24.

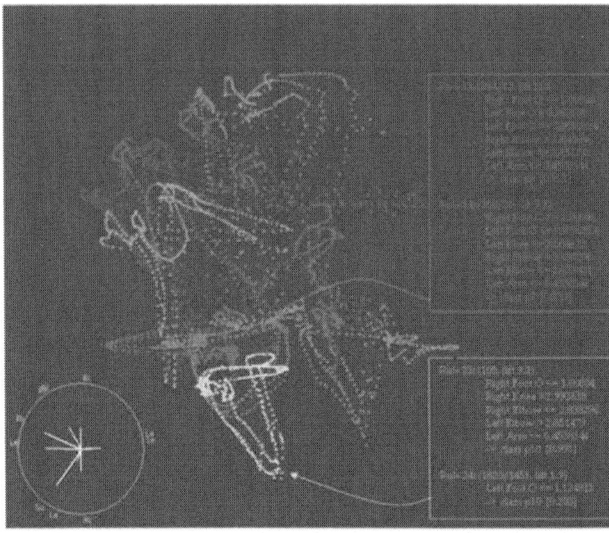

Figure 10: Selective symbolic model rules identifying participants 9 and 10 with a 2D projection of the 12 dimensional featurespace.

The primary aim of this study was to identify a person based on a combination of subtended angles at the feet, knees, thighs, arms, and elbows. In this process 12 features were extracted and using a decision tree and converting this into a rule set classifier 93.2% accuracy was achieved. The participants were 5 males and 5 females between 18 to 40 years of age, indicating that the results obtained were not dependent to specific characteristics of participants. The extracted features could also be used in gender classification, or even different motion classifications. In order to be able to use the described method in a real application, an image processing and computer vision section for data acquisition should be added to the system. The goal in this section is only to test the hypothesis that a plausible signature model to recognize specific individuals could be developed from an appropriate set of features.

SYMBOLIC MODELLING OF DISTINCTIVE MOTION TASKS

This section progresses the development of symbolic modelling to see if it can be used to model various distinctive tasks of human movement skills. As mentioned in Section 3, the CMU Motion Capture Database (2007) offers a significant array of general motions, which would take a considerable period of time to replicate. This data, however, is freely available from the Carnegie-Mellon Motion Capture Database, in the Acclaim ASF/AMC format (CMU Motion Capture Database, 2007).

The data consists of motion capture sequences for various activities such as sports, walking, running, dancing, and nursery rhyme actions. These are captured at a rate of 120 frames per second. For each frame, the optically inferred x, y and z axis rotation for each bone of the body are recorded with respect to the degrees of freedom available for the bone, e.g. the upper arm (humerus) has x, y and z rotations while the forearm (radius) has only x-axis rotation from the elbow.

In total, there are 28 bones in the model as shown in Figure 11, with the 29th bone (root point) representing the rotation and translation of the whole body. This root point serves as the point of origin for the whole skeleton and is situated between the lower back, left hip joint and right hip joint, as illustrated in Figure 11.

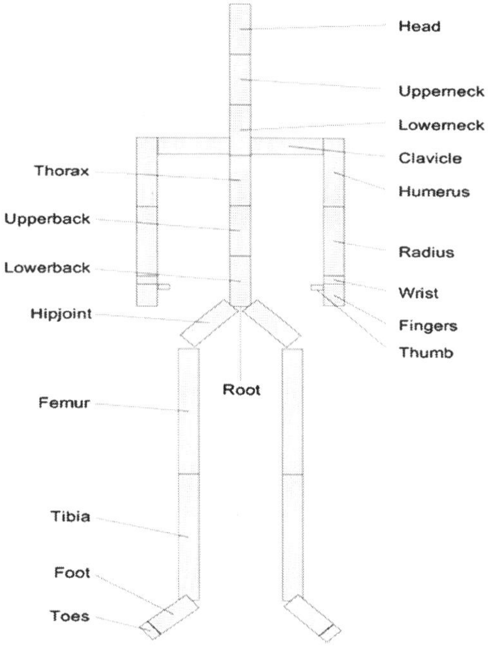

Figure 11: Names and locations of the bones as per the CMU database used in this work.

A plot showing an example of the dataset is shown in Figure 12. In Figure 12, the x-axis represents the frame number of the motion and the y-axis represents the degree of rotation applied to each bone in the skeleton. Figure 12 shows the x, y, and z axis rotation of the lower back bone for two walking motions and a golf swing.

For the purposes of this work, four types of motions consisting of walking, running, golf swing and golf putt were used. The motions were chosen to provide visually similar motions (walking and running), visually dissimilar movements but which utilised a similar set of bones (golf swing and golf putt).

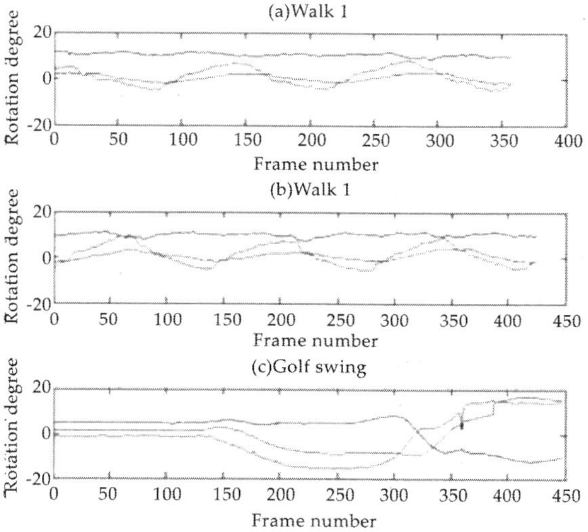

Figure 12: The plots of x, y, and z axis rotations of the lower back bone of two walking motions and a golf swing with different lengths. Each curve represents rotation of the back bones in the skeleton vs. time.

Symbolic Motion Classification using See5

In this section, multiple experiments in developing symbolic models of the motion data using a See5 decision trees were performed where the M value was increased by power of 2 up to 32,768. For each experiment, the size of the decision tree, rule set and the average classification accuracy of each (which was confirmed by 10-fold cross-validation) were recorded. An example of the resulting decision tree for M=8 is shown in Figure 13. In Figure 13, a motion is classified by first looking at the root node of the tree, which contains a threshold decision about the left humerus, x-axis rotation. If the condition is not true, then the next node visited specifies that the left wrist, y-axis rotation be examined. Continuing down the tree to one of the leaf nodes, a data frame of a motion can be classified as a golf swing, golf putt, walk or run motion.

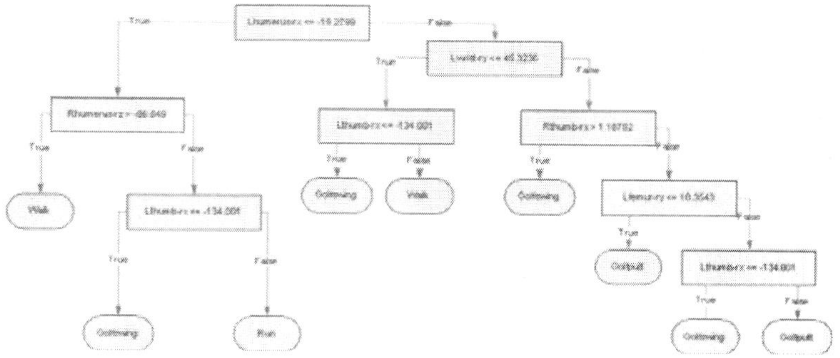

Figure 13: Symbolic motion decision tree for: walk, run, golfswing and golfputt, using M=8.

It can be readily observed in Figure 13 that to in order to classify these four motion classes, only seven bone tracks out of a possible 62 in the motion data base are actually used, and that these seven are the most important features for differentiating between the four motion classes.

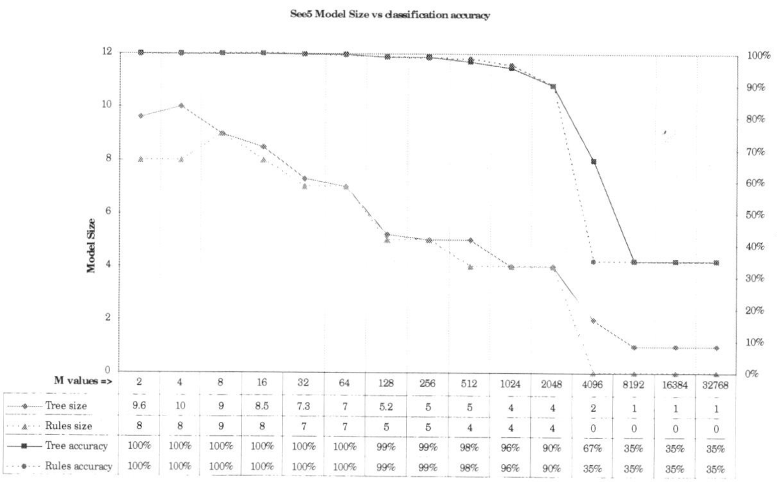

Figure 14: Symbolic Model size and accuracy variations as measured by 10–fold cross validation for four motion classes (walk, run, golfswing, and golfputt).

From the graph presented in Figure 14, the tree in Figure 13 would perform classification with 99.9% accuracy per-frame, which results in

100% accuracy in motion classification. Plots of the M value vs. tree size vs. classification accuracy are shown in Figure 14.

It is evident that in Figure 14 that, there is a knee point in the graph approximately where M=1024, beyond which the classification accuracy begins to decrease significantly i.e., for M=1024 and M=2048, classification accuracies are 96% and 90%, respectively. A typical confusion matrix for such models is illustrated in Figure 15. In Figure 14 there is a further observed knee point at around M=2048, after which for greater values of M the accuracy rate again drops significantly (67% for M=4096 and 35% for M=8192).

(golfswing)	*(golfputt)*	*(walk)*	*(run)*	*<= predicted as*
4463	4			golfswing
1	2507			golfputt
		6616		walk
			1608	run

Figure 15: Typical confusion matrix of the motion model (M=128) for golf-swing, golfputt, walk and run.

It is also of note that parameters of M=2 up to M=32 yields almost 100% classification results. Figure 14 also shows that M=8 for this dataset provides the best classification performance (99.95%), where using smaller M values was not observed to improve classification performance. Using M=8, the resulting decision tree is relatively small with 17 nodes and seven bone motion tracks in total. Hence for the purpose of this work, experiments were performed using decision tree generated with M=8.

Symbolic Modelling of Normal and Abnormal Motion Behaviours

In order to investigate the concept of being able to detect normal and abnormal motion behaviours, a further series of experiments, again involving the Moven inertial motion suit were designed. In this context individuals were asked to carry a back pack with a 5kg weight in it. From these tasks, the same range of features (as used in Section 3) was used again, for the various individuals undertaking the trial.

For this trial, the goal was to clearly identify if a person is carrying a weight or not. However, in addition to this each participant was invited to subtlety disguise their gait on occasions of their choosing, informing the investigators at the end of any recording trial if they had do so. Thus motion data was collected for individual walking gaits that were influenced, or not, by an unfamiliar extraneous weight and also, or not, by a deliberate concealing behaviour of the participant. Again symbolic models of these motion behaviours were induced using the See5 algorithm (RuleQuest, 2007) from the participants using various combinations of subtended joint angles. The algorithm formulates symbolic classification models in the form of decision trees or rule sets, based on a range of several concurrent features or attributes. The model development process followed the same procedure previously discussed in the pervious sections.

For this particular work it was decided to formulate two parallel classifiers to identify both the gender of an individual as well as attempting to deduce if the individual was in fact carrying a weight. The layout of the system is shown in Figure 16.

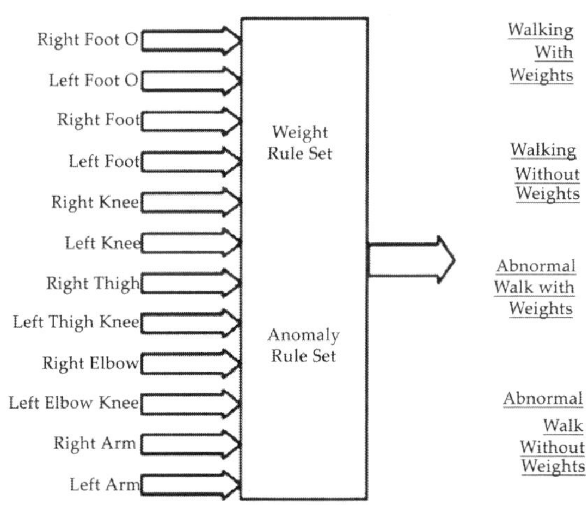

Figure 16: Symbolic model proposal to identify: a weight induced gait anomaly; or an abnormally motion arising from some premeditated disguise.

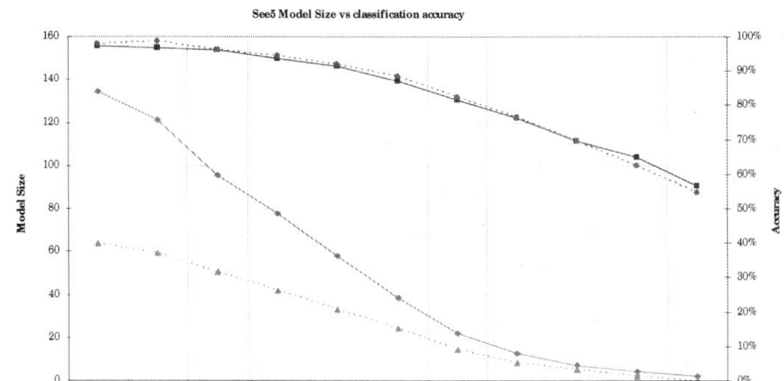

M value =>	2	4	8	16	32	64	128	256	512	1024	2048
Tree size	134.5	120.9	95.5	77.6	57.9	38.4	21.9	12.9	7.2	4.3	2
Rules size	63.7	59.1	50.5	42	33	24.4	14.7	8.3	5.5	2.1	0
Tree accuracy	97.3%	96.7%	96.1%	93.6%	91.3%	86.9%	81.4%	76.3%	69.6%	65.0%	56.5%
Rules accuracy	98%	99%	96%	95%	92%	88%	82%	77%	70%	63%	55%

Figure 17: Symbolic Model size and accuracy variations as measured by 10–fold cross validation for detecting weight induced gait anomalies.

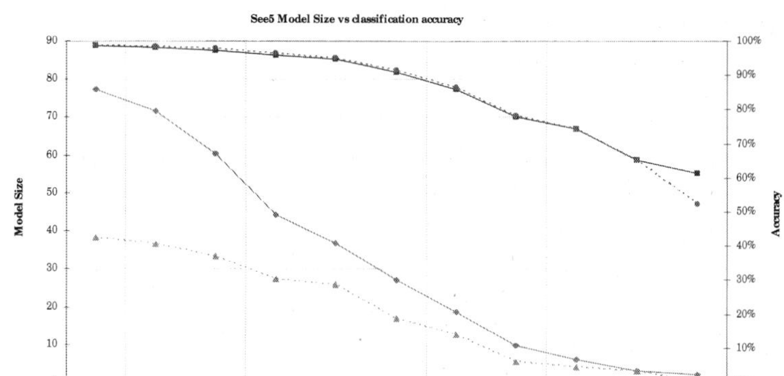

M value =>	2	4	8	16	32	64	128	256	512	1024	2048
Tree size	77.2	71.6	60.5	44.2	36.8	27	18.7	9.6	5.9	3	2
Rules size	38.2	36.5	33.3	27.3	25.8	17	12.8	5.5	3.9	3	0
Tree accuracy	98.5%	98.0%	97.1%	95.8%	94.7%	91.0%	86.0%	78.0%	74.2%	65.3%	61.3%
Rules accuracy	99%	98%	98%	96%	95%	91%	87%	78%	74%	65%	52%

Figure 18: Symbolic Model size and accuracy variations as measured by 10–— fold cross validation for detecting disguised gait related motion behaviours.

Motion data for all 12 subtended joint angles was used in both rule sets in an attempt to classify disguised motion behaviours, and or, individuals that may be carrying an extraneous weight. As in the Sections, 3 and 5, a series of plausible models were firstly analyzed as illustrated in Figure 17 and 18, before their appropriate formal forms were realized as illustrated in Figure 19 and 20.

The participants undertaking these motion experiments were 4 males and 5 females between 18 to 40 years of age. The primary aim of this study was to identify if a person is carrying an object and maybe concealing the object under his clothes based on a combination of the subtended angles at their feet, knees, thighs, arms, and elbows. In this process, again 12 features were extracted and using decision tree and rule set classifier models, more than 87% accuracy was achieved for detecting individuals carrying an extraneous weight, and an accuracy of at least 89% was also achieved in detecting unnatural (pretense) in gait motions.

CONCLUSION

The results from Section 4 and Section 5 clearly support all of the three objectives discussed at the end of Section 2. These being to firstly; develop a plausible model for dynamic finger printing of motion data between individuals. Secondly, investigate a model that could to also identify distinctions between various motion tasks, and finally to formulate a model to identify motion pretence, or acting, as well as normal and (physically induced) abnormal motion behaviours.

Rule 1: (182, lift 2.2) Right Foot O <= 1.282073 Left Foot O <= 1.515643 Right Elbow > 3.057076 => WithWeights [0.995]	Rule 2: (479/ 6, lift 2.2) Right Foot O > 1.282073 Left Arm <= 0.1065039 => WithWeights [0.985]	Rule 3: (216/ 11, lift 2.1) Left Foot O <= 0.2262519 Left Foot <= 2.165244 Right Elbow > 2.876776 Left Elbow > 2.802877 => WithWeights [0.945]	Rule 4: (400/ 21, lift 2.1) Left Foot O <= 1.515643 Right Elbow > 3.027218 Left Elbow <= 3.027386 Right Arm <= 0.2105112 => WithWeights [0.945]
Rule 5: (175/ 26, lift 1.9) Left Elbow <= 2.605653 Left Arm <= 0.3550883 => WithWeights [0.847]	Rule 6: (475/ 74, lift 1.9) Right Foot O <= 0.05478672 => WithWeights [0.843]	Rule 7: (146/ 25, lift 1.8) Right Foot O > 0.05478672 Left Foot O > 1.515643 Left Elbow > 3.02983 Right Arm > 0.1455698 => WithWeights [0.824]	Rule 8: (260/ 49, lift 1.8) Right Foot O <= 1.282073 Right Elbow <= 2.399449 Left Elbow <= 2.802877 => WithWeights [0.809]
Rule 9: (528/ 139, lift 1.6) Left Foot O > 0.3129601 Left Foot O <= 1.515643 Right Knee <= 2.996378 Right Elbow <= 3.057076 Left Elbow <= 2.802877 Left Arm <= 0.3550883 => WithWeights [0.736]	Rule 10: (1092/ 294, lift 1.6) Right Foot O > 1.282073 Left Foot O <= 1.515643 Left Elbow <= 3.027386 Left Arm <= 0.3550883 => WithWeights [0.730]	Rule 11: (297/ 80, lift 1.6) Right Foot O <= 1.282073 Left Foot O > 1.422201 Left Foot O <= 1.515643 Left Elbow > 2.802877 Right Arm <= 0.3095023 => WithWeights [0.729]	Rule 12: (141, lift 1.8) Left Foot O > 1.515643 Left Elbow > 3.02983 Right Arm <= 0.1455698 =>WithOutWeights [0.993]
Rule 13: (259/ 7, lift 1.8) Right Foot O <= 1.282073 Left Foot O > 1.422201 Right Arm > 0.3095023 =>WithOutWeights [0.969]	Rule 14: (267/ 17, lift 1.7) Right Foot O > 1.282073 Right Elbow > 3.027218 Right Elbow <= 3.083863 Right Arm > 0.2105112 Left Arm > 0.1065039 =>WithOutWeights [0.933]	Rule 15: (556/ 77, lift 1.6) Right Foot O > 0.05478672 Right Foot O <= 1.282073 Left Foot O > 1.422201 Right Elbow <= 3.057076 Left Arm > 0.1278452 =>WithOutWeights [0.860]	Rule 16: (6287/ 2659, lift 1.1) Right Foot O > 0.05478672 =>WithOutWeights [0.577]

Default class: WithOutWeights			
Evaluation on training data (6762 motion frames)			
Decision Tree		Rules	
Size	Errors	No	Errors
26	926(13.7%)	16	819(12.1%)

Figure 19: An example motion model for detecting subjects carrying an additional 5kg weight.

Rule 1: (145, lift 2.1) Right Arm > 0.6420745 Left Arm <= 0.432619 => Abnormal [0.993]	Rule 2: (344/ 12, lift 2.0) Right Foot O <= 0.9345771 Left Foot O > 0.206346 Left Foot O <= 1.478905 Right Knee <= 3.012932 Right Arm > 0.2752825 Left Arm <= 0.432619 => Abnormal [0.962]	Rule 3: (298/ 12, lift 2.0) Right Elbow <= 2.419674 Left Arm > 0.432619 => Abnormal [0.957]	Rule 4: (246/ 12, lift 2.0) Right Elbow > 2.905675 Right Elbow <= 3.027221 Left Elbow <= 3.013905 Left Arm <= 0.432619 => Abnormal [0.948]
Rule 5: (395/ 59, lift 1.8) Right Foot O > 1.3187 Right Foot O <= 1.408431 Left Foot O > 0.206346 Right Elbow <= 3.027221 Right Arm > 0.2752825 => Abnormal [0.849]	Rule 6: (4467/ 2094, lift 1.1) Left Foot O > 0.206346 => Abnormal [0.531]	Rule 7: (144, lift 1.9) Left Thigh > 0.5184123 Left Arm <= 0.432619 => Normal [0.993]	Rule 8: (655/ 80, lift 1.7) Left Foot O <= 0.206346 => Normal [0.877]
Rule 9: (371/ 58, lift 1.6) Right Knee > 2.892806 Right Elbow > 2.419674 Left Arm > 0.432619 => Normal [0.842]	Rule 10: (3128/ 1117, lift 1.2) Right Arm > 0.1541919 Left Arm <= 0.432619 => Normal [0.643]	Rule 11: (202/ 73, lift 1.2) Left Foot > 1.776459 Right Elbow <= 3.032689 Right Arm <= 0.1541919 => Normal [0.637]	

Default class: Normal			
Evaluation on training data (5122 motion frames)			
Decision Tree		Rules	
Size	Errors	No	Errors
19	606(11.8%)	11	554(10.8%)

Figure 20: Motion model for detecting subjects manifesting disguised gait motion behaviours.

Motion capture data of human behaviour is necessarily by its nature highly complex and dynamic. Alternative approaches often seek to avoid where ever possible the so-called "curse of dimensionality" (Bellman, 1957) by developing methods to reduce this dimensionality to a tractable lower number of dimensions. Whilst these methods made succeed to various degrees they essentially smother or aggregate out fine detail and various nuances of motion behaviours.

In contrast, the application of symbolic machine learning is able to readily cope with the multidimensional properties of motion data, as evidenced by the example models developed in the previous sections. In effect, an appropriate (symbolic and inductive) DM algorithm will structure and or adjust numerous internal relationships between all of the input features that relate to and support the corresponding output, thereby avoiding, or significantly mitigating, the "curse of dimensionality".

However, whilst such models were often pruned significantly, which may also reduce the domain dimensionality the models address, this process always provides a transparent view of any resultant rules, patterns—often leading to new discovered knowledge. Thus the developer is able to readily critique and further explore various properties and consequences, often through a visualization process, that an individual element of existing or discovered knowledge poses in relation to any reduction in a models resolution (Asheibi, 2009).

Apart from this, motioning the induced symbolic patterns also provides a diagnostic ability guiding the often cyclic and interactive nature of applying machine learning in general. Previous other studies have validated this approach by combining together with unsupervised mixture modelling for gait recognition (Field et al., 2008)(Hesami et al., 2008).

The premise of this proposed work is that all humans have, by the stage of adolescence (or maturity) developed various stylistic signatures or patterns of motion behaviour that can be typically (uniquely) associated with an individual. These become (fundamentally) imprinted as patterns within the central nervous system (CNS) and govern everyday motions such as walking gaits, various gesticulations and other dynamic movements (trunk rotations) of an otherwise static body (Cuntoor et al., 2008). As is obvious, much of these motions can be unconsciously affected or modulated by underlying emotions (Dittrich et al., 1996) or by some conscious intent in order to conceal one's true identity.

In particular, the highly coupled nature of such complex data provides numerous opportunities for the discovery of actionable knowledge patterns, which in turn can be adapted for abnormal motion detection and tracking in two-dimensional (2D) video streams.

It is conjectured that the study of these dynamic (spatiotemporal) multidimensional manifestations will facilitate a new approach to anomaly pattern detection for human motions. By employing (symbolic) machine learning and other related data mining techniques, on a comprehensive range of motion capture trials, it is envisioned that a unique ontology ("structure or science of being", or taxonomy) of such manifest anomaly patterns could be formulated. This would provide a valuable resource structure of (manifest) pattern relationships. Amongst other future goals this research should address is that the motion ontology framework should be utilized to facilitate the derivation of various 2D images and silhouette maps to be subsequently utilized in video pattern analysis for anomaly– identification and ultimately tracking.

REFERENCES

1. Argyle M. 1988 Bodily Communication, Routledge publishing.
2. Asheibi A. Stirling D. Sutanto D. 2009 Analysing Harmonic Monitoring Data using Supervised and Unsupervised Learning, Paper ID: TPWRD-00394-2007. IEEE Trans. On Power Delivery, 24 1 January.
3. Bellman R. 1957 Dynamic Programming, Princeton Univ. Press, Princeton, New Jersey.
4. Braune W. Fischer O. 1904 Der Gang Des Menschen/The Human Gait, Springer, Berlin.
5. Cook D. Swayne D. F. 2007 Interactive and Dynamic Graphics for Data Analysis With R and GGobi, Springer, www.springerlink.com/content/978-0-387-71761-6.
6. Donato G. Bartlett M. S. Hager J. C. Ekman P. Sejnowski T. J. 1999 Classifying facial actions, IEEE Transactions on Pattern Analysis and Machine Intelligence, 21 10 Oct. 1999 Page(s): 974- 989.
7. Ekman P. 1999 Facial Expressions, In T. Dalgleish and T. Power (Eds.), The Handbook of Cognition and Emotion. 301 320 . Sussex, U.K.: John Wiley & Sons, Ltd.
8. Field M. Stirling D. Naghdy F. Pan Z. 2008 Mixture Model Segmentation for Gait Recognition, Symposium on Learning and Adaptive Behaviours for Robotic Systems, LAB-RS'08. ECSIS, 6 8 Aug. 2008 Page(s): 3- 8.
9. Hesami A. Stirling D. Naghdy F. Hill H. 2008 Perception of Human Gestures through Observing Body Movements, Proc. of the International Conference on Intelligent Sensors, Sensor Networks and Information Processing, ISSNIP2008, Sydney, Dec.
10. Johansson G. 1973 Visual Perception of biological motion and a model for its analysis, Percept. Psycophys. 14 1973 210 211 .
11. Mc Neill D. 2005 Gesture and Thought, The University of Chicago.
12. Marey E. J. 1904 Centre Nationale d'Art Moderne, E-J Marey 1830/1904: La Photofraphie Du Movement Paris Centre Georges Pompidou, Musee national d'art moderne, Paris.
13. Mehrabian A. 1971 Silent messages, Wadsworth publishing, Belmont, California.
14. CMU Motion Capture Database. 2007 CMU Graphics Lab Motion Capture Database, http://mocap.cs.cmu.edu/.

15. Cuntoor N. Yegnanarayana B. Chellappa R. 2008 Activity Modeling Using Event Probability Sequences, IEEE Transactions on Image Processing, 17 4 April 2008.

16. Dittrich W. Troscianko T. Lea S. Morgan D. 1996 Perception of emotion from dynamic point-light displays represented in dance. Perception, 25 727 738 .

17. Roetenberg D. Luinge H. Slycke P. 2007 Moven: Full 6DOF Human Motion Tracking Using Miniature Inertial Sensors, Xsen Technologies, December, 2007.

18. RuleQuest Research 2007 Data mining tools, www.rulequest.com.

19. Schmidt K. Cohn J. 2002 Human facial expressions as adaptations: Evolutionary questions in facial expression. Yearbook of Physical Anthropology, 44 3 24 .

20. See5. 2002 See5: An Informal Tutorial, RuleQuest Research, www.rulequest.com.

21. Straker D. 2008 . Changing Minds: in Detail, Published by Syque Press (October 2008).

22. Xsens Technologies 2009 Moven®/MVN®, wireless inertial motion capture www.xsens.com.

Chapter 18

A FRAMEWORK FOR BRIDGING THE GAP BETWEEN SYMBOLIC AND NON-SYMBOLIC AI

Gehan Abouelseoud[1] and Amin Shoukry[2]

[1]Alexandria University

[2]Computer Science and Eng. Dept, Egypt-Japan University of Science and Technology (EJUST), Alexandria, Egypt

INTRODUCTION

FRS (Fuzzy Rule Systems) and ANNs (Artificial Neural Networks) have gained much popularity due to their capabilities in modeling human knowledge and in learning from data, respectively. Fuzzy systems have the advantage of allowing users to incorporate available experts' knowledge directly in the fuzzy model [2]. Thus, decisions made by fuzzy systems are transparent to the user (i.e. the reasons behind the decisions made are clearly understood by tracing the decision and finding out which rules fired and

Citation: Gehan Abouelseoud and Amin Shoukry, "A Framework for Bridging the Gap Between Symbolic and Non-Symbolic AI", IntechOpen Recurrent Neural Networks and Soft Computing, 2012, DOI: 10.5772/38915.

contributed to it). However, there are many parameters whose values are arbitrary. These values have to be "guessed" by a fuzzy system designer, yet they largely influence the system behavior. Thus, a fuzzy system is as good as its programmer.

On the other hand, ANNs have the advantage of being universal functions approximators, requiring only sample data points and no expert knowledge [3]. Despite their advantages, they are essentially black box models. This means that the reasons behind their decisions are concealed in the knowledge acquired in the trained weights. However, these usually have no clear logical interpretation. Thus, their reliability is questionable.

To combine the advantages of both systems while overcoming their disadvantages, two approaches have been proposed in literature. The first is rule extraction from weights of trained ANNs [4]-[7]. However, the proposed approaches often yield some "un-plausible" rules, thus rule pruning and retraining is often required. For examples, some rules may be impossible i.e. their firing depends on conditions that can never occur in reality (impossible antecedents). For example, a rule dictating that a certain action is to be taken in case time is negative. The second approach is ANFIS [8], which attempts to cast the fuzzy system as an ANN with five layers. Although, only two of these layers are adaptable, this model is still more complicated to build and train than a conventional feed-forward ANN for two main reasons. First, the user's expertise is required to choose appropriate consequent (output) membership functions. Second, the desired output needs to be known a priori. This may not be possible for several applications including design problems, inverse problems and high dimensional problems. For example, in a robot path tracking problem, the ANN is required to predict the correct control input. In such application, the desired performance is known but no real-solid rules exist, especially, if the robot is required to be self-adaptive. Similarly, consider a car shape optimization problem. The ANN is required to estimate the shape parameters required to achieve certain air resistance during car motion. Constraints on the shape parameters exist; however, no clear rule database exists relating shape parameters to the desired performance.

The present work proposes a new approach that combines the advantages of fuzzy systems and ANNs through a simple modification of ANN's activation calculations. The proposed approach yields weights that are readily interpretable as logical consistent fuzzy rules because it includes the "semantic" of both input and output variables in the learning/optimization process.

The rest of the chapter is organized as follows. Section II describes the proposed framework. Section III demonstrates its effectiveness through a case study. Section IV shows how it to can be generalized to solve optimization problems. An illustrative example is given for this purpose. Finally, Section V concludes the chapter with a summary of the advantages of the proposed approach.

THE PROPOSED APPROACH

Fig.1 shows a typical feed-forward ANN with a single hidden layer of sigmoid neurons. Conventionally, the output of such an ANN is given by:

$$o_k = \sum_{j=1}^{N_h} w_{jk} sig\left(\sum_{i=1}^{N_i} w_{hij} x_i^p + b_j \right) \tag{1}$$

where $w_{hij}, b_j, w_{jk}, x_i^p, N_h, N_i, o_k$ o are the weight of the connection between the ith input to the j th hidden neuron, the bias of the jth hidden neuron, the weight of the connection between the j th hidden neuron and the kth output neuron, the number of hidden neurons, the number of inputs (elements in each input pattern), the kth output, respectively. The number of output neurons is N_o .

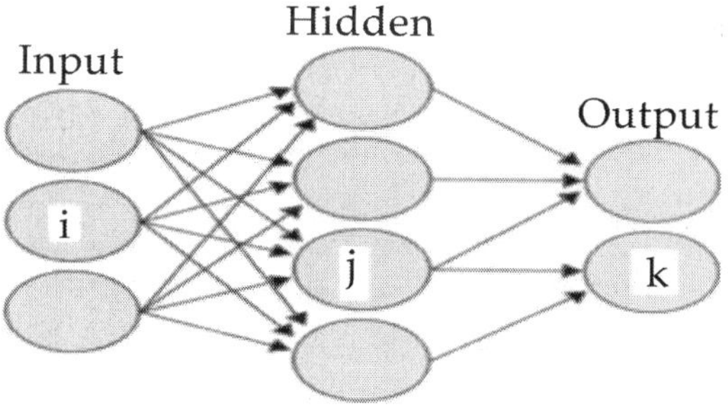

Figure 1: Architecture of a typical feed-forward ANN.

It has been proved in [1] that the sigmoid response to a sum of inputs is equivalent to combining the sigmoid response to each input using the fuzzy logic operator "ior" (interactive or). The truth table of the "ior" operator is shown in table 1. The truth table can be readily generalized to an arbitrary number of inputs. Eq.(1) can, thus, be interpreted as the output of a fuzzy inference system, where the weight w_{jk} is the action recommended by fuzzy rule j. However, this w_{jk} does not contribute directly to the ANN output. Instead, its contribution to the output is weighted by the sigmoid term $sig\left(\sum_{i=1}^{N_i} w_{hij} x_i^p + b_j\right)$.

The sigmoid term corresponds to the degree of firing of the rule, which judges to what extent rule 'j' should participate in the ANN final decision. Moreover, the inference is based on 'ior' rather than the product/'and' fuzzy operator used in ANFIS. It is clear from the 'ior' truth table that the 'ior' operator decides that a rule fully participate in the ANN final decision if all its inputs satisfy their corresponding constraints or if some of them does, while the others are neutral. On the other hand, it decides that the rule should not participate if one or some of the inputs do not satisfy their constraints, while the others are neutral. In the case that some of the inputs completely satisfy the constraints; while others completely violate them, the rule becomes neutral participating by half-weighted recommended action in the final ANN output. The mathematical expression for "ior" is as follows [1]:

$$ior(a_1, a_2, \ldots\ldots\ldots, a_n) = \frac{a_1.a_2 \ldots\ldots\ldots\ldots a_n}{(1-a_1).(1-a_2)\ldots\ldots\ldots(1-a_n) + a_1.a_2\ldots\ldots\ldots a_n} \quad (2)$$

In linguistic terms, an antecedent formed by "ior-ing" several conditions, is equivalent to replacing the conventional phrase: "if A & B & --- then" with "So long as none of the conditions A, B, … are violated --- then". Throughout the paper we will use the Mnemonic "SLANCV" as a shortcut for this phrase. Thus we can say that Eq. 1 can be restated as a set of rules taking the following format:

$$SLANCV \quad x_i^p > -\left(b_j / N_i\right) / w_{hij} \quad then \quad o_{jk} = w_{jk}$$
$$i = 1,2,\ldots,N_i$$

Despite the successful deployment of the "ior" based rule extraction in several applications ([1], [6] and [7]), it has several disadvantages. For example, the weights and biases of a hidden neuron have no direct clear logical interpretation. This makes the incorporation of available knowledge difficult. Such knowledge is of great use in accelerating the ANN training procedure. Besides, leaving weights and biases values unconstrained often lead to some un-plausible rules (rules with impossible antecedent) that need pruning. Therefore, to overcome these disadvantages, our approach is to modify Eq.(1) as follows:

$$o_k = \sum_{j=1}^{N_h} w^c_{jk} sig\left(\sum_{i=1}^{N_i} w^c_{hij}(x^p_i - b^c_{ij}) \right) \tag{3}$$

where, w^c_{hij} are the weights joining input i to hidden neuron j, and w^c_{jk} are the weights joining hidden neuron j to output neuron k.

The superscript 'c' denotes that these weights are constrained. In general, a constrained variable Par^c that directly appears in Eq. (3) is related to its corresponding free optimization variable Par by the following transformation:

$$Par^c = (Parmx - Parmn)\, sig\left(\frac{Par}{\max(Parmx, (Parmn())} \right) + Parmn \tag{4}$$

where, Parmx,Parmn are the maximum and minimum values of the parameter, respectively.

Comparing Eqs. (1) and (3), it is clear that our approach introduces two simple, yet effective, modifications:

- First, in Eq. (3), wh_{ij} is taken as a common factor to the bracket containing the input and bias. Second, there is a bias corresponding to each input (b_{ij}). Using these two modifications, Eq.(3) has a simple direct fuzzy interpretation.

$$SLANCV \quad x^p_i > b^c_{ij}\left(if w^c{}_{hij} > 0\right), x^p_i < b^c_{ij}\left(if w^c{}_{hij} < 0\right) \quad then$$

$$o_{jk} = w^c{}_{jk}, where\ i = 1,2,\ldots,N_i$$

Table 1: Truth Table of the IOR- Operator

First Input	Second Input	IOR Output
0	0	0
0	0.5	0
1	0.5	1
0.5	0.5	0.5
1	0	0.5
0	1	0.5
1	1	1
0.5	1	1
0.5	0	0

This direct interpretation makes it easy for the designer to incorporate available knowledge through appropriate weight initialization; as will be made clear in the adopted case study.

- The weights and biases included in Eq. (3) are constrained according to limits defined by the system designer. This ensures that the deduced rules are logically sound and consistent.

Furthermore, often, the nature of a problem poses constraints on the ANN output. Two possible approaches; are possible; to satisfy this requirement:

- Modifying Eq.(2) by replacing the sigmoid with a normalized sigmoid.

$$o_k = \sum_{j=1}^{N_h} w_{jk}^c \frac{sig\left(\sum_{i=1}^{N_i} w_{hij}^c (x_i^p - b_{ij}^c)\right)}{\sum_{j=1}^{N_h} sig\left(\sum_{i=1}^{N_i} w_{hij}^c (x_i^p - b_{ij}^c)\right)} \tag{5}$$

- Adding a penalty term to the objective function used in the ANN training so as to impose a maximum limit to its output.

In this research, we adopted the first approach.

To apply the proposed approach to a particular design problem, there are essentially three phases

1. Initialization and knowledge incorporation: In this phase, the designer defines the number of rules (hidden neurons) and chooses suitable weights and biases constraints.

2. Training phase.
3. Rule Analysis and Post-Rule-Analysis Processing: The weights are interpreted as fuzzy rules. A suitable method is used to analyse the rules and improve the system performance based on the insight gained from this rule analysis.

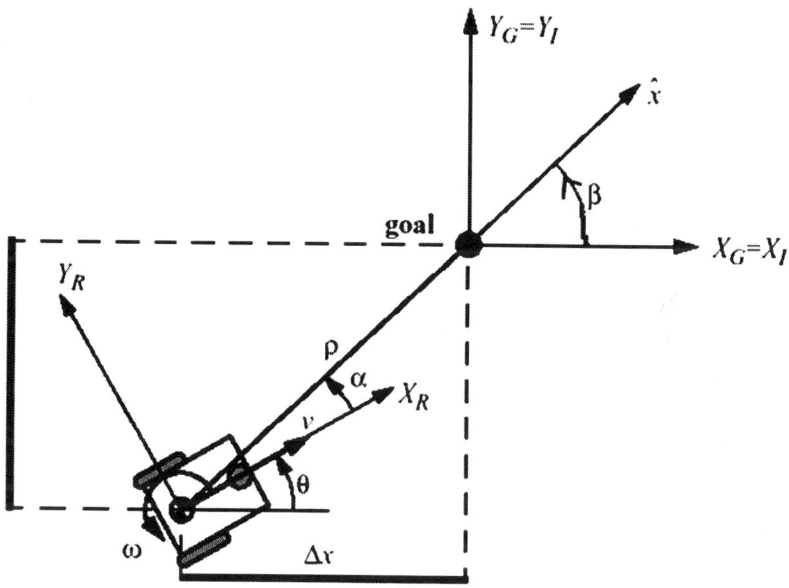

Figure 2. Parameters used to describe the path tracking problem.

CASE STUDY: ROBOT PATH TRACKING

In this example, an ANN is adapted to assist a differential-wheel robot in tracking an arbitrary path. Fig. 2 illustrates the geometry of the problem. The robot kinematic model, the need for a closed loop solution, the choice of a suitable representation for the ANN's inputs and outputs as well as the initialization, training and interpretation of the weights of the obtained ANN, are discussed below.

The Robot Kinematic Model

The kinematic model of the robot is described by the following equations:

$$\dot{X}_R = v \cos\theta$$
$$\dot{Y}_R = v \sin\theta$$
$$\dot{\theta} = \omega \tag{6}$$

Where

$$\dot{X}_R, \dot{Y}_R, \theta, \omega$$

represent the horizontal, vertical components of the robot linear velocity v, its orientation angle and its angular velocity, respectively. Once a suitable control algorithm determines v and ω , it is straightforward to determine the corresponding v_r , v_l values by solving the following 2 simultaneous equations:

$$v = \frac{v_r + v_l}{2} \qquad \omega = \frac{v_r - v_l}{l} \tag{7}$$

where l is the distance between the right and left wheels to yield:

$$v_r = \frac{2\text{v} + \omega l}{2}, v_l = \frac{2\text{v} - \omega l}{2} \tag{8}$$

It is clear from (7) that an average positive (negative) velocity indicates forward (reverse) motion along the robot current axis orientation. Similarly, it is clear from (8), that a positive (negative) difference between v_r and v_l indicates a rotate-left or counter-clockwise (rotateright or clockwise) action. In case both wheels speeds are equal in magnitude and opposite in sign, the robot rotates in-place.

Open vs Closed Loop Solutions of the Path Tracking Problem

Consider a path with known parametric representation (x(t), y(t)) . In this case, the robot reference linear and angular velocities (denoted v_{ref} and ω_{ref}) can be calculated based on the known desired performance. For the robot to follow the desired path closely, its velocity should be tangent to the path. Thus the reference velocities can be computed using:

$$v_{ref} = \sqrt{\dot{x}(t)^2 + \dot{y}(t)^2}$$

$$w_{ref} = \frac{d}{dt}\sin^{-1}\left(\frac{\dot{y}(t)}{v_{ref}}\right) \tag{9}$$

Applying the control inputs (v_{ref} and ω_{ref}) (or equivalently (v_r and v_l)) to the robot would enable it; in the absence of noise and other types of inaccuracies; to follow the required path. This open loop design will be called the "direct forcing case".

However, to assist the ANN in learning the concept of a path (not an instance of a path) as well as making it robust against disturbances, we need to find a closed-loop control law. This is not straightforward because the kinematics model is nonlinear. In what follows, our objective is to show how the proposed ANN-based framework can provide reliable closedloop control; of the form shown in Fig. (3); compared to the direct forcing case. In Fig. (3), the role of the Robot kinematics simulator is to predict the robot location at the current time step given its current control inputs.

For the purpose of illustration, we will restrict our case study to the following family of paths:

$$x(t) = bt; 0 < b < 1 \quad \text{and} \quad y(t) = ct^3; -1 < c < 1$$

where 't' is the time vector= [0 (start time):0.05(time step):1 (final time)].

The ANN is trained on randomly chosen 11 members of this family and tested on different 11 members of the same family. To demonstrate the robustness of the proposed approach an additive disturbance (of uniform distribution) is added to both v and ω . The value of the disturbance can reach up to 200% of v value and 100% of ω value.

Choice of the ANN's Inputs and Outputs

Several possible input-output choices exist. The first has been reported in [11]. The time is the input and the speeds are the output. An alternative choice is to consider the coordinates (x, y) of each point on the path; as inputs; and the corresponding actions (speeds); as output. The third choice is to input the path as a whole as a single input vector and the corresponding sequence of actions (speeds) as a single output vector. All these choices share two fundamental disadvantages. First, it is impossible to interpret the trained weights as fuzzy rules. Furthermore, the ANN does not learn the "concept" of path tracking in general. Instead, it learns to track a single path only. In addition, the first and second choices do not explicitly capture the relation between consecutive path points. To overcome these limitations, we investigated different combinations of ANN inputs and outputs.

Only the two input-output combinations, that produced the best results, are discussed:

i. Case "A":

The inputs to the ANN are chosen to be:

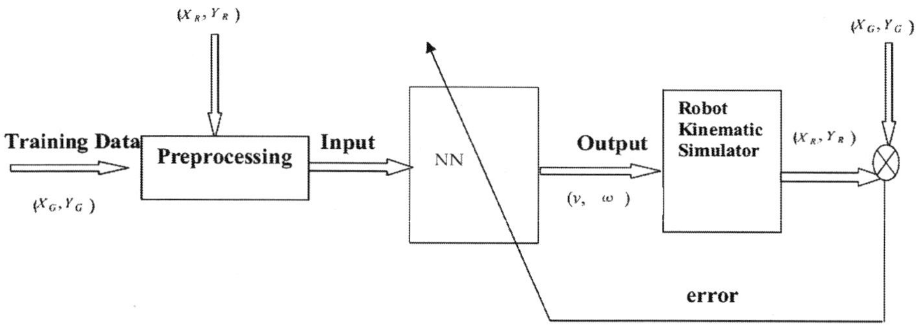

Figure 3: A typical closed loop Block diagram for the flow of information in the Robot Path Tracking Problem.

- the distance ' ρ ' to the goal point, $\rho = \sqrt{(X_G - X_R)^2 + (Y_G - Y_R)^2}$
- the angle 'α ' that the robot needs to rotate to orient itself in the direction of the goal.

Refering to Fig. (2), α is calculated using the following equations:

$$\beta = \tan^{-1}\left(\frac{Y_G - Y_R}{X_G - X_R}\right), \alpha = -\theta + \beta$$

- the deviations between the actual and reference robot linear and angular velocities $(v_{ref} - v_{rob})$ and $(\omega_{ref} - \omega_{rob})$

The outputs are the corrections (increase/ decrease) (v_{inc}, ω_{inc}) needed for (v, ω) to keep following the required path. Fig. 4 shows the block diagram of this system.

ii. Case "B"

The inputs and outputs are as in case A, with the following two additional inputs:

- the previous control inputs (v, ω).

Fig. 5 shows the block diagram of this system.

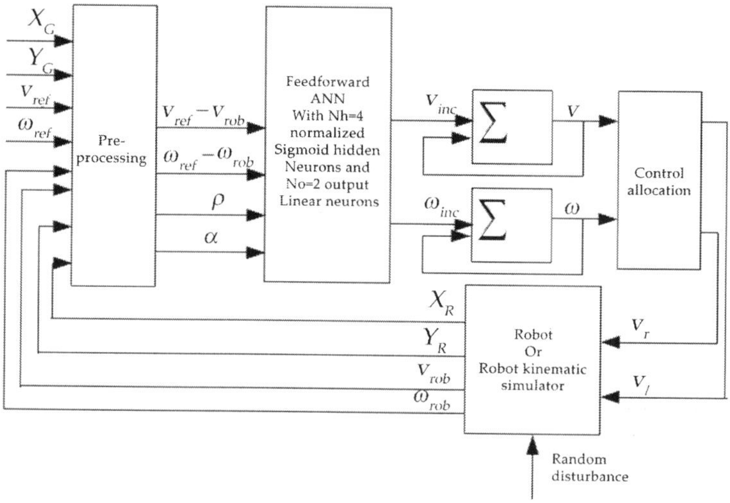

Figure 4: Block diagram of the system in case study case A.

Initialization and Knowledge Incorporation Phase

Knowledge of the range of each input and output variables helps in the proper initialization of both the weights and biases. Recall that N_i is the number of inputs, N_o is the number of outputs, N_h is the number of hidden neurons (which corresponds to the number of rules used by the ANN in decision making). Biases are stored in a matrix B^c of dimensions N_i x N_h : Thus, each column of B^c corresponds to a certain rule. Each entry (row number, column number), in B^c , contains the threshold to which the corresponding input is compared in a particular rule (corresponding to that column). The weights connecting the inputs to the hidden neurons are stored in a matrix Wh^c of dimensions N_ix N_h . As before, each column corresponds to a certain rule. This time, however, the sign of the number in each entry (row number , column number), in Wh^c , controls the comparison operator with the corresponding threshold stored in B^c (negative corresponds to "less than", while positive corresponds to "greater than"). Thus, each column from B^c together with the corresponding column from Wh^c determine the antecedent of a 'SLANCV' rule. The weights connecting the hidden neurons to the output neurons are stored in a matrix Wo^c of dimensions N_h x N_o . The numbers in each row of Wo^c indicate the outputs (consequents) suggested by a certain rule. Such an interpretation of the weights/bias matrices as rules antecedents/ consequents helps in selecting suitable initial values for the ANN's parameters as well as understanding its decisions after training. For example, at initialization, the values in each row of B^c should be constrained to lie within the range of possible values for this input. For instance, if the third input is ρ , then the entries in the third row of B^c should all be positive. Similarly, if the first output is the velocity correction v_{inc} then the first column of Wo^c can take both negative /positive small values to allow the ANN to increase or decrease the robot speed while avoiding instability.

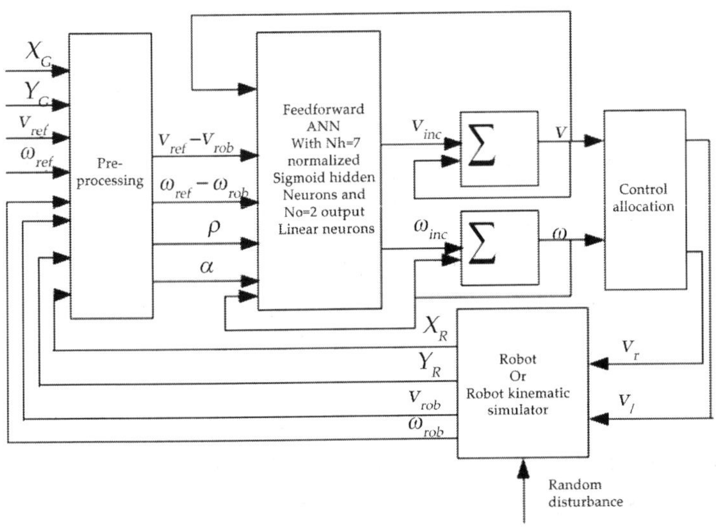

Figure 5: Block diagram of the system in case study case B.

Training Phase

To train the ANN, the following objective function is minimized:

$$objg = \frac{1}{N_p} \sum_{p=1}^{N_p} \sum_{i=1}^{K_i} \rho_i^p \tag{10}$$

where 'ρ_i^p' is the distance between the desired location on the pth path and the corresponding robot's actual location at the ith time instant, K_i is the number of points along the path to be tracked and N_p is the number of paths used in training. Since the objective function is not a closed form function (in the ANN weights and biases) and can only be computed through a simulator, a numerical optimization algorithm is used [10], [11], where gradients are computed using a numerical version of the Broyden–Fletcher–Goldfarb–Shanno (BFGS) optimization algorithm. The gradients are calculated numerically using the following formula:

$$\frac{\partial Objg}{\partial Par_i} = \frac{Objg(Par_i + \varepsilon) - Objg(Par_i - \varepsilon)}{2\varepsilon} \tag{11}$$

where ε is a very small number, typically $\varepsilon = 10^{-11}$. Pari is the ith element of the vector Par For case 'A', the number of adjustable parameters is 2* (4 (inputs) * 4 (hidden neurons/ rules)) + 4 (hidden neurons)* 2 (outputs), which equals 40 parameters. For case B, the number of adjustable parameters is 2* (6 (inputs) * 7 (hidden neurons/ rules)) + 7 (hidden neurons)* 2 (outputs), which equals 98 parameters . Figs. (6 and 7) show the results for cases 'A' and 'B', respectively.

It is clear that case 'B' is more robust to disturbance. However, both cases clearly outperform the open loop (direct forcing) case.

The fact that case 'B' produced better results is to be expected. Certainly, feedback provides memory to the system and helps in accumulating knowledge, which is an essential aspect of learning. We tried two types of feedback; direct feedback in which the ANN learns the link between its own outputs (v_{inc} ,ω_{inc}) and the errors in performance and indirect feedback in which the ANN learns the link between a certain action (v, ω) and the errors in performance. The second memory type produced better results. Appendix A compares these results obtained using our approach with those obtained using conventional ANN [1,6,7] and ANFIS [8].

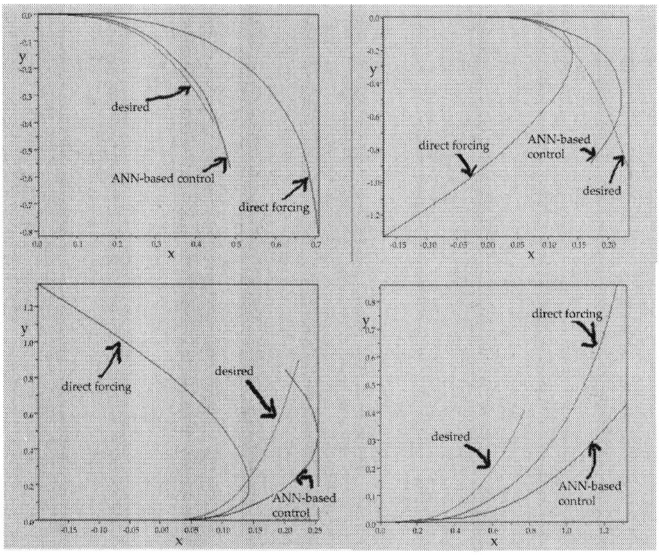

Figure 6: Results for case A. It is clear that closed loop ANN-based control is more robust to disturbances than open-loop direct forcing.

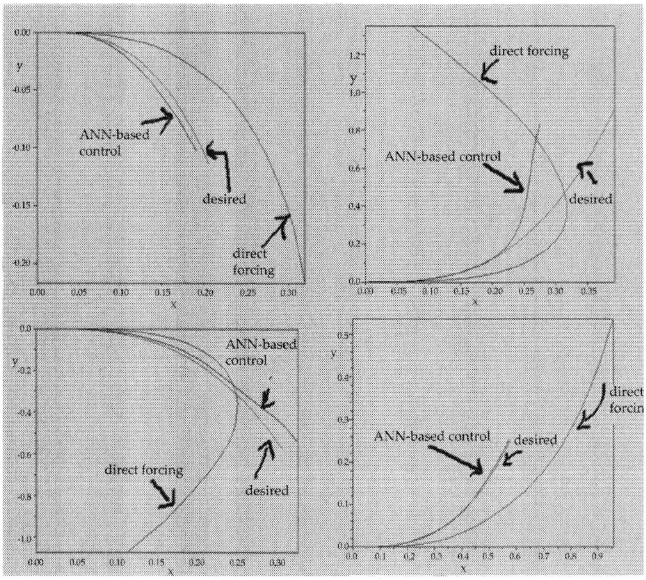

Figure 7: Results for case B. It is clear that closed loop ANN-based control is more robust to disturbances than open-loop direct forcing. It is evident that adding to the ANN inputs the control actions at the previous step improved the results compared to those of case A.

Rule Extraction, Analysis and Post-processing

Following the guidelines given in section 3.4, rule extraction from the trained weights and biases becomes straightforward.

Fig. (8) illustrates how the rules, for case 'A', have been extracted. The discovered rule-base is summarized as follows :

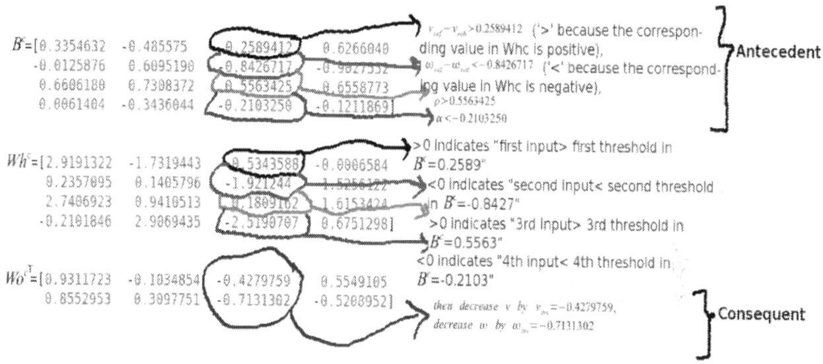

Figure 8. An illustration of the rules extraction from weights and biases matrices.

$$SLANCV \quad v_{ref} - v_{rob} > 0.3354632, \quad \omega_{ref} - \omega_{rob} > -0.0125876, \quad \rho > 0.6606180, \quad \alpha < 0.0061404$$
$$then \quad increase \quad v \quad by \quad v_{inc} = 0.9311723, \quad increase \quad \omega \quad by \quad \omega_{inc} = 0.8552953$$
$$SLANCV \quad v_{ref} - v_{rob} < -0.485575, \quad \omega_{ref} - \omega_{rob} > 0.6095190, \quad \rho > 0.7308372, \quad \alpha > -0.3436044$$
$$then \quad decrease \quad v \quad by \quad v_{inc} = -0.1034854 \quad increase \quad \omega \quad by \quad \omega_{inc} = 0.3097751$$
$$SLANCV \quad v_{ref} - v_{rob} > 0.2589412, \quad \omega_{ref} - \omega_{rob} < -0.8426717, \quad \rho > 0.5563425, \quad \alpha < -0.2103250$$
$$then \quad decrease \quad v \quad by \quad v_{inc} = -0.4279759, \quad decrease \quad \omega \quad by \quad \omega_{inc} = -0.7131302$$
$$SLANCV \quad v_{ref} - v_{rob} < 0.6266040, \quad \omega_{ref} - \omega_{rob} < -0.9027552, \quad \rho > 0.6558773, \quad \alpha > -0.1211869$$
$$then \quad increase \quad v \quad by \quad v_{inc} = 0.5549105, \quad decrease \quad \omega \quad by \quad \omega_{inc} = -0.5208952$$

Similarly, the rule-base for case 'B' is summarized as follows:

$$SLANCV \quad v > 1.1800783, \quad \omega < 0.5797996, \quad v_{ref} - v_{rob} > 0.3283071, \quad \omega_{ref} - \omega_{rob} > -0.1895662,$$
$$\rho > 1.2352848, \quad \alpha > -0.3088343 \quad then$$
$$increase \quad v \quad by \quad v_{inc} = 0.6156011 \quad increase \quad \omega \quad by \quad \omega_{inc} = 0.8038080$$
$$SLANCV \quad v > 1.258573, \quad \omega > 0.817993, \quad v_{ref} - v_{rob} < -0.4339714, \quad \omega_{ref} - \omega_{rob} > 0.0552364,$$
$$\rho > 1.246724, \quad \alpha > 0.2514984 \quad then$$
$$decrease \quad v \quad by \quad v_{inc} = -0.6117282, \quad increase \quad \omega \quad by \quad \omega_{inc} = 0.4437878$$
$$SLANCV \quad v < 0.2235206, \quad \omega > 0.7800814, \quad v_{ref} - v_{rob} < -0.7777185, \quad \omega_{ref} - \omega_{rob} < -0.7120344,$$
$$\rho > 1.5151515, \quad \alpha < -0.0003327 \quad then$$
$$decrease \quad v \quad by \quad v_{inc} = -0.4927100, \quad decrease \quad \omega \quad by \quad \omega_{inc} = -0.0680181$$

$$\text{SLANCV} \quad v > 1.4049258, \quad \omega < -0.3766352, \quad v_{ref} - v_{rob} < 0.0756778, \quad \omega_{ref} - \omega_{rob} < -0.7105109,$$
$$\rho < 0.1169906, \quad a < -0.2358675 \quad \text{then}$$
$$\text{decrease} \quad v \quad \text{by} \quad v_{inc} = -0.2173556, \quad \text{decrease} \quad \omega \quad \text{by} \quad \omega_{inc} = -0.3592529$$

$$\text{SLANCV} \quad v < 0.9930632, \quad \omega > 0.8202830, \quad v_{ref} - v_{rob} > -0.1180958, \quad \omega_{ref} - \omega_{rob} < -0.6506539,$$
$$\rho > 0.1156700, \quad a < 0.2116636 \quad \text{then}$$
$$\text{increase} \quad v \quad \text{by} \quad v_{inc} = 0.3642978 \quad \text{decrease} \quad \omega \quad \text{by} \quad \omega_{inc} = -0.8061992$$

$$\text{SLANCV} \quad v < 0.7807096, \quad > 0.4849204, \quad v_{ref} - v_{rob} > 0.4796375, \quad \omega_{ref} - \omega_{rob} > 0.5922044,$$
$$\rho > 1.2921157, \quad a > 0.2758564 \quad \text{then}$$
$$\text{increase} \quad v \quad \text{by} \quad v_{inc} = 0.3424317, \quad \text{increase} \quad \omega \quad \text{by} \quad \omega_{inc} = 0.8207999$$

$$\text{SLANCV} \quad v < 0.5131913, \quad \omega > -0.2240559, \quad v_{ref} - v_{rob} < -0.4493099, \quad \omega_{ref} - \omega_{rob} > 0.7691425,$$
$$\rho < 1.0904348, \quad a > 0.2174491 \quad \text{then}$$
$$\text{decrease} \quad v \quad \text{by} \quad v_{inc} = -0.5255254, \quad \text{increase} \quad \omega \quad \text{by} \quad \omega_{inc} = 0.8943484$$

In order to improve system performance and remove any inconsistencies, the rules above must be analysed. The following rule analysis procedure is limited to case 'B'. This procedure is assisted by a plot of the DOF (Degree-Of-Firing) of each rule (output of sigmoid) versus time. Such plots help in visualizing which rules the ANN is applying at each time instant and judging the decision/ performance made at this particular time. In particular, it is highly useful to identify dominant rules (rules having relatively high outputs) at the time a wrong decision is made. Correction is then possible by retraining the ANN keeping all rules fixed except the faulty dominant one. For example, as shown in Fig. 9, at the time the deviation from the desired path becomes maximum, rule 5 is dominant followed by rule 4. Accordingly, three different strategies, have been attempted, to retrain the ANN, to improve its performance. In all three strategies, all rules have been kept fixed except:

- for the first strategy: rule 5 (5th column of Wh^c ,5th column of B^c and the 5th row of Wo^c),
- for the second strategy: both the degrees of firing of rules 4 and 5 (4th and 5th column of Wh^c), and their then part (4th and 5th row of Wo^c),
- for the third strategy: the degree of firing of rule 5 and its then part (5th column of Wh^c and the 5th row of Wo^c), with the addition of a new rule (rule 8) whose SLANCV part is the same as that of rule 5 (8th column of B^c is same as its 5th column) but whose DOF (8th column of Wh^c) and consequent part (or then part) (8th row of Wo^c) are to be determined by the training algorithm. The idea is to insert a new rule that co-fires with the malfunctioning rule to provide some correcting action. For our case study, this

last strategy gave us the best results. The results after retraining are shown in Fig. 10.

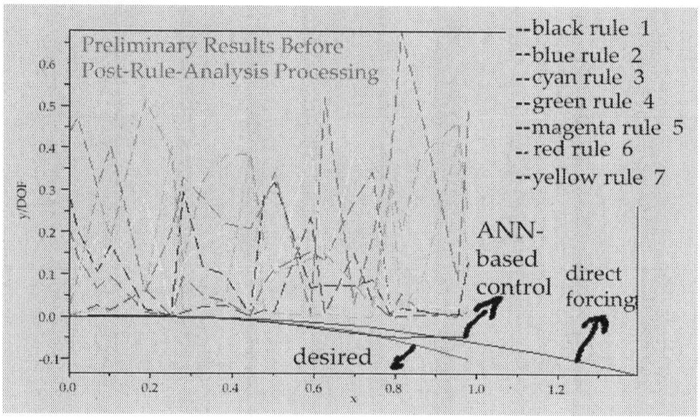

Figure 9: Preliminary results of case B before post-rule analysis processing. The dashed curves represent the degree of firing of each rule with time. It is clear that rule 5 is the dominant rule at the time the deviation from the desired path became maximum.

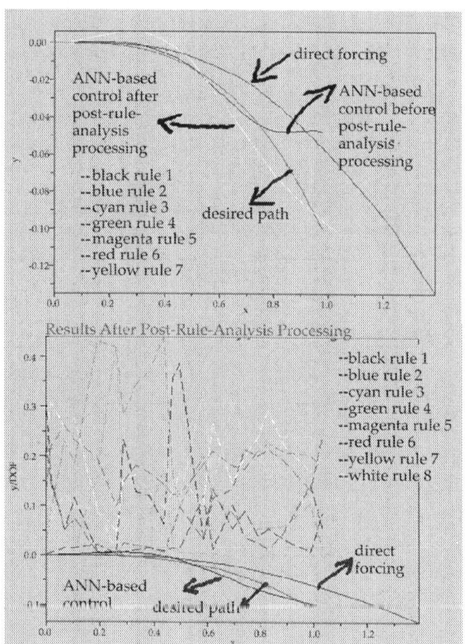

Figure 10: Results of case B after retraining with only the parameters corresponding to rules 5 degree of firing and its "then" part allowed to vary in addition to adding a new rule 8 with the same SLANCV part as rule 5 to co-fire with it and to provide corrective action. It is clear that this retraining improved the results.

DISCUSSION

Although the rule analysis method presented, helped in improving the results, it still needs improvement. In fact, rule analysis is a rather complex task. Understanding the logic behind the rules (why each rule recommends a particular consequent given a certain antecedent) is not simple for the following reasons:

- Rules operate in parallel and collectively. Each rule recommends a certain consequent depending on the context of the neighbouring rules. Collectively, the overall decision helps achieving the desired objective.
- According to the training algorithm, rules recommend certain consequents in order to minimize the desired objective function which is a function of the robot kinematics.
- Rules are derived through batch (off-line) training. Therefore, the overall objective function is minimized over time and not at a particular instant (rules are derived based on a global point of view). Therefore, a rule may not sound reasonable to employ at a certain moment. This what makes it necessary to train an ANN over a certain family of curves. Different families of curves are expected to require different global rules.
- Rule analysis is a trial and error process. Its complexity is proportional to the dimensionality (number of independent variables).

DIRECTIONS FOR FUTURE RESEARCH: SOLVING GENERAL OPTIMIZATION PROBLEMS

The proposed approach can be applied to general optimization problems and not just pathtracking. From an abstract point of view, any optimization problem can be mapped to a goal-tracking problem in which:

- The goal point is the desired performance or desired objective function value.
- The input is the absolute difference between the objective function value at the current solution and the desired objective function value.
- The output is the correction, to the current solution, recommended by the ANN. This is fed-back as an input to the ANN at the next iteration.

- Time evolution corresponds to iterations.

Hence, over the different iterations the ANN is expected to suggest a sequence of corrections that helps in approaching the required objective function value.

This approach to optimization is expected to be less prune to local minima trapping, provides better insight into the nature of the investigated problem and can easily deal with multi-objectives/ errors optimization. For example, in the robot path tracking problem, $\alpha, \rho, v_{ref} - v_{rob}, \omega_{ref} - \omega_{rob}$ were error measures that, ideally are required to be all zeros. In this case, the robot perfectly tracks the desired path and remains tangent to it at all times. This property (being tangent at all times) is desirable, from a practical point of view, because even if the robot path is close to the desired path but with too many frequent changes of orientation, wear out will occur to the robot parts and it will be questionable whether the robot can, practically, makes these moves).

Illustrative Example: Minimizing the Camel Benchmark Function

The camel objective function is defined as follows

$$f_{Sixh}(x_1, x_2) = \left(4 - 2.1x_1^2 + \frac{x_1^4}{3}\right)x_1^2 + x_1 x_2 + \left(-4 + 4x_2^2\right)x_2^2; -3 \le x_1 \le 3, -2 \le x_2 \le 2.$$

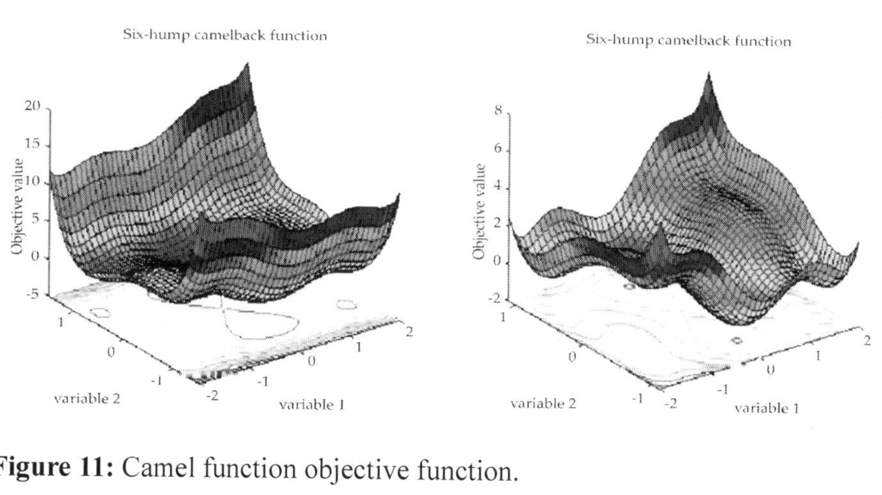

Figure 11: Camel function objective function.

The global minimum is $f(x_1, x_2) = -1.0316; (x_1, x_2) = (-0.0898, 0.7126), (0.0898, -0.7126)$.

Fig. 11 shows a plot of the camel objective function.

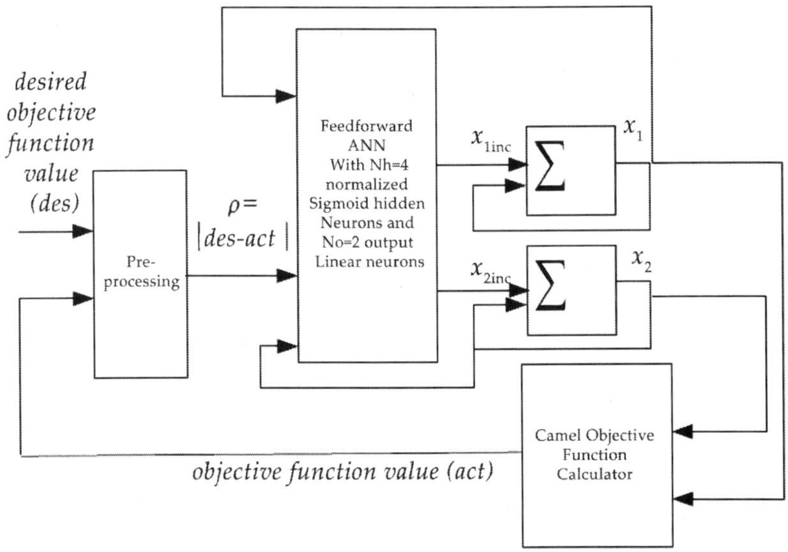

Figure 12: Block diagram of the system used to find the global minimum of the camel objective Function.

To apply the proposed ANN-based approach to find the global minimum of this function, the closed loop system in Fig. 12 has been adopted. The inputs to the ANN are the values of x_1 and x_2 at the previous iteration, the absolute difference between the desired value of the objective function at the current iteration and its current actual value. The outputs of the ANN are the recommended corrections to x_1 and x_2 (increase or decrease). The ANN weights and biases were estimated numerically as before such that the ANN minimize the camel objective function. The desired value for the objective function at any iteration is chosen to be slightly less than its current value (For example, for decrratio=0.85, the desired objective function at a certain iteration =decrratio* the actual objective function value corresponding to the current solution). This idea is borrowed from [13], where it has been recommended to be adopted with any optimization method. The ANN has been trained, as before, using a numerical BFGS algorithm, where the objective function to be minimized is defined as :

$$Objg = \rho_F \tag{12}$$

where, ρ_F is the absolute difference between the desired objective function value and its actual value at the last iteration. The rules extracted from the trained ANN are as follows:

$$SLANCV \quad x_1 > 0.6755704, \quad x_2 > 1.5209156, \quad \rho > 0.0161157,$$
$$decrease \quad x_1 \quad by \quad x_{1inc} = -0.0668506 \quad decrease \quad x_2 \quad by \quad x_{2inc} = -0.0203858$$
$$SLANCV \quad x_1 > 1.4023561, \quad x_2 > 0.4688876, \quad \rho > 0.1058053,$$
$$increase \quad x_1 \quad by \quad x_{1inc} = 0.0617143 \quad decrease \quad x_2 \quad by \quad x_{2inc} = -0.0195689$$
$$SLANCV \quad x_1 > 0.6147782, \quad x_2 < 0.1270187, \quad \rho < 1.504679,$$
$$increase \quad x_1 \quad by \quad x_{1inc} = 0.0301060 \quad decrease \quad x_2 \quad by \quad x_{2inc} = -0.0128413$$
$$SLANCV \quad x_1 > 0.0459426, \quad x_2 > 0.3802873, \quad \rho < 0.5840457,$$
$$decrease \quad x_1 \quad by \quad x_{1inc} = -0.0279751 \quad decrease \quad x_2 \quad by \quad x_{2inc} = -0.0338863$$

Fig. 13 shows the results (the objective function value versus iteration number). Clearly, the ANN-based optimization technique found the global optimal (The initial solution was $x_1 = -2.5, \quad x_2 = 2.5$). When using the BFGS technique to minimize the camel objective function, with the same learning rate, the algorithm completely diverged. For a lower learning rate, direct BFGS reached the global optimal. However, ANN-based optimization offers greater promise for higher dimensions/ multi-objective/ error problems and provides insight into the solution through the analysis of the derived rules. A key reason for the robustness of the ANN-based optimization over direct optimization is that it explores the objective function and derives problem-dependent heuristics (as opposed to meta-heuristics).

An important direction for future investigation is the use of concepts borrowed from "robust optimization" to enhance the quality of the rules extracted from the ANN. Robust optimization concepts can be applied at two different levels:

- At the action level, we can include noise during ANN training. The error in pathtracking can be defined as a function of the difference between the average expected location and the desired location as well as the variance (or standard deviation of this error). This will make the ANN develop a decision strategy that is more prudent and is less likely to cause divergence from the desired path, in case of disturbance.

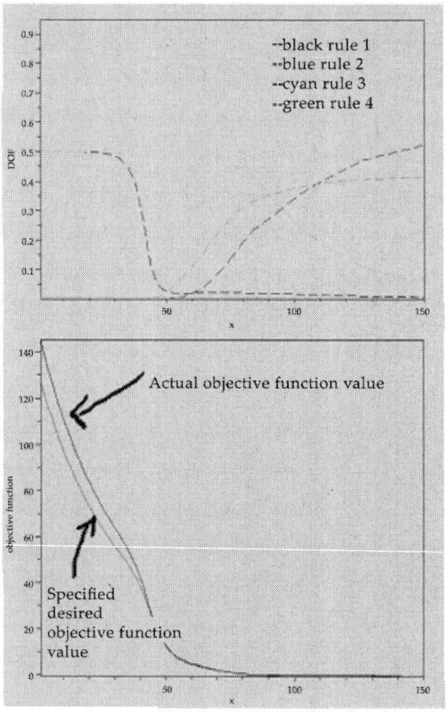

Figure 13. The results of using the proposed approach to find the global minimum of the camel objective function. The top figure shows the degree of firing of the different rules versus iteration number. The bottom figure depicts the objective function value versus iteration number. The approach successfully converged to the global minimum.

- At the weights/biases level, robust optimization can be used to develop robust rules. Rules are robust when they are reachable from different weight initializations (i.e. they are not sensitive to a particular initialization) and lead to acceptable performance when subject to small perturbations.

CONCLUSION

We believe that merging symbolic AI (logic) with non-symbolic AI (ANNs) through our new proposed framework can achieve the following advantages:

- The resulting learning system is transparent to the user and its reliability can be easily assessed. A suitable strategy has been outlined for improving its reliability based on the analysis of the extracted rules.

- The system is robust to noisy inputs and disturbances.
- The logic-based approach to optimization can be less prune to local minima trapping.
- The approach is applicable to a broad class of engineering problems where the corresponding correct output to a certain example input is not necessarily available (but a means of assessing the fitness of the output is available through simulation or practical measurements).

We do not claim that the proposed approach can outperform existing approaches in all problems, however, we can certainly claim that we offered researchers, a framework truly worthy of investigation for complex optimization, control and design problems. The best approach will always remain problem-dependent which is the charm and challenge of engineering optimization.

ACKNOWLEDGEMENT

After thanking Allah almighty for giving the authors the stimulus required to complete this work, the first author would like to thank her students for their valuable inquiries during a summer course on ANNs that she gave. Their eagerness to understand, queries and continuous criticisms helped her a lot to better formulate her thoughts and served as a valuable encouragement, especially during the early part of this work. It is true that, at many occasions, we learn from our students even more than we teach them! The second author acknowledges the support of EJUST.

APPENDIX A

In this appendix, we apply the conventional ANN formulation ([1] ,[6] ,[7]) and ANFIS [8] to case 'B' of the adopted case study. It is noteworthy that strict application of the conventional ANN or ANFIS to this case study is not possible because the desired system output is unknown (it is not possible to evaluate the objective function except by using the robot simulator). Therefore, the numerical BFGS has been used, as before, in the training phase with the same objective function defined in Eq. (9). As said earlier, with the conventional ANN formulation the output of the ANN described in Eq. (1) can be interpreted as fuzzy rules of the form:

$$SLANCV \quad x_i^p > -\left(b_j / N_i\right) / w_{hij} \quad then \quad o_{jk} = w_{jk}$$
$$i = 1,2,\ldots,N_i$$

The fact that the antecedent of a rule depends on both the bias of the corresponding hidden neuron and the weights from inputs-to this hidden neuron makes it difficult the use of known inputs constraints in weights/ biases initialization. Thus, we are forced to use small random weights and biases initially, train the ANN, extract the rules and then re-train in case some of the rules yield un-plausible antecedents. For our case study, 3 out of the 7 rules had un-plausible antecedents. For example, one of the rules stated:

$$SLANCV \quad v < -1.1959689, \ w > 0.1675315, \ v_{ref} - v_{rob} > 0.8878251, \ \omega_{ref} - \omega_{rob} > 0.1353403,$$
$$\rho < -2.7918388, \ \alpha < -0.2060677,$$
$$decrease \ v \ by \ v_{inc} = -6.6637447, \ decrease \ \omega \ by \ \omega_{inc} = -5.8271221$$

Clearly comparing ρ to a negative threshold is not logical.

Fig. 14 illustrates a typical ANFIS architecture for the case of a 2 inputs (x_1, x_2) single output, 2 rules example. A_{ij} is the membership function of the ith input in the jth rule. The DOF of a rule is computed using the 'Product' operator, i.e. it is the product of the output of the membership functions of a certain rule (Layer 2). NORM units (Layer 3) divides the individual DOF of each rule by the sum of DOF of all rules to produce a normalized degree of firing $\overline{w}_j$. Layer 4 computes the consequent of each rule j for each output, f_{jk} as a function of the inputs $f_{jk} = \sum_{i=1}^{Ni} p_{ijk} + r_{jk}$. The overall system output is computed as a weighted sum of the different rules consequents $f_k = \sum_{j=1}^{N_h} \overline{w}_j f_{jk}$ (N_h is the number of rules which equals 2 in Fig. 14). Similarly, to be able to compare our approach to ANFIS [8], we use the same block diagram given in Fig. 5 but replacing the typical feed-forward ANN with an ANFIS. The ANFIS formulation does not impose restrictions on membership function's choices. Therefore, sigmoid membership functions have been chosen, for the purpose of comparison with our approach. In this case, the membership function of the jth rule takes the form:

$$A_{ij} = \text{sig}\left(a_{ij}\left(x_i - c_{ij}\right)\right)$$

Our approach can be viewed as a modified ANFIS system with the 'Product' operator replaced by the 'ior' operator and with $p_{ijk} = 0$. As indicated by the results (Fig. 15), these modifications enhance the performance considerably. ANFIS training involves the estimation of the parameters , p_{ijk}, r_{jk} for each rule contributing to the output as well as the membership functions parameters a_{ij}, c_{ij} of each rule. The extracted rules after training are as follows:

If $v<0.7060525$ and $\omega<-0.2171979$ and $(v_{ref}-v_{rob})>-0.4880470$ and $(\omega_{ref}-\omega_{rob})<-0.3037414$ and $\rho<0.5722985$ and $\alpha<-0.2705797$ then

$$v_{inc}= -0.4977126*v + -0.2252019*\omega + -0.2252019*(v_{ref}-v_{rob}) + 0.6536675*(\omega_{ref}-\omega_{rob}) + -0.9451361*\rho + 0.4929671*\alpha + 0.5642881 + 0.2698283,$$

$$\omega_{inc}= -0.2601446*v + -0.0526525*\omega + -0.0526525*(v_{ref}-v_{rob}) + -0.0256069*(\omega_{ref}-\omega_{rob}) + 0.2929497*\rho + -0.6628894*\alpha + 0.6927095 + 0.5608534$$

If $v>0.9536605$ and $\omega<-0.3640576$ and $(v_{ref}-v_{rob})>0.3635925$ and $(\omega_{ref}-\omega_{rob})>-0.5739730$ and $\rho<1.0224728$ and $\alpha>0.2518629$ then

$$v_{inc}= -0.2252019*v + 0.5305053*\omega + 0.5305053*(v_{ref}-v_{rob}) + 0.5254107*(\omega_{ref}-\omega_{rob}) + 0.6691905*\rho + 0.1321803*\alpha + -0.4524829 + 0.5491040,$$

$$\omega_{inc}= -0.0526525*v + -0.1341857*\omega + -0.1341857*(v_{ref}-v_{rob}) + -0.4909975*(\omega_{ref}-\omega_{rob}) + -0.2141134*\rho + -0.2020394*\alpha + -0.1888051 + 0.0712943$$

If $v<0.7970844$ and $\omega>0.2311402$ and $(v_{ref}-v_{rob})>-0.2606914$ and $(\omega_{ref}-\omega_{rob})>0.3577566$ and $\rho>0.5432450$ and $\alpha>0.3045580$ then

$$v_{inc}= -0.4283727*v + 0.5254107*\omega + 0.5254107*(v_{ref}-v_{rob}) + -0.1505240*(\omega_{ref}-\omega_{rob}) + 0.8359596*\rho + -0.0326892*\alpha + 0.3081844 + -0.0003588,$$

$$\omega_{inc}= 0.3454005*v + -0.4909975*\omega + -0.4909975*(v_{ref}-v_{rob}) + -0.2778212*(\omega_{ref}-\omega_{rob}) + 0.6994976*\rho + 0.4886705*\alpha + 0.2289957 + 0.5789568$$

If $v < 1.0159674$ and $\omega > 0.1276233$ and $(v_{ref} - v_{rob}) < -0.2720805$ and $(\omega_{ref} - \omega_{rob}) > -0.0879307$ and $\rho > 0.7092691$ and $\alpha > -0.1685077$ then

$$v_{inc} = -0.9451361 * v + 0.4569717 * \omega + 0.4569717 * (v_{ref} - v_{rob}) + 0.8359596 * (\omega_{ref} - \omega_{rob}) + 0.6869079 * \rho + 0.7200467 * \alpha + 0.7449960 + 0.8263730,$$

$$\omega_{inc} = 0.2929497 * v + 0.2791020 * \omega + 0.2791020 * (v_{ref} - v_{rob}) + 0.6994976 * (\omega_{ref} - \omega_{rob}) + 0.1821506 * \rho + 0.6257491 * \alpha + 0.4543635 + 0.2199392$$

If $v > 1.205344$ and $\omega > -0.0968983$ and $(v_{ref} - v_{rob}) > 0.0373195$ and $(\omega_{ref} - \omega_{rob}) < 0.1962357$ and $\rho > 1.2538755$ and $\alpha < 0.1218786$ then

$$v_{inc} = 0.5642881 * v + -0.2429051 * \omega + -0.2429051 * (v_{ref} - v_{rob}) + 0.3081844 * (\omega_{ref} - \omega_{rob}) + -0.6377921 * \rho + 0.2372806 * \alpha + 0.3039157 + -0.7144687,$$

$$\omega_{inc} = 0.6927095 * v + 0.7480350 * \omega + 0.7480350 * (v_{ref} - v_{rob}) + 0.2289957 * (\omega_{ref} - \omega_{rob}) + 0.5583952 * \rho + -0.4286680 * \alpha + 0.2080972 + -0.4508324$$

If $v > 1.205344$ and $\omega > -0.0968983$ and $(v_{ref} - v_{rob}) > 0.0373195$ and $(\omega_{ref} - \omega_{rob}) < 0.1962357$ and $\rho > 1.2538755$ and $\alpha < 0.1218786$ then

$$v_{inc} = 0.5642881 * v + -0.2429051 * \omega + -0.2429051 * (v_{ref} - v_{rob}) + 0.3081844 * (\omega_{ref} - \omega_{rob}) + -0.6377921 * \rho + 0.2372806 * \alpha + 0.3039157 + -0.7144687,$$

$$\omega_{inc} = 0.6927095 * v + 0.7480350 * \omega + 0.7480350 * (v_{ref} - v_{rob}) + 0.2289957 * (\omega_{ref} - \omega_{rob}) + 0.5583952 * \rho + -0.4286680 * \alpha + 0.2080972 + -0.4508324$$

If $v > 0.6196452$ and $\omega < 0.2323706$ and $(v_{ref} - v_{rob}) > 0.5647629$ and $(\omega_{ref} - \omega_{rob}) > -0.0327666$ and $\rho > 0.4904380$ and $\alpha > -0.1630150$ then

$$v_{inc} = -0.0378747 * v + -0.3878541 * \omega + -0.3878541 * (v_{ref} - v_{rob}) + 0.1955760 * (\omega_{ref} - \omega_{rob}) + 0.0637325 * \rho + -0.1970121 * \alpha + -0.2060373 + 0.4825770,$$

$$\omega_{inc} = 0.1229261 * v + 0.1073914 * \omega + 0.1073914 * (v_{ref} - v_{rob}) + -0.2504428 * (\omega_{ref} - \omega_{rob}) + 0.6417333 * \rho + 0.2868955 * \alpha + 0.1431067 + 0.1357565$$

It is clear from Fig. 15, that both the conventional ANN and ANFIS produce inferior results to those obtained using the proposed approach (refer to Fig. 7). Thus, the proposed modifications to conventional ANNs succeeded in producing an improved ANFIS system capable of outperforming both conventional ANNs and ANFIS for problems where the desired ANN output is not known a priori (like in the path tracking case study).

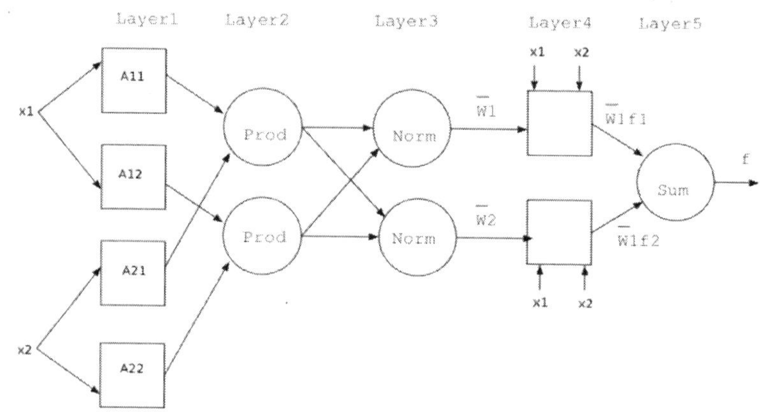

Figure 14: Architecture of a typical ANFIS system for a 2 inputs single output example.

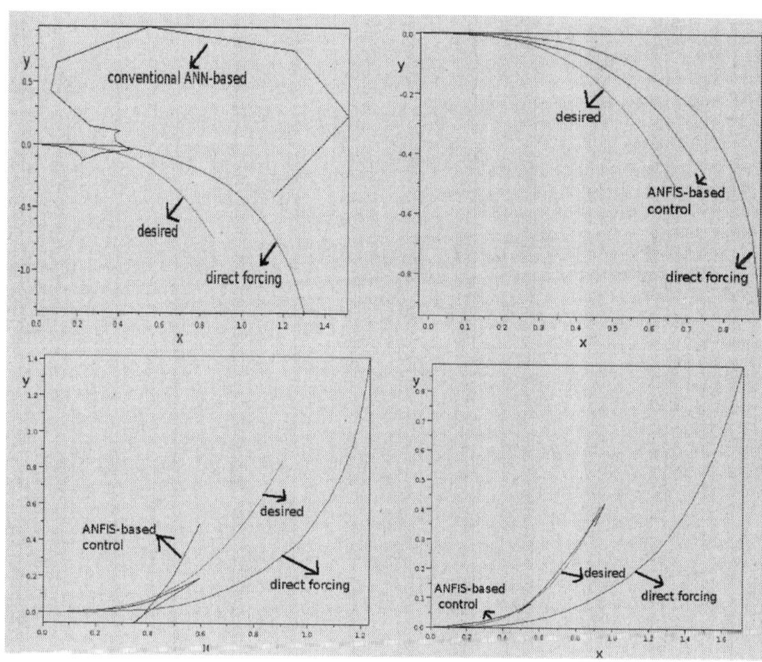

Figure 15: Shows the results for using conventional ANNs and ANFIS instead of the proposed formulation for case B. By comparing these results with those in Fig. 7 , the superiority of the proposed approach-thanks to Allah- is clear.

REFERENCES

1. G J. Benitez, J. Castro, I. Requena, "Are Artificial ANNs Black Boxes?", IEEE Trans. on ANNs, September, 1997.
2. J. Espinosa, J. Vandewalle, V. Wertz, Fuzzy Logic, Identification and Predictive Control (Advances in Industrial Control), Springer-Verlag, London, 2005.
3. Simon Haykin, ANNs and Learning Machines, Prentice Hall, November 2008.
4. R. Setiono, "Extracting M-of-N Rules from Trained ANNs", IEEE Trans. on ANNs, Vol. 11, No.2, January 2000. pp. 510-519.
5. S. Mitra, S. Pal, "Fuzzy Multi-Layer Perceptron, Inferencing and Rule Generation" , IEEE Trans. on ANNs, Vol. 6, No.1, January 1995. pp. 51-63.
6. H. Senousy, M. Abou-El Makarem, "New Reliable Neural-Based Automatic Diesel Fault Diagnosis Systems", International Conference on Mechanical Engineering and Production MDP9, Cairo, Egypt, January 2008.
7. I. Hamid, H. Senousy, M. Abou-Elmakarem, "An Improved Fuzzy Logic Controller For Ship Steering Based on Ior Operator and Neural Rule Extraction", ICCES08, Faculty of Engineering - Ain Shams University, Computer Engineering & Systems Department Cairo, EGYPT, November 25-27, 2008.
8. J.S. R. Jang "ANFIS: Adaptive Network based Fuzzy Inference Systems", IEEE Transactions on Systems, Man, and. Cybernetics, vol. 23, no. 3, (1993) 665–685.
9. http://sourceforge.indices-masivos.com/projects/forallahfacon/
10. S. Rao, Optimization, Theory & Applications 2ed, July 1984, John Wiley & Sons (Asia).
11. R. Gonzalez, ANNs for Variational Problems in Engineering, PhD Thesis, Department of Computer Languages and Systems, Technical University of Catalonia,21 September 2008.
12. G. Dudek, Michael Jenkin, Computational Principles of Mobile Robotics, Cambridge University Press; April 15, 2000.
13. A. El-Bastawesy , A. El-sayed, M. Abdel-Salam, B. Salah, I. Adel, M. Alaa El-laffy, M. Tariq, B. Magdi, O. Fathi, A New Intelligent Strategy for Optimal Design of High Dimensional Systems, 2011 IEEE GCC Conference & Exhibition, Dubai, United Arab Emirates, 2011

INDEX